Understanding Risk

The Theory and Practice
of Financial Risk Management

D0731512

CHAPMAN & HALL/CRC
Financial Mathematics Series

Aims and scope:

The field of financial mathematics forms an ever-expanding slice of the financial sector. This series aims to capture new developments and summarize what is known over the whole spectrum of this field. It will include a broad range of textbooks, reference works and handbooks that are meant to appeal to both academics and practitioners. The inclusion of numerical code and concrete real-world examples is highly encouraged.

Series Editors

M.A.H. Dempster
Centre for Financial Research
Judge Business School
University of Cambridge

Dilip B. Madan
Robert H. Smith School of Business
University of Maryland

Rama Cont
Center for Financial Engineering
Columbia University
New York

Published Titles

American-Style Derivatives; Valuation and Computation, *Jerome Detemple*

Engineering BGM, *Alan Brace*

Financial Modelling with Jump Processes, *Rama Cont and Peter Tankov*

An Introduction to Credit Risk Modeling, *Christian Bluhm, Ludger Overbeck, and Christoph Wagner*

Introduction to Stochastic Calculus Applied to Finance, Second Edition, *Damien Lamberton and Bernard Lapeyre*

Numerical Methods for Finance, *John A. D. Appleby, David C. Edelman, and John J. H. Miller*

Portfolio Optimization and Performance Analysis, *Jean-Luc Prigent*

Robust Libor Modelling and Pricing of Derivative Products, *John Schoenmakers*

Structured Credit Portfolio Analysis, Baskets & CDOs, *Christian Bluhm and Ludger Overbeck*

Understanding Risk: The Theory and Practice of Financial Risk Management, *David Murphy*

Proposals for the series should be submitted to one of the series editors above or directly to:
CRC Press, Taylor and Francis Group
24-25 Blades Court
Deodar Road
London SW15 2NU
UK

CHAPMAN & HALL/CRC FINANCIAL MATHEMATICS SERIES

Understanding Risk

The Theory and Practice
of Financial Risk Management

David Murphy

CRC Press
Taylor & Francis Group
Boca Raton London New York

CRC Press is an imprint of the
Taylor & Francis Group, an informa business

Chapman & Hall/CRC
Taylor & Francis Group
6000 Broken Sound Parkway NW, Suite 300
Boca Raton, FL 33487-2742

International Standard Book Number-13: 978-1-58488-893-2 (Softcover)

Library of Congress Cataloging-in-Publication Data

Murphy, David (David V. J.), 1965-
    Understanding risk : the theory and practice of financial risk management / David Murphy.
      p. cm. -- (Chapman & Hall/CRC financial mathematics series)
    Includes bibliographical references and index.
    ISBN 978-1-58488-893-2 (pbk. : alk. paper)
    1. Financial risk management. 2. Investment analysis. 3. Portfolio management. I. Title. II. Series.

HG4529.5.M87 2007
658.15'5--dc22

                                           2007022602

Visit the Taylor & Francis Web site at
http://www.taylorandfrancis.com

and the CRC Press Web site at
http://www.crcpress.com

For Sue

Contents

PART TWO Economic and Regulatory Capital Models

CHAPTER 3 ■ Capital: Motivation and Provision 163

Introduction

THIS BOOK IS AIMED at those who wish to understand financial risk management better. This includes risk managers; their colleagues in trading, banking and other areas of their firms; regulators, auditors and others supporting or overseeing risk takers; and anyone else with an interest in gaining some intuition into how losses can arise in risk taking and how the severity and frequency of those losses can be controlled.

Good risk managers require both technical and artisanal skills. Some of the attributes in a risk manager which are helpful include:

- *Product and market savvy*, because you cannot understand the risk of a portfolio without understanding what is in it and how those things trade;

- *A measure of expertise in quantitative finance*, because risk is often expressed in the framework of a mathematical model of asset returns, and many products, including most derivatives, are valued and risk managed in a similar* framework;

- *Patience and excellence in handling large amounts of data reliably*, because most risk managers have access to many, many pieces of risk information every day, only a tiny proportion of which have any material bearing on their firm's risk profile;

- *Good social and communications skills*, because risk managers often have to get information from traders, finance and operations professionals, and others, and to persuade management of the veracity of their findings;

- *A hard head and an unbending sense of what is right for their institution*,[†] because there may be times when they are under pressure to permit a trade that should not proceed in its current form, or to hold back trading that should be encouraged.

We hope to give some insight into some of these skills during the course of the book. Often our discussion starts with a sketch of some theory relevant to the problem. This is

---

* Risk managers and traders typically use similar models for looking at risk but often not identical ones. To see this just think of the difference between the assumptions about how the yield curve moves in your swaptions model and your value at risk (VAR) model.

† Michael Lewis' *Liar's Poker*, Frank Portnoy's *F.I.A.S.C.O.: Guns, Booze and Bloodlust—The Truth About High Finance*, Emanuel Derman's *My Life as a Quant* and especially Nassim Taleb's *Fooled by Randomness: The Hidden Role of Chance in Life and the Markets* are all good, easy-to-read books which give insights into how market participants can sometimes think and behave.

then contextualised and some issues around its use are explained, with the hope that the reader will gain some intuition into both practical and theoretical issues.

Our perspective of risk management is sometimes as much a craft as a science. This is because risk managers may be more interested in tools that are effective in a particular limited application rather than in more general but less immediately useful frameworks. That is not to dismiss theoretical considerations, simply to suggest that theory is sometimes only helpful to the practitioner insofar as it provides insight into real problems in a short space of time and with a small investment in its development. Most risk managers have to prioritise between competing issues, so they tend to value approaches which give a lot of insight for their time and money.

PREREQUISITES

A reasonable familiarity with at least the basics of financial mathematics and the theory of options is assumed. Most terms are explained in passing, but the majority of the discussion is in the middle ground between an elementary account and a thorough quantitative approach, with the emphasis on the intuition behind models rather than detailed mathematics.

One reason for keeping the technical level relatively low is that many traders do not know much more, and if they can run their books without knowing what a Borel σ space is, then their risk manager should be able to operate at a similar level if need be. Another is that there are a number of excellent books on mathematical finance and it would add little to cover the same territory.*

The only exception to the elementary level of mathematics assumed is an occasional dip into the stochastic calculus. This is not necessary for understanding most of the book, but some background here might be helpful.†

We do assume basic knowledge of at least one area of the financial markets, partly because the author believes that when all is said and done market prices are the result of nothing more than trades between two willing parties—and sometimes these parties behave in ways that theory does not predict. So some exposure to the irrationality of markets would be helpful. The treatment of one market should, though, be comprehensible to professionals who have a good grounding in another: the sections on interest rate risk should be accessible to an equity specialist, for instance, and *vice versa*.

REFERENCES, QUESTIONS AND DIALOGUES

References to other books, articles and (inevitably faster dating) web pages are given as footnotes. Cross-references within the book are given in [square brackets].

---

* Good references here include the latest edition of John Hull's *Options, Futures and Other Derivatives*, Riccardo Rebonato's *Volatility and Correlation* or *Paul Wilmott on Quantitative Finance*.
† Wilmott's book or Mark Joshi's *The Concepts and Practice of Mathematical Finance* gives good background; something more advanced, like Karatzas' *Brownian Motion and Stochastic Calculus*, gives a more complete picture.

The text contains a number of exercises: you might want to read through these and mentally sketch out an answer. If you have time and the topic is interesting, it may be worth writing one out.

There are a number of dialogues scattered around the book where the author wants to comment on a situation without taking a position.

ACKNOWLEDGMENTS

The author would like to thank Andrew Street for sharing his insight over the years; Gary Carlin, for leading by example; and John Lambert, for the wealth of his knowledge and commitment.

This book was written in the autumn of 2006 and spring of 2007; efforts have been made to ensure that material is up to date as of the time of writing, but the markets will inevitably move quickly and make fools of us all. Any errors and omissions remain, of course, the sole responsibility of the author.

David Murphy
Spitalfields, Spring 2007

Part One

Risk Management and the Behaviour of Products

CHAPTER 1

Markets, Risks and Risk Management in Context

INTRODUCTION

This chapter provides some of the context for risk management: we discuss markets, their history and their (mis)behaviour; financial institutions of various kinds; and the external constraints within which these firms operate dictated by the regulatory and accounting environment.

We look at the various kinds of risk financial institutions are susceptible to. This leads to a discussion of the aims of risk management; to some simple risk reporting; and to some basic controls that are often used in market risk management. We also consider cultural and organisational issues which may help or hinder a risk manager.

1.1 FINANCIAL MARKETS OVERVIEW

1.1.1 Introducing the Markets

The financial markets are sometimes portrayed as huge arenas of money filled with voracious traders making and losing fortunes each day. This cinematic conceit sees every market as fast moving and massively volatile with enormous volumes being traded. What is being traded hardly matters for the film maker's purposes: usually we are led to believe that it is equities, but it could equally well be bonds, pork belly futures or gold.

Of course, the stereotype has an element of truth. If we take the London Stock Exchange (LSE) as a representative equity market, it has over 2,500 companies listed on it with a combined net worth of more than £3 trillion. The daily turnover is typically billions of shares. The global spot foreign exchange (FX) market is even more impressive: daily volume there is often in excess of $1.75 trillion, much of it in one of only three crosses: USD/JPY, USD/EUR or USD/GBP. The bond markets hold their end up well too: there are estimated to be over $25 trillion of U.S. dollars (USD)-denominated bonds outstanding, with U.S. Treasury securities comprising about 20% of that. Treasury trading volume alone is around half a trillion dollars per day.

The fortunes-made-and-lost part of the stereotype has some element of truth too, at least if we consider institutions: to pick one example from many, Goldman Sachs' trading and investment banking revenue in the 3 months to 26 May 2006 was $2.31B. Compare that with the cost of a new school in England—very roughly £25M—and you can see that Goldman's trading and investment banking business made enough to build a new school every two days.

These enormous numbers and rather jejune comparisons can distract us from some basic questions:

- What is a market?
- What is a market price?
- How do firms make money trading?

1.1.1.1 Markets
Markets are places where buyers and sellers meet. Like any other collection of people, they have diverse needs, wants, and views of the world.

1.1.1.2 Market prices
There is nothing magical about a market price. It is simply the result of a transaction between a willing buyer and a willing seller. If there are more sellers than buyers, the price they trade at tends to go down.

This process of matching buyers and sellers may be intermediated by technology—as in a *screen market*—or by a *broker*—but fundamentally it is a sociological phenomenon. Price discovery is only possible if there are sufficiently many buyers and sellers, and it only comes about because of their actions. Thus, we distinguish a *market price* for an asset from its *value*:

- An asset may have a long-term value to its holder without being saleable, or only saleable at a much lower price than its value to the holder;
- On the other hand, we may well believe that market prices are too high and do not reflect the real value of an asset. For instance recently someone paid $135M for Gustav Klimt's portrait of Adele Bloch-Bauer when he could have had a great Rembrandt for $40M, but that does not mean that everyone thinks it was a good trade.

Value, then, is *context dependent*: how valuable something is to you may depend on many factors including your holding period, your accounting policy, your funding costs, your capital allocation strategy and how desperately you want a particular painting on your wall. In contrast, a market price is a single idealised number representing 'free market' transactions in an asset at some point in time.

Market prices are often quoted with a *two way* price: a lower price at which we can sell, the *bid*, and a higher one at which we can buy, the *offer*. *Mid market* is half way between the bid and offer price, and the difference between them is known as the *bid/offer spread*.

1.1.1.3 Financial assets and their markets

There are two types of assets: financial assets and real assets. For the most part, we will be concerned only with financial assets, which we further subdivide into *direct investments* (investments where the holder takes a direct ownership interest in the asset, as in a corporate bond); and *indirect investments* (where there is indirect ownership, as in most mutual funds, exchange-traded funds and real estate investment trusts). In the next few sections we introduce the main classes of assets and how they trade.

1.1.2 Securities

Many direct investments are made using *securities*. A security is an agreement between an *issuer* and a *holder* entitling the holder to certain payments under certain conditions, and possibly granting other rights.

- These benefits are often documented via a *prospectus* which sets out the nature of the agreement.
- The payments may be fixed, as in a *fixed rate bond*; variable on the basis of some interest rate or other index, as in a *floating rate note* (FRN); or at the issuer's discretion, as in the dividend payments on a *stock*.
- The rights may include *voting rights*, as in an equity holder's right to vote at a company AGM or a bondholder's vote in insolvency proceedings, or *covenants* which constrain the issuer's freedom of action in some way to the presumed benefit of the security holder.
- The duration of the agreement can be fixed, as in a typical bond, or *perpetual*, as for most equities.

The fundamental analysis of a security involves the assessment of the value of the future payments promised and rights granted versus the likelihood of those payments materialising or those rights not being exercisable.

Typically, a security is issued for *funding* purposes. That is, the issuer needs cash for some purpose—for instance to cover operating expenses, to engage in a new project, to increase its buffer against possible future losses, or to acquire another company. This cash is the sum investors pay for a security when it is first offered: they expect some return on this investment, either via the price appreciation of the security or through payments made to the holders of the security, or both. A debt security can therefore be thought of as a form of *loan*: the initial buyer hands over cash for the right to receive interest payments and an eventual return of principal. An equity security is different in that there is no return of principal: instead, the issuer grants the equity holder an ownership interest (which entitles them to a share of the issuer's profits) and voting rights.

A security is therefore an *asset* for the holder: they have the right to the benefits of the security. On the contrary, it is a *liability* to the issuer: they have to meet the obligation imposed by the security.

1.1.2.1 Primary versus secondary markets

Primary markets are where securities are initially sold. In this market, the issuer receives the proceeds from a sale of securities. This sale is typically managed by an *investment bank* that may *underwrite* some or all of the issue, that is, undertake to buy the new securities if there is not sufficient primary market interest. Primary market issues may be either in the form of an *initial primary offering* (the first time a company's stock is offered to the public markets) or a *rights issue* (where an offer is made by a company to its existing shareholders to enable them to buy new shares in the company, usually at a discount).

Once the securities have been sold into the primary market, they begin trading in the *secondary market*. Here holders buy and sell securities for their own benefit. The existence of the secondary markets allows the primary markets to function more efficiently: then holders know (or at least believe) that they will be able to sell their securities if they need to; frequent issuers have clear price benchmarks for new issues; and underwriting is less risky.

Exercise. How would you decide on the price you would be willing to pay for a security if you knew that you could never sell it?

1.1.2.2 Equity, debt and leverage

One key difference between equity and debt is that the obligation to pay interest on debt is often binding: if a firm cannot meet interest or principal payments, it may be forced to *default*. On the contrary, equity represents a *residual claim* on a firm after all other obligations have been met, and any payments on equity are at the issuer's discretion.

Typically, at maturity the best thing that can happen to a bondholder is that they are paid the expected amount, whereas an equity price can go up arbitrarily far if the corporation's profits increase faster than its costs.

This means that an equity investment in a firm is a higher-risk, higher-return investment, whereas a debt obligation from the same firm is usually thought to be lower risk. Put another way, an equity holder only gets a return once obligations on debt have been met, but they have a claim on *any* excess return.

Equity is also said to be *loss absorbing*: the funds raised by issuing equity are available to absorb losses since they never have to be repaid.

The ratio between a firm's total value and its equity is known as *leverage*. The higher the leverage, the riskier the firm is since there is less equity available to absorb losses.

1.1.2.3 Ratings agencies

A *ratings agency* is an independent company that assesses the credit quality of debt securities and assigns a score to them based on that assessment. Typically the ratings agencies make their assessment based on:

- An obligator's ability and willingness to repay the debt;
- Their history and strategy of borrowing and repayment;
- Their current leverage and the extent and volatility of their assets and liabilities;

- The consequences for the holder of the instrument of default or other failure to pay by the obligator.

Typically, a ratings agency assessment will be on a fixed scale. This contains 10–20 buckets divided into two parts: higher credit quality or *investment-grade* ratings and lower quality, speculative, or *junk* ratings. Ratings are often stated using a scale from AAA to BBB+ (the investment-grade ratings) then on down from BBB to C (the junk ratings) ending in D (default). This scale was popularised by Standard & Poor's rating agency.* The lower on the scale a security is, the more likely the agency thinks it is to default on the security, i.e., fail to meet the promises made to the securities' holders.

1.1.2.4 Bankruptcy and seniority
When a company defaults, it usually enters into the *bankruptcy* process. The precise details of the process vary from jurisdiction to jurisdiction, but typically there are two choices: either the firm is *reorganised*, or it is *liquidated*. In either case, an independent party is appointed to protect the interests of creditors.

Once the bankruptcy process has been completed through reorganisation, the sale of the firm as a going concern, or its liquidation, a certain amount of money is available to pay creditors. Imagine a queue of people waiting to be paid on a first come, first served basis. *Seniority* is the term for where in the queue you stand: this is a property of the kind of obligation held. The most senior creditor presents their claim first and is paid in full (or as fully as the funds available allow). Then the next most senior creditor is repaid, and so on. Typically, most or all bondholders (aka *senior debt* holders) rank at the same level in the queue: next come any *subordinated* creditors, and finally the equity holders. If you rank at the same level in the queue as another creditor, you are said to be *pari passu* with them.

Obviously, the more senior your claim, the more chance you have of getting paid in the event of bankruptcy. Thus, senior debt holders often recover some of the amount they are owed—known as the senior debt *recovery* value—whereas equity holders often get little or nothing.

1.1.3 Other Instruments
A security documents the granting of rights in exchange for cash. As such they are *funded* transactions: the buyer pays for the rights now and hopes for a future return. Other market transactions of interest include those where currencies are exchanged or where instruments whose value depends on other financial variables are traded.

1.1.3.1 Foreign exchange
FX markets allow participants either to exchange amounts of one currency for another immediately—the *spot* market—or to agree to such an exchange at some point in the future—the *forward* market.

---

* See www.standardandpoors.com, www.moodyskmv.com or www.fitchratings.com for more details of three of the better known ratings agencies.

1.1.3.2 Derivatives

One major impetus for the development of derivatives markets was the securities market. Once securities became freely traded, market participants started to look for other ways of taking a view than simply exchanging cash for securities immediately. One obvious idea is to agree to the exchange but not to consummate it until some agreed future date: I want to buy something, you want to sell it, but I do not want to pay the entire purchase price (or maybe you do not have the security). So we agree to trade in the *future* at a price we fix today. Alternatively, perhaps I think that a security is going up in price, but I cannot afford to buy it; what I really want is the ability to buy it at today's price should I do desire at some point in the future: the *option* to buy.

Derivatives are available on a range of underlyings including equity and debt securities, currencies, commodities and interest rates.

1.1.3.3 Physical versus OTC markets

Physical markets are those that have a location where trading takes place: most major equity markets, for instance, are physical. Physical markets are often based on an *exchange*: this is a body which provides a forum for trading together with other infrastructure such as *settlement* systems. Trading is usually regulated, with market participants being obliged to adhere to certain standards of *conduct of business*.

OTC or *over the counter* markets in contrast consist of a network of dealers and trading here may be more lightly regulated or totally unregulated. The distinction between physical and OTC markets used to be much clearer before the days of electronic trading: now many physical markets also trade on electronic platforms, whereas some OTC markets (such as the NASDAQ) are characterised by regulatory features, such as best execution rules, more typical of a physical market.

Understanding the precise nature of the market—how it is regulated, the nature of participants' obligations, whether trades must be reported—is crucial in understanding what a market price means. To see this, compare and contrast two idealisations:

- *Market maker* style. All trades have as one counterparty one of a small number of market makers. The market makers have an obligation to make firm prices on both bid (to buy) and offer (to sell) in normal market size while the market is open. All trades are reported within a short period of execution.
- *OTC inter-professional* style. Some market participants make one- or two-way screen prices, but these are not firm. Trade reporting is not compulsory; limited price discovery is possible, perhaps via a proprietary market data system.

Exercise. Which style of market would you prefer for:

— ordinary hedging activity where you have to buy less than 1% of average daily market volume of a security on a typical day;

— block trading, where you have to sell a large position of more than 20% of average daily market volume? [See section 9.1.3 for a further discussion of block trading.]

1.1.3.4 Exchange-traded derivatives

Exchange traded derivatives are instruments which trade on a recognised derivatives exchange. This may be a securities exchange or a specialist body which only facilitates the trading of derivatives.

Exchange traded derivatives are often standardised instruments. They can be very liquid with many market participants trading them, but many are much less well traded.

Many exchanges have features which mitigate risk for participants:

- Trades between market participants are *intermediated* by the exchange. This means that participants are not exposed to the risk of each other's default, but only to the (presumably unlikely) default of the exchange;

- In exchange for this protection, the exchanges demand that participants post *margin*. This is a sum which provides some *collateral* against the risk of their default.

1.1.3.5 OTC derivatives and ISDA

If an exchange contract does not meet your needs, you can negotiate directly with another market participant. These *over the counter* or OTC derivatives markets are rather flexible since an OTC derivative is a private bilateral contract between two parties and hence capable of considerable contractual freedom. The price paid for this freedom is firstly the potential for *credit risk*—since any situation where someone might owe you money in the future under a contract has the potential for credit risk—and secondly possible *illiquidity*—in that you might have found the only person in the world who is willing to enter into that form of contract.

Like the OTC securities market, the OTC derivatives market is a network of dealers rather than a single organised market. Due to the tailored nature of OTC derivatives, trades must be documented in detail.

The *International Swap Dealers Association* (ISDA)* acts as a trade association for derivatives dealers. One of its functions is to provide.

- Standard *documentation* for derivatives which can be readily modified to capture a wide range of OTC transactions; together with

- Legal agreements which define the nature of derivatives trading relationships so that both parties to a transaction have *legal certainty*;

- And moreover which permit various forms of credit risk mitigation such as a *bilateral collateral agreement* to be agreed. [See section 10.1.1 for more details.]

We now turn to the individual markets for equity securities, debt securities, and various other instruments.

1.1.4 Equity Markets

An equity represents an *ownership interest* in a corporation. The holders of shares are entitled to the assets of a company once other liabilities have been met. This *capital* is

---

* See www.isda.org for more details.

originally provided by issue of shares in the primary market. Many equities are traded on a stock exchange [but see the discussion of private equity later on in this section]: we begin our discussion of various markets with these instruments and the markets they trade on.

1.1.4.1 Equity and capital

The terms *stock*, *shares* and *equity* tend to be used interchangeably. Although there are a few shares which are not *equity-like* (notably some *fixed rate preference shares*), unless specifically stated otherwise it is a reasonably safe assumption that all three terms represent the lowest tier in a corporation's capital structure, that is, the back of the queue of seniority.

The term *capital* refers both to equity and to certain *equity-like* instruments: its key feature is that it has loss absorption capability. If a firm is too highly leveraged only a small fall in earnings may leave it unable to meet its commitments on its debt. Therefore firms require sufficient capital to support earnings volatility. For a fixed amount of risk, the more capital a firm has, the less likely it is to default. Many financial services firms are regulated, and a key feature of this regulation is the requirement to keep sufficient capital, i.e., to limit the likelihood of default. Earnings volatility is caused by risk, so the amount of capital needed is a function of the risks being run. Most firms calculate both their own estimates of the capital they need to support the risks they are running—an *economic capital* requirement—and the capital they are required to have by supervisors—a *regulatory capital* requirement.

1.1.4.2 Ownership and dividends

Shareholders elect the board of directors of a public company to run the company for them. Or, at least, that is the theory. In practice it seems that, in many cases, the board of directors has a certain amount of staying power even in the face of shareholder hostility.*

If a company is sufficiently profitable, the directors may pay some or all of the profits back to shareholders via a *dividend*. The right to dividends and the right to elect directors form two of the chief benefits of owning equity.

1.1.4.3 U.S. stock markets

In the United States, there are three major stock exchanges where securities are traded:

- The New York Stock Exchange (NYSE)
- The NASDAQ
- The American Stock Exchange, or Amex

Additionally, U.S. (and many other) equities trade on electronic communications networks (ECNs) such as Archipelago, Instinet or Island.

The U.S. stock markets are particularly important because together they constitute a substantial fraction of global equity trading. This makes them very liquid, at least for the

---

* The voting rights that come with most equity bring an interesting risk management problem: how does a firm decide how to vote the stock it holds? Note firstly that this stock may be held in different places within the firm—as proprietary positions, as hedges against derivatives or in asset management—and having different desks voting differently might be embarrassing. Not voting at all is not really a good alternative as it may be important reputationally to be an active shareholder.

larger stocks, and hence they give good *price discovery*. Furthermore, a large number of non-U.S. companies are listed on U.S. markets.

> *Exercise.* Find out the 20 most liquid stocks on the NYSE in some convenient period. How many of them are U.S. companies?

More recently, changes in the corporate governance requirements for listing in the United States under the *Sarbanes–Oxley Act* had decreased the attractions of U.S. markets for foreign companies. [This is discussed further in section 5.5.1.]

1.1.4.4 European stock exchanges
The largest European stock exchanges are:

- In London, the *London Stock Exchange* or LSE;
- In Germany, the *Frankfurt Stock Exchange*, operated by Deutsche Börse;
- In France, the Netherlands and Belgium, *Euronext*, formed by a merger of national markets.

Most European countries have a national stock exchange: some have regional ones too.

> *Exercise.* The consolidation of stock exchanges has been much in the news recently. What is the story and why is it important to market participants?

1.1.4.5 Stock market indices
A stock market index is a measure of the performance of the broad market. It is typically some average of the prices of all or some stocks traded on the market, perhaps weighted by market capitalisation. There are thousands of different indices calculated by the exchanges, various information providers, and investment banks. Some of the best known ones (and the only ones we will use for examples) are:

- The *Dow Jones Industrial Average* (DJIA) is a price-weighted average of the 30 largest and most widely held public companies in the U.S., calculated by Dow Jones Indices. One of the reasons the DJIA is interesting is that it has a long history: it was first published in 1896 (albeit only with 12 stocks in it), and so there is a large amount of historical information available for studying its behaviour;
- The *S&P 500* is a market-value-weighted average of the 500 most important U.S. stocks chosen for their market size, liquidity and industry grouping. It represents 70% of the capitalisation of U.S. publicly traded companies;
- The *FTSE 100* index, colloquially the 'footsie', is an index of the 100 largest companies listed on the LSE meeting certain criteria of liquidity etc. As with many indices, the composition is decided by a committee based on broad rules. In the case of the FTSE,

this committee meets quarterly. Depending on its precise actions and market events, the FTSE may at any time contain slightly more or less than 100 companies;

- The *CAC 40* or just CAC is a float-weighted index of the 40 largest and most liquid stocks traded on the Paris Bourse;
- The *DAX* is an index of 30 major German companies trading on the Frankfurt Stock Exchange. Unlike all the others mentioned above, it is a *total return index*: both dividends and price changes are factored into the index calculation, rather than just price changes.

The major stock indices of large European countries used to be important, but they are now fading a little, with some liquidity moving to pan-European or pan-Eurozone indices, notably the EuroSTOXX 50.

The diversity of means of calculation, number of stocks and total capitalisation of the indices above is worth noting. The DAX with only 30 stocks and its total return nature is a very different beast from the S&P with 500 stocks.

1.1.5 Interest Rate Instruments

The terms *interest rate instrument* and *fixed income instrument* typically refer to some instruments which are sensitive to the level of interest rates—such as fixed rate bonds— and also to some which aren't, or at least aren't much—such as floating rate notes.* They include debt securities and derivatives which depend largely on the value of interest rates or other interest rate instruments.

1.1.5.1 Debt securities
Debt securities are typically structured as:

- An initial payment by an investor to the issuer, which can be thought of as like the funding of a loan;
- Periodic interest payments on the loan, known as *coupons*, from the issuer to the holder of the security;
- A final repayment of *principal*.

The agreement to make these payments and the granting of certain rights to the holder are packaged up in a *bond* or *note*. This can often be traded. The key features of a bond therefore include:

- The issuer (sometimes known as the *obligator*);
- The method of calculation of the coupons, their frequency, details of how interest is calculated, and in what currency payments are made;
- The repayment schedule or other details of the circumstances under which the principal will be repaid;
- The *seniority* of the security and any collateral or other protection for the holder;

---

* One important reference for the bond markets is Frank Fabozzi's compendious *Handbook of Fixed Income Securities*.

- *Covenants* made by the issuer;
- Details of the transferability of the security and any conditions of sale;
- Any *optionality* (for instance the issuer's ability to force us to sell the bond back to him or her early—an *issuer call,*—our ability to force them to buy it back—a *holder put,*—the issuer's ability to *defer* the payment of coupons under some circumstances, possible changes in the coupon, etc.).

Exercise for you.   Find and read the prospectus for a range of different bonds.

1.1.5.2 Interest rate and credit derivatives

Two important innovations in the OTC derivatives market came with the development first of instruments to speculate or hedge on the evolution of interest rates (*interest rate derivatives* in the 1970s and 1980s) and then of derivatives based on the creditworthiness of a bond issuer or loan counterparty (*credit derivatives* in the 1990s).* These latter agreements allow two market participants to trade exposure to the underlying issuer without having to trade one of the issuer's bonds (and thus find the cash to pay for it, deal with its coupons and so on). This convenient packaging of the exposure required and only that exposure is a feature of derivatives product development.

1.1.5.3 Interest rates and bond prices

We assume a basic familiarity with the basic ideas of bond mathematics, summarising them to fix notation and to highlight some issues as we go.

Consider a typical fixed rate bond with semi-annual coupons paying c% annually. The market price P of a bond is the *present value* (PV) of the coupon cashflows and the face F, discounted at the appropriate interest rate, so if the coupon flows happen at times t_i with $i = 1 \ldots n$ and PV_t is a function discounting a cashflow at t back to today

$$P = \sum_{i=1}^{n} PV_{t_i}\left(\frac{c}{2}\right) + PV_{t_n}(F)$$

Suppose we own a 6% annual pay bond which paid its penultimate coupon yesterday. As there is only one cashflow left, we know that $P = PV_{1\ year}(106)$. Usually, the situation is not as simple as this: in particular, if a bond is traded between coupon dates, we have to allocate the next coupon between the buyer and the seller. *Accrued interest* on the bond is the amount that has accumulated at a given period between coupons: this belongs to the seller, the remainder of the coupon being due to the buyer. Bond prices are often quoted *clean* (that is, without the accrued interest) but traded *dirty* (including accrued).

1.1.5.4 Interest conventions

The accrued interest payable by a bond buyer to a bond seller between coupon dates is determined by the coupon rate, the dates and a *day count convention*. This tells us how to

---

* [We discuss credit derivatives more extensively in section 2.5.] A comprehensive account can be found in Philipp Schonbucher's *Credit Derivatives Pricing Models: Models, Pricing and Implementation* or John Gregory's *Credit Derivatives: A Definitive Guide.*

calculate the number of days between a coupon payment date and trade date. The interest accrued in a period less than a whole coupon period is $F \times$ rate $\times$ day count.

There are a number of different day count conventions. The main ones are as follows:

- *Actual/360*, or Act/360. This is the most common convention and is used by bonds denominated in many currencies including the USD and the euro. Each month is treated normally and a year is assumed to be 360 days long.

- *30/360*. Each month is treated as having 30 days and the year is again assumed to be 360 days long.

- *Actual/Actual*. Here each month is treated normally, and the year has the usual number of days, with leap years affecting the result.

There is also Actual/365 (unfortunately common in bonds denominated in GBP) and a slightly different version of Act/Act for U.K. government bonds or *gilts*.

Bond coupons are usually quoted in terms of annualised equivalents. But what that means exactly depends on a *compounding convention*. If we have a 6% bond, it might pay 6% of notional once a year, 3% every 6 months, or something else entirely. If we have compounding at frequency n times per year, and an annual rate r quoted in terms of that compounding, the interest for a period of d days less than a compounding period is $F \times r/n \times$ day count(d), whereas the interest for a whole number m of compounding periods is $F \times (1 + r/n)^m$.

Finally, note that if a payment date falls on a holiday, it is made on another business day determined by a *business day convention*.

One reason that we have outlined this material rather than skipping it or going straight to instantaneously compounded rates is that it is occasionally a source of error. Bonds differ in their day count conventions; different currencies have different holiday schedules; different instruments in the same currency have different coupon frequencies. If we want to model interest rate instruments successfully, we have to get these (tedious and arbitrary) details right.

1.1.5.5 Deriving curves

Once we have got all of the above sorted out for a given currency, we can derive the effective government rates from government bond prices, *bootstrapping the curve*. Conceptually, the process starts with a bond with a single cashflow, and we find the rate such that if we discount at that rate, we recover its price. That gives us (assuming the bond is liquid and does not have too high or low a coupon) the government yield at that maturity. We carry on with a bond with two remaining cashflows, one discounted at the known rate, and so on. This allows us to build the *zero coupon yield curve*, our first basic tool in understanding rates.*

---

* There is some discussion of building yield curves in most mathematical finance textbooks, and Yolanda Stander's *Yield Curve Modelling* goes into some detail. It is also worth obtaining documentation on how a firm's proprietary interest rate derivatives systems build the yield curve, if you can.

1.1.5.6 Present value

The value of a cashflow in the future at an earlier date depends on the size of the cashflow, the dates, the currency and the curve we are discounting it along. The term yield curve builder is slightly misleading, then, in that it suggests something rather geometrical. What we really want is a function giving *present values*

$$PV(\text{start date, end date, currency, issuer})$$

such that if we apply PV(now, coupon date, $, govy) to the coupons of any on-the-run U.S. government bond and add up the results, we get its price, and similarly for other currencies, and other issuers. [We discuss some more desirable properties of a curve builder in section 2.5.2.]

1.1.5.7 Yield

A bond's *yield to maturity* is that return a holder would obtain if:

- The bond is held to its stated maturity date;
- The bond pays all the promised cashflows; and
- All payments made are reinvested at this yield.

1.1.5.8 Premium and discount

In many ways, bonds are messy things: all those coupons complicate understanding the behaviour of the instrument. It is much easier to understand a single cashflow than the series of those cashflows that together form a bond. Part of the problem comes with *premium* and *discount*: a bond priced at a premium has a coupon rate above its yield to maturity. We are paying more than 100 now to get something that pays 100 in the future, together with above-market coupons. In a discount bond, we have below-market coupons, made up for by a purchase price less than 100. These considerations become particularly important if there is any risk we might not get our 100 back.

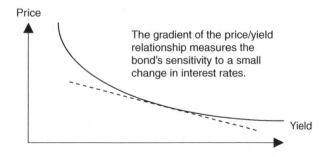

The gradient of the price/yield relationship measures the bond's sensitivity to a small change in interest rates.

1.1.5.9 Price and yield

There is an inverse relationship between price and yield for fixed rate bonds as in the illustration. To a first approximation, as interest rates rise, bond prices fall proportionally. However, this relation is not linear: if rates rise a long way, bond prices fall less fast than a straight-line relationship would suggest. This phenomenon is known as *convexity*.

The gradient of the price/yield relationship for the current level of yields is sometimes known as duration. Thus, if a bond falls in price by 50 bps for a 10 bps rise in interest rates, it is said to have a duration of 5. [See section 2.1.7 for a further discussion of duration.]

1.1.5.10 Bonds on the curve
It is worth noting:

- The government yield curve does not give the price of new government borrowing. That comes from a *primary* issue of new bonds, often by some kind of auction process. This would typically give a clearing yield close to the curve, but not necessarily exactly on it.
- The yield curve certainly does not give the price of anyone else's borrowing either: at least for AAA governments, all non government issuers in a currency issue and trade in the secondary market at a premium to the government curve. For a given maturity, this *credit spread* reflects the extra risk investors take by buying a bond not issued by the government, so issuers' curves sit above the government curve. In EUR, the various governments are perceived by the market as having different levels of risk: the Italian government typically trades at a few basis points spread over Germany for instance, so here some governments have a credit spread.

> *Exercise.* Consider medium-term bonds issued by Deutsche Telecom. These have traded anywhere from less than 100 to more than 400 bps over the German government curve over the last few years. If Deutsche Telecom were to cease operation, the majority of Germans might not be able to make or receive telephone calls. Do you believe the German government would let that happen? If not, would any government intervention offer some measure of protection to bondholders? Extend your argument to national champions in other industries.

The *swap* or *Libor curve* (dashed) sits over the government curve (solid), with the difference between them known as the *swap spread*. The Libor curve is simply the curve that we derive from an analysis of interest rate swap (IRS) prices and related instruments rather than bond prices. Since many financial institutions fund at levels around Libor, the Libor curve is in some ways more fundamental to banks than the government curve.

There are a number of standard ways of understanding size of the swap spread. One common one is to say that the Libor curve represents the cost of unsecured funding for a high-credit-quality bank and thus to explain the swap spread as compensation for credit risk. This suggestion sidesteps the fact that most swaps are done on a collateralised basis and the experienced losses on swaps between international banks are very small so the credit risk involved is miniscule. Instead perhaps the swap spread represents a *convenience or liquidity yield*, and/or it may include the return on the *regulatory capital* required against a swap. In the end, the swap spread is just given by the level at which market participants are willing to engage in swaps versus that at which they will trade government bonds. As such, it is *volatile*: the 10-year swap spread in dollars has moved by more than 50 bps over the course of a month in the past.

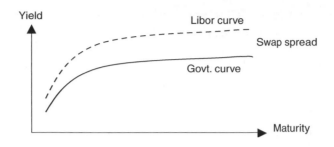

> *Exercise for you.* Check that you remember how to build the zero coupon curve for governments, Libor and credit risky issuers.

1.1.5.11 How bonds return principal
The most common bond structure is a return of 100% of the principal value at maturity: this is known as a *bullet* structure. The alternative is an *amortising* bond where principal is returned along with interest during the life of the bond.

1.1.5.12 The universe of debt instruments
A huge diversity of debt securities has been issued, so any survey of them is necessarily either sketchy or extremely lengthy. Here we give a high-level overview.

One estimate is that there are over 500,000 different debt securities in issue. Most of those are small and illiquid offerings. There are a number of well-known names with very liquid bonds: the G10 governments, U.S. agencies, international banks and some other frequent borrowers. But these are a small percentage of the total number of issuers if not the total market turnover: many bonds are obscure and infrequently traded.

1.1.5.13 Money market instruments
The money market comprises high-credit-quality, short-term (less than 1 year), and large-denomination interest rate instruments.

Many of these instruments are quoted on a *yield basis* including

- *Money market deposits*, or depos. These are ordinary bank deposits, usually in large size, and deposited for a fixed term or on a rolling overnight basis.

  In many countries, deposit taking institutions—banks—have a special status: they are typically regulated, for instance, both in terms of their conduct of business and their capital requirements. Part of the reason for this is that deposits are often *insured*. That is, if a bank defaults, some monies deposited at the bank may be partially or fully protected by either the government or a government agency. Deposit insurance may be capped (for instance at the time of writing the cap is $100,000 in the U.S.) or restricted to certain classes of customer.

- *Certificates of Deposit* or CDs. These are bearer instruments representing (typically large) deposits at a bank (or occasionally other deposit taking institutions such as a building society). Jumbo CDs have a reasonably liquid secondary market. Unlike depos, CDs are typically not insured.

Money market securities are sold on a *discount* basis: they do not pay interest but instead are sold at a discount to face value. They include

- *Government bills.* These are used to provide short-term liquidity for governments. They are typically issued in a range of original maturities, e.g., for the United States, 4, 13 and 26 weeks.
- *Bankers' acceptances.* These are bank obligations payable on some future date, often created by exporters during the process of international trade.
- *Commercial paper* (CP). This is high-quality, unsecured and short-term corporate debt. Original maturities range from 30 to 270 days, with much paper issued at 30 days. Typically, CP is issued as part of a rolling programme. Corporates that issue CP often pay a fee to a bank for a *CP backup line*: this is a source of cash that they can draw on if they are not able to roll their CP. As the CP market is both a cheap source of funds and very credit sensitive, these facilities are only likely to be used when the CP issuer is in distress.
- There are about 2000 companies that issue CP, but about 75% of volume is paper issued by financial institutions.

1.1.5.14 Medium-term notes

The term *bond* tends to be used to refer to longer-term paper, that is, securities with an original maturity of 2 years or more, whereas *note* tends to refer to shorter-term paper. There is, however, no precise definition and some practitioners refer to all debt securities as bonds regardless of their maturity. The term medium-term note (MTN) typically denotes a security with an original maturity typically between 1 and 5 years, often offered as part of a continuous programme of issuance.

1.1.5.15 Treasury notes and bonds

Nearly all governments issue some form of intermediate and long-term government securities. There is massive liquidity in many of these government bond markets, especially the United States (the *Treasury bond market*), the better credit quality Eurozone issuers (including French OATs, German bunds and Italian BTPs) and the United Kingdom (gilts).

Exercise. Review the following list of high-credit-quality countries: Australia, Austria, Bermuda, Canada, Denmark, Finland, France, Germany, Iceland, Ireland, Isle of Man, Liechtenstein, Luxembourg, Monaco, the Netherlands, New Zealand, Singapore, Spain, Sweden, Switzerland, the United Kingdom and the United States.

Classify their bond markets into very liquid, fairly liquid and illiquid. What criteria did you use? Are there any countries on the list that surprise you? How does credit quality relate to liquidity?

1.1.5.16 Repo

Originally in the government bond market (and soon thereafter in other bond markets), it became possible to *finance* a bond purchase using a *repo* or repurchase agreement. Repos allow market participants to get the economic benefit of the possible returns on a bond without having to find the entire purchase price immediately. The idea is that if the bond has good enough credit quality, it can act as the collateral for a loan, therefore allowing most of its purchase price to be funded. In a repo:

- The seller agrees to sell a bond now and repurchase it at the agreed repurchase price at the repurchase date;
- The buyer agrees to buy now and sell at the repo price at the repo date. They have made a loan to the seller with the bond as collateral;
- The intermediate cash-flows on the bond during the repo (e.g., coupons) go to the original holder. The seller therefore has the benefit of coupons and (when they buy the bond back) price changes in the security, but they do not own the bond and it is not on their balance sheet;
- If the seller defaults, the buyer can *perfect* (gain ownership of) the bond and sell it giving them a good measure of protection against the seller's credit risk.

Typically in a repo *fungible** collateral is returned. Overnight, term or open repos are often all available.

The repo price and duration determine the *repo rate*. This is usually lower than Libor in the currency concerned, although of course it depends on the quality of the collateral chosen.

The most liquid and highest quality bonds in a currency typically all repo at the same rate, and this is known as *general collateral* (GC). In USD, the Libor/GC spread is typically some teens of basis points, but it can be fairly volatile. Bonds which repo at a different level from GC are said to be trading *special*.

In a repo, the full value of the collateral is not lent: for instance, in a government bond trade, we might lend 98% of the value of the bond, giving a 2% *repo haircut*. The size of the haircut is intended to reflect the volatility of the price of the collateral and its credit quality, so for riskier collateral 10% or even 25% haircuts are not uncommon.

The bond's repo haircut must be supported by other borrowing by the repo seller: in effect the haircut constrains the leverage available by repo as the seller has to find at least the haircut amount in cash. A 2% haircut therefore corresponds to a 50 times leveraged position.

1.1.5.17 Corporate bonds

Many corporations issue bonds to fund themselves. Typically, long-term funding is obtained by issuing unsecured senior bonds of maturity between 5 and 30 years: these often form a

---

* Two securities are fungible just when either is deliverable for the other. Thus if we are expecting 100 shares of DaimlerChrysler, we do not care which 100 they are, as long as they are the right class of share. Similarly, we do not care which 5% OAT of 25th October 2016 we get back on a repo as long as the face value is correct.

major part of a firm's funding. There may well be one or more *benchmark* bonds: relatively large and liquid issues from which the firm's credit spread can easily be derived.

In addition to senior unsecured debt many other forms of bond can be issued including;

- *Subordinated bonds.* These are unsecured debts with a lower priority claim than senior bonds. They are often issued either by firms which need a more equity like capital instrument for regulatory purposes, such as banks or insurance companies, or by those that need them for ratings agency purposes. There can also be tax advantages to certain subordinated instruments.

- *Mortgage or other secured bonds.* Here the bond is secured by a claim over specific collateral owned by a corporation, such as a building.

Currently [for reasons we shall see in section 11.1.6], there is a vogue for very long term instruments, with some issuance at fifty and even one hundred years.

1.1.5.18 Eurobonds

The term *Eurobond* refers to the largest unsecured corporate bond market. The origins of the term are historical: in the 1960s, it became expensive for tax and regulatory reasons for U.S. corporations to issue debt in their home market, so they turned to Europe. Hence, internationally issued bonds became known as Eurobonds, and the Euromarket (which is by no means entirely within Europe) became the dominant liquidity pool for corporate issuers. This process was further aided by the establishment of two efficient electronic settlement providers for the Euromarkets: Euroclear and Clearstream. Today Eurobond secondary market liquidity ranges from excellent (for the benchmark issues of some frequent borrowers) to completely non-existent.

1.1.5.19 Floating rate notes

FRNs are bonds that have variable rates of interest through their life. The coupon usually pays a fixed spread over a reference index, such as 3- or 6-month Libor.

1.1.5.20 Asset-backed securities

Many different types of bonds fall into the category of asset-backed securities or ABS.* In each case, the promise to pay on the security is backed by a claim on some asset, that is some *collateral* for the security. The process of gathering this collateral and issueing a security is known as a *securitisation*. Common collateral types include collections or *pools* of:

- Retail or commercial mortgages;
- Credit card receivables;
- Trade receivables;

---

* The ABS markets are too fast moving and too fragmented for there to be a book which covers everything and is up-to-date. Lakhbir Hayre's *Salomon Smith Barney Guide to Mortgage-Backed and Asset-Backed Securities* is useful for MBS, and John Deacon's *Global Securitisation and CDOs* is another possible reference.

- Lease or equipment finance receivables;
- Bank loans or auto loans;
- Corporate bonds;
- ABS (that is, ABS of ABS, or *eating your own lunch*);
- Cashflows from an entire business, that is, *whole business securitisation* [of which more in section 10.3.3].

The first step in understanding an ABS is to understand its collateral and the cashflows generated from that. There may be for instance *prepayment risk* if the underlying collateral can prepay: we might get our money back early. Second, the nature of the securities issued must be understood and their rights to cashflows generated from the collateral should be analysed.

[The technology of securitisation used to produce ABS is discussed in Chapter 5 and in Chapter 10 we look at particular asset backed securities in more detail.]

1.1.6 Foreign Exchange Markets

Over a hundred currencies in the world, there are three which are highly important for international commerce and finance: U.S. Dollar (USD), Japanese yen (JPY) and the Euro (EUR). A few more are of secondary importance including GBP, CHF, CAD, SEK, HKD and AUD. Each currency has a three-letter code. In addition, market practitioners are often concerned with assets in various emerging market currencies including ARS, BRL, IDR, INR, MXN, RUB, THB, TRY, TWD and ZAR.

A *cross* is a pair of currencies for which an exchange rate is quoted. Thus USD/GBP is a cross where at the time of writing you can sell USD and buy GBP at the rate of $1.8647 per pound purchased (in sizes of $10M) or buy USD/sell GBP at $1.8644 per pound sold.

Most currencies trade via the dollar, so for instance a Swedish exporter wishing to change Hong Kong dollar (HKD) receipts into his home currency would not find much activity directly in SEK/HKD; instead the HKD would first be quoted in USD and then those USD in SEK. This of course reduces the problem of quoting on 100 currencies from providing $100 \times 100 = 10,000$ quotes (or 4950 exploiting symmetry) to that of providing 99, a considerable simplification. The exception to the rule of going via USD is certain EUR crosses: EUR/JPY and to a lesser extent EUR/Scandi, EUR/Swissy and EUR/emerging Europe are all fairly liquid.

Exercise. What FX cross is known as cable? In what way is it different from all the other crosses?

1.1.6.1 Characteristics of the spot FX market
Typically we find the following:

- Massive turnover in the major crosses;
- Very tight bid/offer spread and the ability to trade tens or hundreds of millions of dollars spot very quickly;

- The market is both a screen and a broker market. There is a simultaneous auction on both bid and offer sides, so the best counterparty to buy USD sell CHF is typically not the best counterparty for the reverse trade;
- Bid/offer spreads are so tight that many market participants believe that few institutions make much money in spot FX: the exception is emerging market crosses where spreads are wider and there are fewer market makers;
- The FX market is a global, with good liquidity 24 hours a day in the major crosses. This means that global books are common, with institutions passing trading from their office in Tokyo, Singapore or Hong Kong to London, then to New York, then back to Asia.

1.1.6.2 Trading activity

It is estimated that less than 3% of the total flow of the FX market comes from global trade. The rest is either proprietary position taking or liability or asset hedging. (Remember that if a risk position goes from one party to another via five hedgers, it might generate six times the original notional in hedge activity.)

It seems to be a feature of some FX crosses, notably USD/JPY, that the spot rate is relatively stable for long periods of time then large moves happen in a few days or less. This may partly be due to central bank intervention. In any case, the reality is often not a random walk but rather a lengthy snooze followed by a short period of panic.

1.1.6.3 Yen carry trade

A famous example of a trade exploiting FX spot rate stability is the *yen carry trade*. This is a position many hedge funds and others have held for some time.

The basic trade is:

- Borrow in yen at the low JPY Libor (say 1%);
- Turn the JPY into a currency with higher rates such as USD in spot market;
- Invest the proceeds in assets with a good yield (perhaps using leverage as in a repo) for a fixed term;
- At the end, turn the USD back into JPY to repay the borrowing.

This strategy makes money providing that the USD/JPY spot rate at the point of repayment has not moved materially against the carry trader.

1.1.7 Derivatives Markets

A derivative is a financial instrument whose value is determined by, or derived from, the value of one or more other instruments. These are called the *underlying(s)* of the derivative. There are four major classes of derivative:

- *Forwards/futures*: these are agreements to buy or sell the underlying in the future at a fixed price;

- *Options*: the right but not the obligation to buy or sell the underlying at a fixed price in the future [as discussed further in section 2.1.6];
- *Swaps*: an agreement to exchange one series of cashflows for another;
- *Structured products*: combinations of derivatives possibly with securities or other assets.

Derivatives have been known for centuries: in the seventeenth century, for instance, options on tulips were traded in the Netherlands, and rice has formed a derivative underlying in Japan for hundreds of years. Prominent contemporary examples include the following:

- *Equity derivatives* whose underlyings are stocks or the level of equity indices. All the equity indices we have discussed, for instance, enjoy a range of exchange traded equity derivatives;
- *FX derivatives* which depend on currency rates, such as the highly liquid exchange traded FX futures and options traded on all the major crosses;
- *Commodity derivatives* which depend on the prices of standard commodity contracts such as the futures and options available on Brent crude, base and precious metals, and agricultural commodities;
- *Interest rate derivatives* which depend either on the prices of bonds or the level of interest rates. These include government bond futures, interest rate futures, and interest rate swaps;
- *Credit derivatives* which depend on the occurrence of a credit event such as a default on a particular bond or loan.

Many other more complex examples are possible

1.1.7.1 Organised, standardised markets

Many asset exchanges, including most stock exchanges and some commodity exchanges, have developed ET derivatives alongside the cash* market. Thus, for instance, futures and options on the CAC trade on Euronext.

ET contracts are typically standardised to a small selection of maturities (and, for options, strikes). For futures, the shortest maturity contract, or *front month*, is typically the most liquid. This standardisation can bring liquidity, as in many *equity index futures*, or it can result in market participants being unable to find a good hedge to their exposure. Typically, exchange traded products are most useful where a standardised product is established which meets most participants' needs: where the market began with OTC trading or where considerable customisation is important in meeting hedging or risk taking needs, OTC products tend to dominate. The table below gives some examples of common derivatives types and underlyings.

---

* The term 'cash' here refers to the securities market, as opposed to the derivatives market.

Underlying	Contract Type				
	ET Future	**ET Option**	**OTC Swap**	**OTC fwd**	**OTC Option**
Equity Index	S&P 500 future	Option on S&P 500	S&P 500 equity swap	OTC S&P 500 forward	OTC S&P 500 option
Single Stock	Single Stock future	Single Stock option	Single Stock equity swap	Equity repo	OTC stock option
Government Bond	Bond future	Bond option	N/A	Repo	OTC bond option
Money Market/Libor	Eurodollar future	Option on Eurodollar future	Interest Rate Swap	Forward Rate Agreement	Interest rate cap, floor, swaption
FX	FX future	Option on FX future	Currency swap	OTC FX forward	OTC FX option
Credit	N/A	N/A	Credit default swap, total return swap	N/A	Risky bond option

[Many of these derivatives will be discussed in Chapter 2.]

Often it takes some time for an exchange to hit on the right design of contract to maximise liquidity: the recurrent failure of ET inflation futures is a good example of the difficulty of finding a contract design many market participants find attractive.

1.1.7.2 Example

A good example of an ET futures contract is the *long gilt future*.* This is an agreement to buy or sell a *notional* underlying 10-year U.K. government bond or gilt, listed on Euronext-Liffe. (The 'long' in the contract name refers to the duration of the underlying.)

The contract is intended to offer some flexibility, so a range of actual gilts can be delivered into it with maturities between 8¾ and 13 years. (Deliverable gilts must also have certain other properties such as sufficient liquidity.) The contract is for £100,000 of gilt, and expirations are every March, June, September and December.

If the future is held to maturity, the holder will receive an amount of one of the deliverable gilts, weighted to give a reasonable proxy for 6% 10-year gilt according to a formula set by the exchange. Of the range of bonds that can be delivered, the cheapest one is known as the *cheapest to deliver*. Depending on interest rates and the coupons of the available gilts within the permitted range which gilt is the cheapest to deliver will vary from time to time. The value of the option to select from a range of deliverable bonds is typically fairly small, but some market participants monitor the value of this *cheapest to deliver option* in case a good opportunity arises.

> *Exercise.* How would you decide if an ET future offered a good price discovery mechanism for valuing an OTC derivatives contract?

---

* See the *Long Gilt Futures Contract Specification* available from www.euronext.com.

1.1.7.3 Why trade derivatives?

The usual reasons given are:

- *Leverage.* A *leveraged position* is one where a change in a risk factor does not always produce a proportional change in P/L: if you are long £1000 of a single stock and it moves 2%, you make £20 because you are not leveraged. With derivatives, a £1000 investment could produce an investment that changes by any amount between nothing and thousands of pounds for a 2% move.*

- *Customisation.* It is sometimes possible to get exactly the risks you want and not the ones you do not.

- Making money from *flow, position taking,* or *arbitrage.* Like any market, the derivatives markets offer the potential for profit either by acting as a market maker, by taking a position, or by exploiting market anomalies.

Many derivatives trades are also driven by:

- *Regulatory arbitrage.* The capital required to take a position via a derivative may be significantly different from that via another route, or it may be possible to pass on an insignificant amount of risk via a derivative and yet make a significant change to the capital required. [See Chapter 7 for more details of this.]

- *Tax optimisation.* Derivatives may have a different tax treatment from other investments or they may permit tax liabilities to be transformed or relocated, enhancing the user's tax position.

- *Accounting or perception arbitrage.* Derivatives may permit a different accounting treatment for a risk, and they may reduce earnings volatility, or otherwise enhances the perceived attractiveness of an investment to third parties.

- *Funding arbitrage.* Derivatives may permit risks to be taken without balance sheet being used, they may allow off-balance-sheet funding, or they may permit both.

1.1.8 Principal Investment and Private Equity

Private equity or PE is the provision of medium to long-term financing to a company in exchange for an equity stake, usually in an unlisted and so hard-to-trade company. In the past typical PE, targets were young, high-growth companies, but this has changed recently as more innovative financing mechanisms are used together with PE.

The motivation for private equity is that the public equity market sometimes finds it difficult to assess new companies' growth prospects. This is a classic *agency problem*: most investors need to do significant due diligence on these situations, but it is too expensive for everyone to do this separately. As there is no trusted source investors can all go for this information, no investor ends up buying the stock, or those that do put a significant

---

* A leveraged position is similar to a leveraged company in that some level of losses will cause default. For a company it happens when the losses exceed the capital available to absorb them: for a position when the value of the position declines below the amount borrowed to fund it.

discount on it for the information they do not have. A single experienced owner of a large stake can afford to do the necessary due diligence and hence to pay more for the stock than a multitude of smaller players.

The advantages and disadvantages of PE are shown in the table below.

Advantages	Disadvantages
Equity finance, so flexible	PE players are highly return oriented and are looking
No fixed debt service requirements	for early and profitable exits perhaps at the expense of
PE sponsor often cannot force bankruptcy	laying the foundations for long-term growth
Refinance is often possible via a rights issue	Refinance can be difficult if project growth has not been achieved
The better PE players have a lot of experience at helping the management grow the company	The original management can lose effective control to the PE investors
PE as a broad asset class has had excellent returns over the past few years	There are arguably too much money chasing and too few opportunities in PE at the moment

Venture capital developed originally to fund high growth, high risk opportunities in unlisted companies. Typical situations in the early days of private equity were:

- New technology companies;
- Firms with new marketing concepts;
- Spin offs or start ups designed to exploit a new product; or
- Potentially high growth spin offs of physical assets, brands or ideas from existing corporates.

1.1.8.1 Methodology
PE investors typically take a significant, although not necessarily controlling equity stake. They typically prefer situations with:

- Good existing staff. As the industry phrase goes, '*bet the jockey, not the horse*';
- Products or processes which have passed through at least the early prototype stage and are adequately protected by patents or copyrights;
- The potential of an exit within a few years via either an initial public offering or a trade sale;
- The opportunity for the venture capitalist to make a value added contribution to the management and/or funding of the company.

1.1.8.2 Private equity terminology
- Some financial institutions use the term *principal investment* for their PE activities, as the bank's capital is committed (sometimes alongside the bank's clients). One motivation here is to lock in investment banking fees once the private company is taken public or sold. Principal investment activities were very profitable in the high-tech boom of the late 1990s, but this form of proprietary risk taking is going out of favour as the new Basel II regulatory capital rules [discussed in Chapter 7] considerably increase the capital required for them.

- The stages of PE are typically called *seed, early stage, development* and *buyout*. A seed stage business is often just an idea. Many PE players are reluctant to invest at this stage: funding instead typically comes from the management team, bank loan finance, and private individuals (aka angels). An early-stage business is a company that has been in business for a short period of time, but may not have a product ready for market and probably will not have a sales history. Some PE players will get involved at this stage, but most prefer to wait until the business has proved its concept and needs further funding for development or marketing, that is, the development phase. Finally, buyout-stage firms are situations where an existing firm, often a struggling one, is taken private. Sometimes this happens because the management team or others believe that a restructuring is best achieved away from the scrutiny of the public market, sometimes because this is the easiest way of obtaining funding for corporate development or restructuring and sometimes because a dominant owner believes that the public market does not properly value the firm.

- In Europe, the term *venture capital* is used to cover all stages of PE investment. In the United States in contrast, 'venture capital' refers only to investments in seed, early and development stages.

1.1.9 (A Short Section on) Commodities Markets

The principal physical commodities of broad interest to the financial markets are:

- *Base metals*, including copper, aluminium, zinc and lead;
- *Precious metals*, including silver, platinum and palladium (but not gold: technically and for historical reasons, this is a currency, not a commodity);
- *Energy-related commodities and consumables*, including Brent crude and related oil products, natural gas and electricity;
- *Agricultural and food commodities*, including soy beans, pork bellies, and coffee.

A typical commodities market is often based around an exchange which in turn defines a contract specification for futures and options. For instance, the Brent crude oil future trading on the IPE is based on the delivery of 1,000 barrels of crude oil with certain chemical characteristics (of sulphur content, etc.) at Sullom Voe* with an option to cash settle. The front-month Brent crude contract provides a price reference for the prices of oil delivered elsewhere, and for various other oil fractions such as gas oil which typically trade at a spread to Brent.

Markets where the commodity cannot be stored easily—such as electricity—have significantly different dynamics to those where storage is possible. If electricity is cheap,

---

* Sullom Voe is an inlet on the Shetland islands in Scotland. One of Europe's largest oil terminals is situated there at the terminus of several pipelines from the North Sea oil fields. The Sullom Voe terminal can accommodate large oil tankers, and at its peak served nearly 700 tankers per day.

perhaps because it is 3 a.m. and more is being generated than used, we cannot buy it and hope to deliver it four or five hours later when demand is stronger. Equally, if we own the gas coming out of a pipeline now but we do not possess (expensive) physical storage capability for it, we might have to pay someone to take it off our hands. If more gas is delivered than the market expects, this can (and has recently) resulted in spot gas prices going negative.

As you have probably already gathered from the above, there can be a lot of detail to master in the typical commodities market. One needs to understand not only the financial characteristics of the market—the forward commodity curve, option volatilities and so on—but also the physical, political and economic factors which can influence the market. Just taking oil as an example, we might be interested in the weather, since hurricanes can severely disrupt oil production in the Gulf of Mexico, the status of the main Middle Eastern, Latin American and Russian oil fields, the political situation in those countries, the economic situation in the main consuming countries and so on.

A large market participant might also be monitoring the cheapest-to-deliver option on the major contracts. Since the Brent contract has the option to exchange for physical, this involves a knowledge of the whereabouts and ownership of the tankers which could take delivery at Sullom Voe, the precise chemical composition of the available crude and how the price of the contract is adjusted to take into account that composition, and so on.* This large corpus of knowledge means that participants in the commodities markets are often rather older and more experienced that those in purely financial markets: it is not uncommon in London base metal trading, for instance, to meet someone who has been in the markets for 20 years or more.

1.2 TRADING AND MARKET BEHAVIOUR

One of the problems with managing any kind of risk is that crises are infrequent. That might sound perverse, but it is true in the sense that we do not know that something—a trading strategy, a risk system or a spacecraft part[†]—really works until we have experienced the full range of conditions under which it might have to operate. So the sooner we see the full range of conditions, the better: just because things have gone well so far under a limited range of market conditions does not mean that they will continue to do so. Therefore, some experience of the unpredictability of markets and of past events can be useful for the risk manager: it gives some insight into how assumptions and predictions have been challenged in the past.

---

* The diversity of chemical composition of physical commodities means that derivatives contracts are often based on an idealised underlying. If the contract is settled by physical delivery, a series of standard price adjustments is applied on the basis of differences between the actual commodity delivered and the standard underlying.

† I am grateful to Louise Pryor for pointing out to me that pieces of foam frequently fell off during space shuttle launches before the Columbia disaster. It was only when the circumstances changed slightly that the risk this posed was fully and tragically understood. This is a commonplace in engineering situations: things work for years, so we assume they will continue to do so. Then apparent success suddenly becomes failure due to a small change in conditions. One response to this is to have a measure of scepticism about whether any construction— physical or intellectual—can perform well in *all* circumstances.

1.2.1 The Difficulty of Forecasting Market Levels

Economic forecasting is not easy: one only has to look at the performance of even the best analysts in predicting the future value of any financial variable to see how difficult it is. There are many reasons why this is the case:

- The use of old, bad, or no-longer-applicable data, assumptions or models in forecasting;
- Unexpected shocks (the September 11th tragedy is a perfect example);
- The tendency of some analysts to follow trends, perhaps because they believe that it is better to be wrong in a crowd than wrong alone (which in career terms is almost certainly correct).

Exercise. With this in mind, it is worth trying to predict the future in any market you find yourself covering. Reading other people's research and studying the fundamentals will give you an appreciation of how the participants think about the market dynamics. Forming your own view will force you to commit in the same way that some traders have to. Pick a horizon such as 3 or 6 months and try to predict what value key market variables such as price level and volatility will take.

Then come back and see how you have done. This makes a good exercise for any risk group before the summer holidays, with a prize at Christmas for the best performance.

1.2.2 The Anatomy of a Market Crisis

The Brazilian crisis of 1999 provides a good insight into a disrupted market. The local context in the years before the crisis was as follows:

- The Brazilian currency, the REAL, was *managed*: it did not float freely, but rather had a level set daily by the government on the basis of a constant rate of decline against the dollar. This was defended by high local interest rates.
- The country had a large and growing current account deficit, but a combination of privatisation and foreign direct investment was expected to cover more than 30% of current account deficit. Brazil had been relatively successful at attracting foreign investment ($13B in 1996, $16B in 1997); however, much of this was highly mobile capital.
- The country was politically stable.
- There had been an expansion of credit domestically, with individuals and corporations both taking on substantial amounts of debt.

1.2.2.1 What went wrong?

There were two main factors which determined the character of the crisis:

- Domestically, there were adverse price shocks between 1997 and early 1999, with exported product prices falling 16%.

- There was a general decline in risk appetite for emerging markets amongst international investors following the Russian default. As this is important background, we turn to it next.

1.2.2.2 International context: Russia

The backdrop to the events in Brazil is formed by the Russian crisis in late 1998. In the early 1990s following the break-up of the Soviet Union, Russia inherited a substantial amount of debt from the old Soviet Union and its satellite states based on the old artificial exchange rate.* Tax collection was poor, and the Russian state started to have difficulty finding enough cash to service its debt. International confidence in Russia began to fall, and the Russian Central Bank disagreed with the Duma (parliament) over how to proceed. This caused further jitters, and the market began to demand a higher and higher yield on Russian government bonds. The Rouble fell, and an attempt by the Central Bank to defend it resulted in considerable depletion of its foreign currency reserves. The stage was set for a crisis and one duly arrived: on August 13th 1998 the Russian stock, bond, and FX markets collapsed.

There was little choice at this point: on the 17th, the government floated the rouble (resulting in a substantial devaluation), defaulted on its foreign debt, halted payment on rouble-denominated debt (primarily short-term securities known as GKOs), and declared a 90-day moratorium on payment by commercial banks to foreign creditors.

1.2.2.3 Back to Brazil

It was in this context that investors started to look at Brazil afresh.

The Brazilian economic strategy assumed that international investors would finance its public account deficit while it made economic adjustments necessary to move into surplus (a similar assumption applies in the United States today). Both the public debt and the budget deficit increased steadily from 1996 to 1998, and as a result, the current account deficit increased by more than 50%, placing increasing stress on that assumption.

In late 1998 after the Russian event and with concerns about the predicted current account deficit, foreign investors started to withdraw capital from Brazil. A temporary respite came in August 1998 when Telebras, the national telephone company, was split up and privatised, generating funds for the government. However, the speed of capital flight was so great that Brazilian foreign currency reserves declined from almost \$75B after the privatisation to less than \$50B in late September 1998.

---

* This phenomenon—long-dated debt issued in times when political circumstances were rather different—is a good indicator of political risk. Governments tend to be less willing to repudiate debt they themselves have issued: debt issued by their predecessors, especially those of a different political persuasion, may be less likely to be honoured.

It was also presidential election season. In October, a few weeks before the first round of elections, the Government officially announced that it was negotiating with the IMF for a bail-out package:

- A tight monetary policy was agreed, and interest rates were increased to approximately 40% in mid-September;
- An aid package from the IMF, multilateral organisations and G7 countries totalling more than $40B was pledged to Brazil;
- Brazil also agreed to maintenance of the pegged exchange rate policy.

Despite the obvious unpopularity of these measures President Cardoso won the first round of the elections, and the situation appears to be going, if not well then at least, more smoothly. Interest rates were even allowed to decline slightly in November.

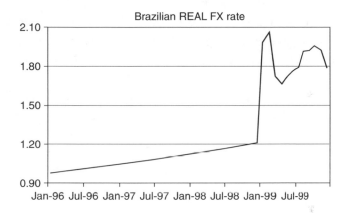

Then, perhaps with an eye to the elections, the Brazilian Congress rejected the proposed IMF fiscal adjustment. Soon afterwards in January 1999, the governor of Minas Gerais (one of Brazil's more important states) threatened that his state would default on its local currency debt. This move may have been motivated by political considerations or perhaps by personal jealousy between the state governor and Cardoso. It had little direct effect since not much state debt was held internationally. However, foreign investors saw it as indicative of increased risk, more capital left Brazil, and foreign currency reserves started to drop at around $1B a day.

1.2.2.4 A devaluation becomes inevitable

This speed of decline was unsustainable. On 15 January 1999, the Brazilian REAL was allowed to float and the classic pattern of overshoot was observed: the currency fell by nearly 50% then recovered significantly.

A sovereign default was averted, and the Brazilian economy went on to recover, with strong GDP growth in late 1999 and 2000.

Exercise. You might find it helpful to examine the detailed history of this crisis, or any other market upheaval that interests you in more detail, both macroeconomically and in terms of the market moves experienced and their impact on participants. Then try to answer the following questions:

— Was the crisis inevitable?
— Did the international capital markets over-react?
— Was the IMF helpful to Brazil in retrospect?
— Was the Russian crisis an important element in Brazil's predicament or would things have worked out similarly even if Russia had not defaulted beforehand?
— Should Brazil have defaulted on its sovereign debt?
— Was it obvious that Brazilian assets were a screaming buy after the crisis?

1.2.3 Current and Past Markets

As I write, the past few years have been relatively benign in the capital markets. Since the market falls of 2000–2001, we have seen rising equity markets, a low and stable interest rate environment, and tight credit spreads. There are some signs that market conditions are becoming more choppy with a minor equity market correction, rising default rates on some types of mortgage and increasing concerns about inflation (at least from some commentators).

This relatively calm environment should probably give us pause for thought. Although the future never replays the past exactly, it is worth having a view to history since calm usually precedes the storm. We will review the history of equity markets in some detail and touch upon interest rate market history.

1.2.3.1 Equity market history

The claim for the first equity market, like the claim to be the oldest pub in London, is shrouded in historical controversy. It is certainly true that by the seventeenth century, the idea of the joint-stock company with publicly issued shares traded on a secondary market was well established: before this, the trail is harder to follow.

The industrial revolution gave a major boost to the development of the markets, and many of the major stock exchanges were established in the decades around 1800. Securities trading developed during the nineteenth century and stock broker became an established profession.

The early twentieth century was a time of considerable volatility for equities. In the United States, for instance, stocks entered the 1900s trending higher and rose 35% in less than 3 years. They swung wildly, peaking around 1909 and then fell heavily. By late 1921, the market was about 40% lower. Then in the bull market of the 1920s, stocks returned more than 400%, a situation that ended in the great market plunge of 1929. The size of the crash is well known: the DJIA went from 381 in early September 1929 to a low of 41 in July 1932. It did not recover its pre-crash levels until November 1954.

Perhaps less famous than the big ups and downs of the market are its doldrums. During the immediate post war period, for instance, the Dow did very little. The market did have a big surge from the early 50s to the mid-1960s, but there was then another stagnant phase.

By 1979, stock prices were extremely cheap: the average P/E ratio of the S&P 500 was around 7. This set the market up nicely for a bull market through to 1987. The pattern should be familiar by now: this 8-year rise was followed by a sharp correction. Markets around the world fell on Black Monday, 19 October 1987: the FTSE 100, for instance, was down 23%.

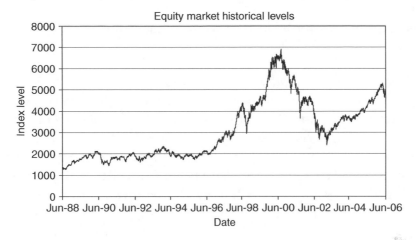

More recently, most major equity markets show broadly the pattern above. Again, we see typical equity market features of long periods of range bond markets, then rapid growth followed by a correction. In the illustration we have:

- A rather boring period from 1990 to 1996 where the market hardly moved at all;
- Rises from 1996 to 1998, accelerating into the
- 'Irrational exuberance' (to use Greenspan's resonant phrase) of 1998–2000;
- Followed by sharp falls with partial recoveries 2000–2002;
- Then a slow but accelerating rise to mid-2006;
- With a very recent minor correction.

Several other equity market trends are worth bearing in mind:

- Over the very long term, so far, equity investments have outperformed fixed-income ones. However, that term needed for outperformance has sometimes been quite extended—if you bought equity at the wrong point, you might have had to wait more than 20 years—and the folkloric outperformance of equity might be challenged at some point in the future.
- Equity market correlation has been broadly increasing over time. That is, globally equity markets tend to move together more than they used to. This makes sense when we think of increasingly globalised investment practices and the growth of dominant equity market liquidity providers (that tend to be global firms). But it is troubling: *diversification* across markets seems to work less well than it used to.
- There is a fairly good correlation of volume with market level: when markets are going up, people tend to buy more, increasing activity. Retail investors in particular stop trading during periods of falling or level markets.

> *Exercise.* Consider a simple client facilitation equity brokerage desk. The desk has no positions at the end of each day.
>
> — Why is its P/L nevertheless strongly dependent on the performance of the market?
> — What does that tell you about the beta* of an equity broker's stock?

1.2.3.2 Volatility and the VIX index

If we look at the daily price returns of an equity market, we find variability. Some days, the index makes money: some days, it has a negative return. *Volatility* measures how big these swings are on average. Suppose we plot the distribution of returns (or, better, since we should compare an asset's returns to those of the risk-free asset which grows exponentially, the natural logarithm of the returns), we find a bell shape. Volatility measures the width of the bell: the higher the volatility, the more likely large positive or large negative returns are.

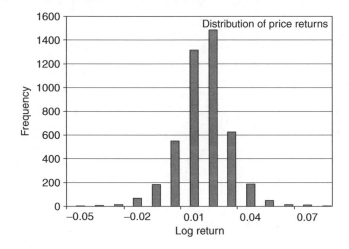

The exact shape of the picture formed by plotting the log returns depends on when we look and for how long: if we were to go back to 1987, for instance, we would see an event right out in the tail of the distribution corresponding to the big fall of that year, whereas the boring early 1990s would produce a tighter bell than the abandon of the latter half of that decade.

The CBOE volatility index (VIX) is a key measure of U.S. equity market expectations of near-term volatility (as measured by S&P index option prices). Since option prices tell us about the market's expectation of the cost of hedging, they measure expected future market volatility. [We discuss this further in section 2.1.4.]

The alternative name of the VIX is the *fear gauge*. This is because spikes in volatility tend to occur when the market falls. Therefore, the VIX is anti-correlated with most other indices: for instance, the correlation with S&P tends to be in the range −0.5 to −0.7.

---

* See Mark Grinblatt's *Financial Markets and Corporate Strategy* for details of beta factors, the CAPM and other standard topics in the theory of corporate finance.

After a choppy period at the turn of the century, the VIX has been below 15% for most of 2005 and into early 2006. However, recently spikes have been seen, perhaps indicating a return of fear to the market.

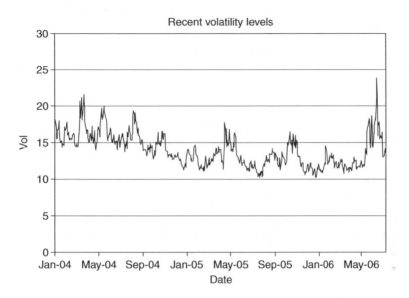

1.2.3.3 Equity derivatives innovations

One of the earliest derivatives was the *equity warrant*. This is an equity call option, packaged as a security, and traded on a stock exchange. One of the first examples was a warrant on ATT, listed on the NYSE in 1970. There are two common forms of warrant: in both cases the holder has the right but not the obligation to buy stock at a fixed price.

- *Corporate warrants* are issued by the underlying company, and if exercised new shares are issued by the company;
- *Covered warrants* are issued by an investment bank. Here the issuer would typically have to purchase some shares in the secondary market to hedge the warrant (and indeed some exchanges require the issuer to own these shares before issuance is permitted): these shares are known as the *cover*.

Warrants with a variety of other structures such as put warrants have also been issued.

The trading of warrants was revolutionised by the development of the Black–Scholes formula in 1973. This put it on a (pseudo)scientific footing and gave participants the comfort needed to extend their derivatives trading activities.

In the 1980s, the Tokyo stock market went through a period of dramatic growth, and Japanese equity derivatives trading grew at a similar pace thanks in part to extensive warrant issuance. The 1989 Nikkei crash left a huge volume of call options worthless, but by that point the genie was out of the bottle and the equity derivatives markets continued to develop rapidly despite the setback to the Japanese market.

Another big spurt of innovation in equity-linked structures was driven by retail note issuance in the mid- to late 1990s. Many retail investors were attracted to investments in the equity by rising prices. However, novice investors were reluctant to invest directly in shares given the risk, the complexity of equity investment and the relatively large capital outlay needed to obtain a diversified portfolio. Investment banks began to structure long term products which would guarantee to return the investor's principal and offer some equity market participation. These went under many names, including the *Guaranteed Equity Bond* (GEB) [discussed further in section 2.3.3].

Meanwhile, the development of the single-stock equity derivatives market was assisted by factors including the following:

- *Corporate bond arbitrage.* This became an established investment strategy in the 1990s. CB had become commonplace in the 1980s, and market players became adept at pulling these instruments apart into credit and equity components. It was quickly realised that many CBs offered a cheap way of buying equity options, which could then be either hedged or sold on in the OTC market;
- *Corporate finance* professionals started to use equity derivatives in mergers and acquisitions as a cheaper and more leveraged alternative to cash positions. These situations often produced very large derivatives positions which, while offering a challenge to dealers, also developed their appetite for large deals.

1.2.3.4 Funding innovations

From the issuer's point of view, how do you choose between equity financing and a debt financing? Until the 1970s, a company would usually pick one extreme:

- Common equity; or
- Fixed rate debt.

In debt financing, one could choose different maturities and priorities among secured, unsecured and possibly subordinated debt, but that was essentially the only choice that was available. More debt increases leverage with a lower weighted average cost of capital but no capability to absorb losses: more equity increases loss absorption capability but at the expense of a higher cost of capital and a lower return for shareholders.

A surge in inflation in the late 1970s and an increase in interest rate volatility spurred innovation in financing instruments, filling the gap between these alternatives.

- *Convertible bonds* or CBs were developed. These allowed the holder to choose between receiving their principal in either cash or converting the bond into a fixed number of shares. CBs therefore package up a corporate bond and an equity derivative.
- More complex *subordinated bond structures* were developed, for instance perpetual bonds which could be *called*—repaid on some specific date or dates—by the issuer. These issuer calls often coincided with a stepup in the coupon rate paid, giving the issuer a strong incentive to call the bond. Thus, the structure allowed the issuer to extend the term of the funding if really necessary—giving some flexibility—at the cost of an increase in coupon.

Other developments were driven by tax needs or investors' desire for more tailored debt instruments:

- Some subordinated bond structures were driven by tax optimisation: the aim was to produce something that was debt-like for tax purposes (because interest payments are typically a pre-tax expense, but equity dividends are paid from post-tax income) but otherwise as equity-like as possible.
- *Dual currency* features were introduced into bonds where coupons were paid in one currency and principal in another, and floating rate structures were introduced where the floating rate was not Libor or a prime rate but instead some currency, equity or commodity rate.
- We have already mentioned *securitisation* technology, where a bond is backed by a concrete asset or pool of cashflows rather than, or as well as, a promise to pay from a recognised issuer.

1.2.3.5 Bond markets and the economic cycle

The history of the bond markets is partly the history of the economic cycle since bond prices depend on interest rates, and these are typically controlled by a central bank. There has been broad agreement by central bank policymakers for some years that economic stability can be guaranteed and inflation controlled by the management of interest rates. This conventional wisdom is summarised in the following table:

Point in the Economic Cycle	Inflation Expectations	Central Bank Action	Bond Market Conditions
Overheating	Low, increasing	Increase rates	Bad
Cooling	Overshooting	Rate increases slowing	Improving
Recession	High, decreasing	Rate cuts	Good
Recovery	Undershooting	Rate cuts slowing	Declining

There is no doubt that, by good luck or good judgement, central banks have been adept in most of the major economies at promoting stability. However, as ex-FED chairman Alan Greenspan himself said:

> In perhaps what must be the greatest irony of economic policymaking, success at stabilization carries its own risks. Monetary policy ... will reduce economic variability and, hence, perceived credit risk and interest rate term premiums.*

In other words, the longer the things are stable for, the less risky the market perceives things to be. And this means a large potential for a blow-up when adverse news does arrive.

---

* See *Remarks by Alan Greenspan to the National Association for Business Economics 2005 Annual Meeting*, available on www.federalreserve.gov.

Exercise. Where is the U.S. economy in the cycle currently? Does a rising rate environment imply a falling or rising currency, other things being equal? At what points in the cycle should bonds outperform equities? Examine the recent history of USD rates, EUR rates, the USD/EUR cross-rate and representative U.S. and EU equity markets. Does this history support this conventional wisdom?

1.2.3.6 Credit market conditions

As we might expect from the previous section, credit spreads at the start of 2007 were close to historic lows in some markets, with benign experienced default rates.

During early 2007 there was spread widening in some markets. Doubtless the broad credit market will experience a *credit crunch*—dramatic spread widening and much less freely available credit—at some point. But what will cause the cycle to turn in less clear. One possible candidate is the U.S. mortgage backed security market: in early 2007 concerns were being raised about the quality of some mortgages, and if these jitters contaminate the wider market, the current era of tight credit spreads may end rapidly. [See section 10.2 for further information on mortgage backed securities.]

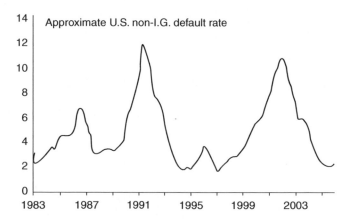

1.2.3.7 Being early versus being wrong

A well-known exchange in financial circles goes:

What's the difference between being late and being wrong?

There is no difference between being late and being wrong. In other words, there *will* be another crisis some time. But if you sell out or go short in expectation of it and it does not happen soon, you lose valuable opportunities and your equity holders will wonder why you have not made as much money as your competitors. Predicting a crisis is easy: predicting when the crisis will hit is the hard part.

Exercise. What looming issue is your favourite candidate for the cause of the next crisis?

1.3 BASIC IDEAS IN RISK MANAGEMENT

Risk means the danger of loss. If I have an exposure to a risk, it means that there are some circumstances under which I can lose money, some hypothetical future loss. So the very concept of risk assumes that I know what the things that I own now are worth: it implies *a concept of valuation*.

Some authors prefer a broader definition of risk: another definition is '*any phenomenon which could affect our ability to meet our objectives*'. But in the end, this usually comes down to money—lost clients, lost reputation, lost freedom to do business as we wish—these all impact our ability to make money. Cash, then, is the common currency of financial risk management.

1.3.1 Types of Risk

There are myriad ways to lose money. We will look at a few.

1.3.1.1 Market risk

The prices of financial assets change. For instance, suppose I own 2000 shares of DuPont. The close of DD last night was $42.58: tonight it is $42.41 (I am still at my desk for the New York close). Therefore, my position has changed in value from $85,160 to $84,820 and I have a (mark-to-market) loss of $340. This assumes that all other things are equal: I am not paying funding on the stock, I do not have margin to worry about, and so on.

Market risk, then, is the risk of loss caused by movements in the prices of traded assets. So far, so obvious. But I am British, so I account in GBP. Therefore, I have to translate those prices in dollars into pounds at the current exchange rate of 0.534445. There is another source of risk in my position: *FX risk*. This occurs whenever any aspect of a position is denominated in a currency that is not the holder's accounting currency.

> *Exercise.* In it better to use the mid market FX rate for measuring FX risk, or bid/offer? The difference is tiny, but it is worth a moment's thought.

1.3.1.2 Aside: funding

The mention of funding in the paragraph above might have struck you as odd. How does funding matter? To see the point, suppose you are a trader and you want to buy a stock. The stock costs money: your firm has to pay out cash to buy the equity. This has to come from somewhere, either from depositors (if you work in a bank) or from the firm's own borrowing. The cost of borrowing this money, then, determines the cost of your position:* before the firm makes a profit on a position, it must first pay its cost of funding. This is a function of interest rates, the bank's borrowing credit spread, etc.

---

* Of course, the firm's systems may not be set up to charge you this cost on every trade, but it is real, and it makes the ability of a position to raise secured funding, for instance via repo, key in determining its all in cost. [We discuss this further in Chapters 8 and 9.]

1.3.1.3 Simple measures of market risk

All of the markets we have discussed—equity, FX, interest rate, credit, commodity—are sources of risk. Within them are many individual risk factors. Moreover, since the pricing of derivatives depends on not only market levels but also volatilities, each underlying's *volatility* may be a risk factor too. Products whose performance depends on more than one asset are becoming increasingly commonplace, and here asset *comovement** matters.

The simplest, if most long winded, way of setting out the risk of a position is simply to list each of the risk factors it has and how much money we would make or lose if they moved. Thus[†]

Position			2000 Shares DuPont	
Risk factor	Equity (DD U.S.)	Sensitivity	$851.60 for 1% move
Risk factor	FX (USD/GBP)	Sensitivity	$455.13 for 1% move

1.3.1.4 Complexity of risk taking

There are many instruments which have complex and non-linear risk characteristics. Their value can change quickly or unexpectedly or both in certain circumstances. For instance, they may have:

- *Highly asymmetric payouts.* If things go well, we make a little, but if an unlikely bad event happens, we lose everything. This is the characteristic of some kinds of options trading;

- *Contingent and higher-order risks.* Here one risk depends on another. For instance, if we swap a callable bond, we may be exposed to interest rate risk if the bond is called;

- *Embedded optionality.* The instrument may appear simple, but in fact it contains features which sometimes make it behave like a derivative.

This means that our job as a risk manager is typically not to diligently find, record and limit *all* the risk factors there may be in a position. Rather we need to understand the important ones, manage those and let the myriad unimportant others go their own way. Thus, we will need to have techniques for aggregating risk, simplifying risk measures and generally being able to see the wood for the trees: these are the subject of the next chapter. Of course, we must also have techniques for determining when those immaterial risk factors become material.

---

* We prefer the term 'comovement' to correlation as it emphasises that we are interested in how things move together in general, not just in a specific model of that collective movement.

[†] It is important to be clear when stating a sensitivity what the size of the move generating the stated P/L is. Here we are using a 1% relative move in stock price and FX rate. There is also a possibility for misunderstanding when the underlying factor is itself quoted in percentage terms: is a 1% move in a 30% volatility from 30 to 31% or from 30 to 30.3%?

> *Exercise.* Select a single risk taker you are familiar with. Try to outline *all* of the risk factors his or her position is sensitive to.

1.3.1.5 Credit risk

Credit risk arises whenever a positive cashflow is expected in the future: if we are expecting to receive cash from someone, there is a chance they might not perform on that obligation, in which case we would lose money because we have to replace that missing cashflow. *Credit risk*, then, is the risk of loss from the failure of the counterparty to fulfil its contractual obligations, perhaps because they have defaulted.

The magnitude of the loss can be gauged by the free market cost of replacing the lost cashflow or cashflows.

Credit risk occurs in a lot of settings:

- *Loans*, where we lend a corporation money on a bilateral basis, expecting them to make payments of interest and to repay principal;
- *Contractual agreements* such as IRSs or purchased options where our counterparty either certainly has to make payments in the future (as in the swap) or may have to (if we exercise an option they have written);
- *Receivables*, where goods are delivered or services performed before they have been paid for.

> *Exercise.* What is the nature of the credit risk you bear to your employer on your salary, pension arrangements, bonus, etc.?

1.3.1.6 Convention: specific risk versus credit risk

We will make the choice that failure to pay on a security is market risk rather than credit risk. This is partly because we are concerned with the effect of creditworthiness on the market price of bonds and credit derivatives, and partly because bank regulators have enforced this (rather artificial) distinction. Thus, *specific risk* (SR) will be used to refer to risks relating to specific issuers of securities such as the default risk of a bond, and *credit risk* will be used to refer to the risk of non-performance on contractual arrangements such as loans that are not packaged up as a security.

1.3.1.7 A taxonomy of credit risks

The extent of credit risk borne depends on how large the cashflow or cashflows at risk are, the probability of being deprived of them and the amount we are likely to receive, if any, if our counterparty does not perform.

The varieties of credit risk include the following:

- *Direct exposure.* We have lent someone some money: will they pay it back?
- *Settlement risk.* Suppose we have traded with the counterparty, perhaps by buying an equity or engaging in a spot FX transaction with them, but there is not a simultaneous

exchange of asset for cash. If we give them something before they give us the other side of the bargain, there is settlement risk because they may default before we get what they have contracted to give us.

A well-known example of this is the failure of the German bank Herstatt in 1974. Herstatt engaged in (among other things) spot FX transactions. One day in June 1974 (as on many other days), the bank sold USD and bought European currencies. As was the convention then, its counterparties paid their side of the trade during the European day. At the end of that day, German supervisors closed down the bank. The expected USD legs owed by the bank were not due until several hours later during the New York business day. Since the supervisor had closed the bank by then, they could not be paid, and the bank's spot FX counterparties were left holding an unsecured claim against the by-now-insolvent Herstatt's assets. Since the failure of Herstatt, there have been a number of initiatives to reduce settlement risk culminating in the development of gross real-time settlement services recently.

- *Pre-settlement risk* occurs when we engage in a transaction where the counterparty may default prior to settlement giving rise to a potential loss. For instance, in an FX forward transaction, we may undertake to exchange $10,000,000 for £5,273,550 in a year's time. In 9 months' time, suppose our counterparty defaults: we still have an obligation to them. Typically, FX forward contracts net on default, so our exposure will be limited to the cost of replacing what is by the time of default a 3-month forward. Thus if 3-month USD/GBP is 0.53 at that point, our $10,000,000 is worth more than their £5,273,550 and we will suffer a loss on their failure to perform under the contract.

- Credit risk can also be *mitigated* by the use of various techniques such as pledging *collateral* or the provision of *guarantees* by third parties. Although this often helps considerably, it does not usually remove credit risk: the risk is then to the joint default of the counterparty and the failure of the credit mitigation mechanism. Thus, if we have an exposure to Firm F whose performance is guaranteed by Bank B, we are concerned with the risk that F and B both default on their obligations: F to pay us and B to make good that payment. Clearly, if the defaults of F and B are unrelated, then this risk is small, but if for instance they are both situated in the same emerging market country, then their *default correlation* may well be high. Since we are talking about two credit events happening, though, which can be wider than default, we prefer the term *credit event correlation*.

- Finally, credit risk can occur in either *funded* or *unfunded* form. In a funded exposure, we have already paid out cash that we are expecting back, as in a loan. In an unfunded exposure, in contrast, we are expecting a cashflow in the future either with certainty or if some event happens.

- That brings us to our final distinction: *contingent* cashflows are those whose presence or size depends on some other market factor; *fixed* cashflows are known with certainty in terms of their size and timing.

The table below shows examples of credit exposures in each part of the taxonomy.

	Funded Exposure	Unfunded Exposure
Unmitigated fixed	Ordinary loan	Bought protection on credit derivatives
Unmitigated variable	Prepaid swap	FX forward without collateral
Mitigated fixed	Mortgage	Back to back credit derivatives
Mitigated variable	Prepaid swap with collateral	FX forward with collateral

1.3.1.8 Current replacement cost

Clearly, if we have an unmitigated fixed cashflow that is at risk, the cost of replacing it is just its PV. What if it is a contingent cashflow? Then to see how much it might cost us to replace the cashflow, we need to analyse

- How big is the expected cashflow today? This is clearly our starting point: the mark-to-market of the instrument under which the cashflow arises.

- How does the size of the cashflow vary? Clearly, in our FX forward example, the size of the difference between the $10,000,000 we are paying and the £5,273,550 they are paying depends on the *volatility* of the USD/GBP FX rate. The more volatility we have, the more the instrument can move during its life, and hence the more we might be owed by the time our counterparty defaults.

- The period during which the cashflow can change. In general, we have exposure for some time interval during which the volatility of the exposure can act. This period includes both the duration of the actual exposure and how long it would take us to replace the exposure. For instance, consider a 2-year FX forward GBP versus INR on £50M. Since INR is a relatively illiquid currency with a managed float, this will be a non-deliverable forward (NDF),* and since the size is fairly large and the maturity fairly long, at least for an INR NDF, it may take us some time to find a market counterparty to replace this trade with. This is especially the case since presumably the default of our counterparty will cause some market disruption, especially as there are only a rather limited number of players in this market. Therefore, we are faced with the possibility of having to replace our NDF in a volatile market with challenging liquidity: in this situation, it might take a week or more to get the trade done.

The quantification of credit exposures is discussed in much more detail in Chapter 5. Now we return to our survey of risk types.

1.3.1.9 Liquidity risk

Liquidity is the ability to meet expected and unexpected demands for cash. *Liquidity risk*, therefore, is the risk that we will not be able to do that—that we will face the requirement to pay cash and be unable to do so.

---

* An NDF is one that is cash settled on the difference between the agreed rate and the spot rate at maturity of the forward rather than by the exchange of gross amounts. Typically, it trades in USD or EUR versus an emerging market currency, often one with limited convertibility.

This does not necessarily mean that the firm is insolvent: liquidity risk can occur when we have more assets than liabilities, but when we are unable to liquidate those assets in time. This is an important risk class for many financial institutions precisely because they often have illiquid assets and more liquid liabilities.

The classic example is a *run* on a bank. Consider a simple commercial bank: it has some equity capital and some retail deposits as liabilities and some long-term loans as assets. If depositors lose confidence in the bank, perhaps because of adverse publicity, and demand their deposits back, then the bank may be unable to liquidate its loans in time to meet their claims, and thus fail. The actions of some depositors thus cause a bank which might well be solvent to default.

Bank runs used to happen with depressing frequency and their occurrence was one of the motivations for bank capital requirements. [Chapter 9 examines liquidity risk and its management in more detail.]

1.3.1.10 Operational risk

Operational risk is one of the less well-defined risk classes. One starting point is a regulator's definition:*

> The risk of direct or indirect loss resulting from inadequate or failed internal processes, people, and systems or from external events . . . strategic and reputation risks are not included . . . but legal risk is.

Faced with this definition, one can sympathise with the earlier term *other risks* for this risk class. Operational risk loss can be categorised under the following (overlapping) categories:

- *Internal and external fraud.* Some person or persons either inside the organisation or outside it, or both, have broken regulations, laws or company policies and losses resulted. Insider trading and theft typically come under this category.

- *Employee practices and workplace safety.* These are losses arising from failure to implement required employment practices and include losses under discrimination suits and workers' compensation.

- *Loss of or damage to physical assets.* Natural disaster, act of God and terrorism losses come under this category. Some events in this category—such as fire damage—may be insurable.

- *Clients, products and business lines.* Here losses arise from failure to engage in correct business practice, for instance, via unsuitable sales to clients, money laundering or market manipulation.

- *Business disruption, system or control failures.* These include all hardware-, software-, telecom- and utility-failure-related losses.

---

* See the operational risk section of www.bis.org/bcbs/.

- *Execution, delivery and process management*. This is a wide category including data entry issues, collateral management, failure to make correct or timely regulatory or legal disclosures, and negligent damage to client assets.

Clearly, both within operational risk and between it and other risk classes, there are, however, many definitional ambiguities. For instance, are fraudulently obtained loans that subsequently default operational risk or credit risk? [Chapter 6 discusses operational risk definition, reporting and management in more detail.]

1.3.1.11 The ubiquity of operational risk
The history of large losses in financial services firms serves to emphasise the importance of operational risk. For instance, consider the following events:

- Barings' $1.2B loss due to the activities of Nick Leeson;
- Metallgesellschaft's $1.3B loss on oil derivatives;
- LTCM's $4B loss on a number of different areas of investment management [discussed in section 6.2.2].

In just three examples, we have come up with losses roughly equal to the GDP of Namibia, and all of them involve some measure of operational risk, mostly centring around inadequate systems or controls, sometimes combined with fraud (as in the case of Barings).

1.3.1.12 Beyond operational risk: strategic and reputational risk
Strategic risk is the risk that the selection or implementation of the wrong strategy will lead to loss of money. This is typically what shareholders pay the board to manage.

A firm's *reputation* is a collection of opinions, past and present, about a firm which are held by stakeholders and others: hopefully, it will include perceptions of integrity, fair dealing and social responsibility. *Reputational risk is the risk that the bank's actions will be perceived by clients, regulators or others to damage its reputation and hence lead to regulatory action, diminution of franchise or other adverse effects.

Although reputational risk is hard to quantify, it can lead to massive losses, often opportunity losses. As an example, consider the recent events relating to private equity opportunities in the United Kingdom.

- First, the context: private equity had become a well-publicised and profitable investment class, particularly following the ground-breaking trades of Guy Hands in the late 1990s [discussed in section 10.3.3].
- A number of investment banks had pursued PE investments on their own account and for their clients. Many of the same firms were also leading corporate brokers and corporate finance advisers to U.K. corporates.
- *The proposed bid*. Early in 2006, an investment bank was leading a PE consortium to buy the U.K. pub operator, Mitchells & Butlers. It was reported that the firm approached M&B with an offer but were rebuffed, the chairman supposedly calling the offer '*hostile and inappropriate*'. The chairman was a prominent figure in U.K.

corporate circles and his remarks were interpreted as damaging to the investment bank's U.K. corporate finance activities since it is difficult not to have a conflict of interest if you are acting for both acquirers and their targets.

- The investment bank subsequently withdrew from the bidding consortium, and it was reported in the press that the senior management at the firm had instructed that the bank's funds should no longer be used for hostile takeovers. Certainly, this withdrawal highlights the need to balance proprietary trading opportunities against perceptions from clients that one might act in a hostile manner towards them. One could see this as a challenge in reputational risk management.

Exercise. A trader in a financial institution buys a corporate bond. The trader pays USD for the bond, holds it for a week and then sells it, again in USD, to a client. The bank accounts and funds in EUR. Outline all the operational and reputational risks you can think of in the transaction.

1.3.2 The Aims of Risk Management

We have seen that extreme market movements happen with some regularity and that financial risk can be taken in many ways, some of them rather complex. There is a long and inglorious history of financial losses resulting from failing to manage these financial risks. Shareholders, regulators and other stakeholders have very little tolerance for bad news. At its broadest, then, risk management is a process to ensure that undesirable events do not occur.

Good risk management requires:

- An understanding of the risks being taken;
- A comprehensive definition of the firm's risk appetite;
- Allowing opportunities to be exploited within the risk appetite;
- But ensuring that risks outside it are not taken.

So specifically there are three components:

- *Risk Measurement*: discovering what risks the organisation is running;
- *Action*, if required;
- *Culture* to ensure that the process works.

Often institutions suffer risk management problems when only the first of these receives sufficient attention.

1.3.2.1 Risk information

For many markets, risks positions are valued every day: they are *marked to market*. In this context, it is important to understand:

- What factors actually drive the daily P/L?
- What could cause a large negative P/L?

This potentially requires measuring a lot of risk factors, only to throw most of them away: you do not know something is irrelevant until you have checked it is.

1.3.3 Sensitivities

A basic measure of market risk is the *sensitivity*. Sensitivities capture the amount of P/L generated by a movement in a risk factor. For instance, the duration of a bond mentioned earlier is a risk factor: it captures the change in a bond's price for a small movement in interest rates.

More generally, we want to know how much the P/L will move if there is a small move in any risk factor. A sensitivity gives us the gradient of the function that relates the P/L V to some market factor S at a point, so technically it is a *partial derivative*

$$\frac{\partial V}{\partial S} \text{ or a finite approximation to one } \frac{\Delta V}{\Delta S}$$

Thus, ΔV is the change in the value of a position if the risk factor S moves by ΔS.

If we experience an actual market move from S_1 to S_2, then the approximate P/L resulting from that change can be estimated to be

$$\frac{\partial V}{\partial S}(S_2 - S_1)$$

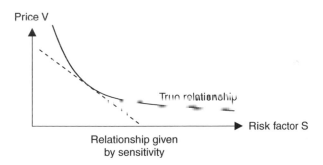

How good an approximation this is depends on

- How big the move is;
- And how far from a straight line the real P/L versus market factor curve is. If the price versus risk factor graph is not (almost) a straight line, it is said to display *convexity*.

1.3.4 Daily Risk Controls: P/L, Limits and P/L Explanation

We have seen that the idea of risk is inextricably intertwined with a concept of valuation, so one set of daily risk controls will ensure that we know the P/L and how it has arisen. Others control various aspects of taking risk.

1.3.4.1 Daily P/L

The *daily P/L* produced under a valuation paradigm is a key risk control. It should not just be a single figure giving the profit or loss on a portfolio. Rather we should be able to find out:

- The contribution to the P/L of each instrument in the book* in their native currencies;
- Together with P/L caused by funding;
- FX movements;
- And cash received or paid.

1.3.4.2 P/L explanation

Once we have the P/L broken down into these components and we know

- The sensitivities of each instrument in a portfolio; and
- How the underlying risk factors moved overnight

we should be able to *explain* the P/L. That is, we can reconcile

- The *Actual* P/L on each instrument with
- The *Predicted* P/L based on the sensitivities and the observed moves in the risk factors plus
- Any *New Deal* P/L resulting from putting on or taking off positions away from mid market.†

This is important because it proves that our risk information is consistent with our P/L. It does not show that it is correct, but it does at least demonstrate that we are capturing the right factors.

If we are missing a risk factor, the reconciliation between the predicted and the actual P/L will not be possible and alarm bells should start to ring.

> *Exercise.* How would you determine what is an acceptable unexplained P/L for a large trading book?

1.3.4.3 Limits

A risk limit is a constraint on a risk taker which expresses his or her firm's risk appetite. For example, suppose a pension fund hires an outside investment manager to invest some of its assets. It wishes these assets to be invested in corporate bonds subject to a

---

* History has given us the term *book* for a portfolio of instruments recorded together for management purposes. This comes from the same pre-modern era as *front office* for the trading area of the firm: many years ago traders sat in the first office in the firm, with support staff further from the entrance in the *middle* or *back* offices.
† It is a serious warning sign if the new deal P/L is often negative, or if positions are often cancelled or sold below their marks.

risk appetite approved by its board. First, then, it constrains the type of risks that can be taken:

- Only senior unsecured debt can be bought: equities, government bonds, ABS, etc. are not permitted.
- All bonds must be denominated in G4 currencies and from G4-domiciled issuers.
- All FX risk is to be hedged exactly on a cashflow-by-cashflow basis.

Within that area of investments, it further demands that nearly all its cash be invested in high-quality bonds and just bonds:

- No more than 5% by value in cash at any time;
- No leverage, no use of derivatives;
- All investments to have a credit rating of BBB+ or better from Moody's or S&P. In the case of a split rating, the lower will apply.

> *Exercise.* Find the highest spread corporate bond you can meeting these criteria.

When an organisation authorises a risk limit for risk-taking activities, it specifies:

- A risk factor;
- A sensitivity;
- A value for the risk metric that is not to be exceeded, the limit;
- Who or what it applies to.

For example

Risk factor	FTSE 100 Index Risk
Sensitivity measure	FTSE 100 Futures Equivalent
Limit	$\perp$ 100 Futures
Application	European Equity Trading Desk (Mr. V. Risky, Desk Head)

Another kind of limit is the *stop-loss*. This is triggered by a given level of cumulative loss on a position. In a liquidation stop-loss, if the limit is hit, the position must be closed, whereas in a consultation stop-loss, the management must be informed and a positive decision made to continue running the position.

1.3.4.4 Limit utilisation and limit breaches

The actual amount of risk being taken as quantified by the sensitivity is called the *limit utilisation*. Any instance where utilisation exceeds the risk limit is called a *limit violation* or a *limit breach*. There are two common types of risk limit in financial institutions:

- *Hard.* These must never be breached.
- *Soft.* The risk taker must seek authorisation, for instance, from a risk manager or a risk committee, before taking more risk.

Occasionally, firms have both sets: an inner, soft limit and an outer, hard one.

A hard-limit breach should be a serious disciplinary offence, typically leading to dismissal unless there is a very good reason for it; a soft-limit breach without authorisation should similarly be treated harshly. If management flunks this challenge it is difficult to see how they can be said to be meaningfully in charge.

1.3.4.5 Setting risk limits

How do we decide on a limit? For a single factor without convexity, it is conceptually straightforward: the senior management decides how much the firm can afford to lose. The historical and possible future moves in the market factor are examined and, on the basis of a conservative estimate of the possible size of a big move, a limit is set such that the big move gives a loss smaller than the permitted one. This firm-level limit is then allocated down to the businesses on the basis of their mandates and budgets.

For instance, on the basis of historical analysis, we conclude that a 1-day move of more than 30% is unlikely in an equity index. The management tells us that the threshold of pain is $10M. Therefore, we recommend a limit of $300K for a 1% move.

For multiple factors, the problem is comovement. Suppose we want to lose no more than $10M for a move in risk factors S and T. We need to know how movements in S are associated with movements in T. For instance, if a 10% move-down in S is usually associated with a big fall in T, we will need smaller S and T limits than if T rises when S falls. But many financial variables have rather little association with each other in ordinary markets; yet when a market crisis occurs, many things tend to fall together. We will face this problem again and again in trying to understand the behaviour of portfolios.

Risk limits can be divided into the following categories:

- Ordinary, book or desk level constraints on one or a closely related set of market variable, designed to express the firm's risk appetite during ordinary trading such as the FTSE 100 example given above;
- *Aggregate risk limits* designed to express the risk appetite of larger parts of the business; and
- *Stress limits*, designed to constrain losses in a crisis situation.

Exercise.   Find a limit structure for a large firm if you can. Are there any circumstances under which a high-level limit could be broken without any lower levels being breached?

1.3.5 What Do You Own? The Deal Review Process

Most businesses rely on one or more *systems* for P/L and balance sheet, risk information, trade settlement and so on. Typically the system *is* the reality:

- If a position is not in the system, no one knows about it.
- If it is not correctly represented in the system, one or more business functions may have bad information.

This makes it critical to ensure that the contractual arrangement you have as a legal matter matches what is in the system. The control that ensures this is *deal review*: positions booked in the system are independently checked against legal documents (*trade confirmations* in the case of derivatives, *custody records* in the case of securities). If we can reconcile derivatives and securities positions plus cash in the bank versus expected cash movements every day, then we have some reason to believe that the system records are consistent with reality.

Another key control is the *trade recognition process*. If a counterparty fails to recognise your trade, that is a potential sign you might have a mis-booking; rejected or failed trades and unsigned confirms after some period need to be followed up.

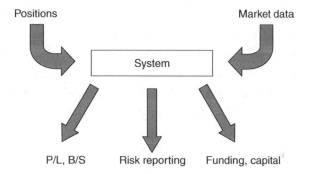

In an ideal world, once we know that something is correctly represented in the system, all business functions will use that data. The more separate copies or different representations there are of a piece of information, the more chance that they will not agree. Thus, though having a single system responsible for position keeping, risk management, record keeping, funding and capital calculation may seem utopian, the price of not having one may ultimately be higher.

1.3.6 Trader Mandates

Organisations usually function best if people are clear about what is and what is not their job. A *trader mandate* specifies the risks that a trader is and is not expected to take. It might be quite narrow for a flow trader in a particular market who is expected to keep his book tightly hedged, or very wide for a cross-market proprietary trader.

For example, for a cash equity desk

Trader mandate	Mr. V. Risky (European equity desk)
Mandated products	Stocks listed on the FTSE 100, CAC 40, DAX, Eurostoxx 50, IBEX, MIB and SMI indices; index futures; equity swaps; and stock loans
Risk	As per limits

whereas for the equity prop desk

Trader mandate	Ms. Al Safe (European equity prop trading)
Mandated products	Stocks, CBs and warrants listed on the London, Frankfurt, Euronext, Milan or Zurich stock exchanges; index futures and option; plain vanilla, Asian, barrier, quanto, and cliquet OTC equity derivatives on mandated underlyings or baskets of underlyings; interest rate futures and swaps (as hedges); equity swaps; stock loans; dividend swaps; CB asset swaps and CB options on mandated underlyings; plain vanilla credit derivatives (as hedges)
Risk	As per limits

1.4 CULTURE AND ORGANISATION

In this section, we review a number of different arrangements for the organisation of risk management and the allocation of responsibilities between it and other groups within financial institutions. Different firms' attitudes towards the management of risk are expressed not only by organisational structure but also in a firm's culture. Although there are various approaches that can be effective, failure to set an appropriate tone here can compromise the effectiveness of an institution's risk management.

There is a wide diversity of financial institutions, from banks, through broker dealers, hedge funds and other investment managers, to insurance companies and others. We map out some of this landscape as a prelude to an analysis of the differences in how they approach risk management. Finally, to give insight into the consequences of failing to get the basics right, we look at several historical risk management failures.

1.4.1 Risk Management in the Broader Institution

There are two basic paradigms for the risk management function:

- *Broad risk management.* In this model, risk management is a large group with responsibility for all risk reporting, marking some or all of the bank's trading books and risk infrastructure. This form of organisation often has fairly formalised reporting structures and procedures.

- *Narrow risk management.* Firms organised this way have smaller risk management groups, typically comprising more experienced individuals. They often rely on other areas, such as finance, for some risk reporting. Individual risk managers are usually product professionals with good market knowledge, and they often make decisions within a relatively lightweight procedural framework.

In the author's opinion, at least, there is no 'best' organisational structure. Rather it is a question of selecting something that is appropriate for the institution concerned.

1.4.1.1 Aside: adding value

Apart from keeping regulators happy, risk management is only truly useful if it adds value to the firm. An incremental extra pound spent on risk management is not spent on developing the business, so it must be justified. More risk management is not necessarily better risk management.

In particular, there is a balance between

Costs	vs.	Quality of staff, systems, data
Timeliness of action	vs.	Committee-based decision making and formalised procedures
Closeness to the business	vs.	Influenced by the business
Saying yes to the right trades	vs.	Saying no to the wrong ones, or 'yes but'
Taking risk with a reasonable expectation of return	vs.	Taking the wrong risks

1.4.1.2 Dialogue

The head of risk management, Dr. R. Careful, is having his weekly meeting with his boss, Finance Director Mr. P. Pincher.

'Don't sit down, Careful, we don't have long. I just want to go over your department budget with you. I've reviewed your estimates for next year, and you need to cut £4M'.

'I can't do that without compromising control'.

'How do you know?'

'We haven't had a big blow-up in 5 years. That's because we have the right infrastructure. Finally'.

'Do you have an elephant protector?'

'What?'

'An elephant protector—something that stops elephants chasing you'. The FD picks up what appears to be an ordinary umbrella and waves it at his employee. 'Ever since I started carrying this I haven't had any problems with elephants.'

'But there aren't any elephants in this town.'

'Exactly.'

1.4.1.3 Organisational structure in large financial services groups

In many large financial services groups, the corporate centre provides central functions to business groups, including risk management. Typically, there will be a board member assigned responsibility for risk issues, often one who also has responsibility for financial and legal functions. The head of risk will report to the board member, and have beneath them a number of functions.

Risk management itself often has a hub-and-spoke structure. The hub contains central functions, with the spokes supporting individual businesses. Central risk functions may include:

- Risk data and risk technology;
- Model verification;
- Country risk;
- Firmwide market, credit and operational risk-reporting groups, etc.

In addition, credit risk management is typically organised into

- Counterparty specialists by type (e.g., hedge funds, banks, large corporates);
- Portfolio management functions, possibly organised by sub-portfolio.

Market risk may be organised into market specialists, etc.

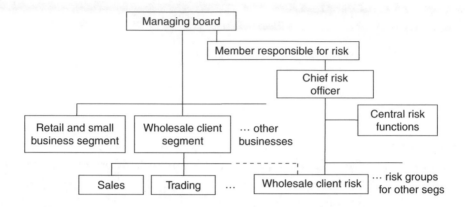

The dotted line shown between a risk sub-group responsible for a particular segment of business and the head of that business is a vexed issue. Obviously if a risk manager is too influenced by the business head he or she supports then their independence may be compromised: on the other hand a risk manager should have some responsibility towards the business they support. Whether the dotted line is there or not, keeping a balance between independence and assisting the business is important for a successful risk group.

1.4.2 Cultural Issues

The term *risk culture* refers to issues around

- The *free flow of information*. Is information on positions and risks freely available to everyone who needs it, or is the process of finding out what the desk is doing like having teeth extracted?
- *Open discussion* of risk issues by all involved parties regardless of status. Can a junior risk manager (or finance professional or a deal lawyer or anyone else) point out that the emperor has no clothes?
- The *incentive structures* within the organisation around risk data, valuation and deal commitment. If a limit violation, a serious data quality issue or a trader violating their mandate is escalated to the senior management, do they support the control function or the trader?
- *Scepticism* about risk and about the comprehensiveness of any risk aggregation process. Does the firm suffer from a single pervasive view of risk (sometimes known as *group think*) or does it embrace a diversity of views?

1.4.2.1 Organisational principles

The first fundamental principle of organisation in financial institutions is *segregation of duties*: there should be a clear distinction between risk takers; staff engaged in controlling activities and information gathering; and those with an oversight or management role.

> *Exercise.* Select a trading business you are familiar with. Suppose you are a criminal intent on stealing the maximum possible from your employer. You can pick any single job you like in trading, finance, operations or the legal department.
>
> What could you do alone? Now select an accomplice. Now how best could you and your partner enrich yourselves at the expense of your employer?

The next equally important principle is *respect for others*. In any organisation where some staff can earn a hundred or more times the salary of others, there is the danger of individuals with higher bonuses or status—or simply those with more forceful personalities— riding roughshod over the rest. Not only does this pose reputational risk in many countries because of harassment, bullying and discrimination legislation, it also impedes the free flow of information, demotivates the staff and can worsen group think. Once an organisation acts as if solely controlled by a single domineering member, control may well have been lost and meaningful risk management is very difficult.

You may also wish to consider

- The assumption that the senior management have all the relevant information and know what they are doing can be dangerous. Unless you question assumptions, it is difficult to find out that they have incorrect, out-of-date or partial information or understanding.

- Ivory tower jobs are dangerous. Risk managers typically need to be close to the market to be effective. If they are writing academic papers or implementing the latest stochastic volatility Markov market rate model, how do they know what their business is doing? That is not to say that there is no place for quants in a risk group: just that there is no place for a risk group *only* staffed by quants.

1.4.2.2 Key features of a successful risk culture

The corporate cultures of firms with a successful risk culture diverge widely. However, they often share certain common features:

- There should be a clear alignment of each individual's interests with those of the organisation. If people are compensated for doing the wrong thing, danger inevitably follows.

- Independence, competence and authority of all functions, not just the front office.

- Trust and respect between different functions.

- Active senior management involvement in the definition of the firm's risk appetite and in ensuring that the risk taken remains inside it.

- Enough risk is taken. This is a Goldilocks business: not too little, not too much, just enough.
- Passing the newspaper test. If you are not embarrassed to have the details of what you are doing published in a respected financial newspaper, then either you are deluded about what is going on or things cannot be too bad.

Risk management is not about ensuring that money is never lost: it is about ensuring that when money is lost, it happens because of risks the firm understands, wanted to take and continuously monitors.

> *Exercise.* Review some negative P/Ls you are familiar with. Do the causes of all of them pass the newspaper test? If not, were organisational or cultural issues partly to blame?

1.4.2.3 Committee structure

Firms often use committees to ensure that decisions are taken by all relevant personnel with common information, for instance

- The *risk committee* acts as a forum for the discussion of risk issues, the setting of the firm's risk appetite, high-level-limit definition, the approval of major changes in risk measurement methodology and (most importantly) is the final court of appeal in any decision on risk management decision. There may be additional risk sub-committees by product or risk class in larger firms.
- *New product approval committee* ensures that new products are only traded with management cognisance and approval of issues from all functional areas.
- *Audit and finance committee* deals with accounting policy issues, recognition of P/L, reserving, and adverse audit reports on businesses or processes.

The key point is that the management must be managing (and be seen to be managing): a risk committee that simply reviews reports, rubber stamps limit increase requests and generally facilitates traders doing whatever they want is worse than useless. Worse because without it, at least the traders would be responsible: with a complicit risk committee, the management is responsible without being in control.

1.4.3 A Taxonomy of Financial Institutions

Financial institutions can be classified along a number of dimensions:

- *Legal status.* Is the institution legally a bank, an insurance company, a broker/dealer, an 'ordinary' company, a partnership or something else? Are there material limits on the contracts it can enter into as a result of its status? (Legal entities which are not insurance companies cannot write insurance in many jurisdictions, for instance: non-banks sometimes cannot take deposits.)

- *Domicile, tax.* Where is the legal entity based, and to whom does it pay tax and on what basis?
- *Regulation.* Who regulates the legal entity for its conduct of business and what regulatory capital requirements does it have? Does it have any preferential features as a consequence of its status such as access to the central bank window?
- *Accounting.* What accounting principles are used for the various risk-taking books?
- *Corporate structure.* Many financial institutions have a complex legal entity structure. Banks may organise their activities around branches, whereas insurance companies may have both on-shore and off-shore insurance and reinsurance companies within their group. Some groups even contain banks, non-bank financial services companies and insurance companies under a single holding company.

The next few sections discuss some of the major types of institution.

1.4.3.1 Non-banks
Non-bank financial services companies include

- *Mortgage companies and building societies.* These are typically specialists in marketing and underwriting retail mortgages, usually funded via securitisation. They are not always regulated in a similar fashion to their bank competitors.
- *Consumer finance firms, savings and loans associations.* These also specialise in retail products, typically unsecured personal loans or credit cards. Regulation depends on jurisdiction.
- *Broker/Dealers.* These are U.S. legal entities existing as a result of the distinction between banking and brokerage brought in after the great crash of 1929 and enforced via the *Glass–Steagall Act.* Although many of the Glass–Steagall distinctions have been abolished in the past 10 years, broker/dealers remain regulated differently from banks (by the SEC rather than the FED). Thus, some investment banks are not, in fact, regulated for much of their business as banks but rather as broker/dealers.
- *Industrial loan corporations.* Another feature of U.S. financial services, ILCs are non-banks which are permitted to engage in activities which are bank like, including taking deposits and making loans.
- *Corporate finance advisors.* These are firms with small balance sheets providing advice and consultancy to corporates. In an investment bank, these services would be provided alongside the issuance of primary securities and secondary trading activities: advisors specialise in counsel alone.

Note that some large non-banks such as GE or General Motors may well have significant financial services activity within their groups conducted in either non-bank or sometimes even bank subsidiaries. In the latter case, the subsidiary is regulated as a bank, but obviously its financial stability may depend to some degree on the health of its ultimate parent. A related situation happens when a corporate sets up a *captive* insurance company to provide insurance services to group companies.

1.4.3.2 Banks

- *Pure retail banks.* Most specialise in deposit accounts, a limited range of savings/investment products, credit cards, perhaps mortgages and some retail lending.
- *Retail plus small- and medium-sized enterprises (SME).* Similar to the above, but in addition, these banks provide some business banking facilities, usually to small or medium-sized companies.
- *Retail, SME and corporate.* Serving large corporates typically in addition requires the capability to make much larger loans and the ability to provide a wider range of products and services to clients.
- *Wholesale/investment banks.* These are banks without retail facilities participating in the capital markets as principals and agents for their clients.
- *Full service banks.* These offer full range of services to clients from retail to wholesale including corporate banking.
- *High net worth specialists.* They provide wealth management services to rich individuals.

In each case, the bank's *trading book* contains positions taken with trading intent. This book is typically dominated by market risk. The *banking book*, or loan book, contains often less liquid credit risk positions, although these may nevertheless be actively managed [as discussed in section 5.4.4].

1.4.3.3 Aside: traditional banks

Let us step back 30 years and look at a simplified balance sheet for a traditional retail bank* from those days. The bank takes deposits, makes loans and manages its liquidity on the interbank market. It has assets and liabilities as shown in the following table.

Assets		Liabilities	
Loans	£900M	Retail deposits	£800M
Securities	£300M	Issued debt	£100M
		Interbank borrowing	£200M
		Shareholder's funds	
		Equity	£100M
	£1200M		**£1200M**

Note the bank's leverage. It has £100M of equity capital. Risky loans held are £900M, so its rough leverage (assuming the securities are comparatively much less risky that the loans) is 9:1. A bank like this would concentrate on:

- Managing its cost of funds (how much it has to pay depositors or the interbank market to get cash);

---

* Barbara Casu's *Introduction to Banking* offers a much more comprehensive introduction to the business of banking.

- Managing its loan spreads (the spread to Libor it can charge investors for providing loans);
- Efficient cash management (so it had enough cash to pay any depositors who wished to withdraw their money, but not so much that the yield from longer-term less liquid investment was being sacrificed).

Much of the bank's liquidity is provided by retail deposits, so its reliance on borrowing to fund its activities is small: if it can avoid a loss of confidence by depositors and manages the credit quality of its loan book properly, it can remain profitable at relatively low risk [see Chapter 8 for more details].

1.4.3.4 Investment managers

The varieties of investment managers providing indirect investments include:

- *Hedge funds.* Often structured as an off-shore partnership for tax reasons, the hedge fund is often an opaque risk-taking vehicle.
- *Private equity.* These funds take unquoted equity stakes in a range of PE situations. They may be structured as off-shore partnerships or on-shore companies.
- Traditional mutual funds (U.S.) or funds under the UCITS* directive (EU) including directional funds, tilt funds, funds of fund managers and index funds. Retail fund management vehicles are structured in various ways, primarily driven by tax and regulation. In Europe for instance, they may well be SICAVs (investment companies with variable capital). Specialist funds in various areas might instead enjoy a particular legal structure, such as REITs in the United States.

1.4.3.5 Insurance companies

An insurance company consists of several functions including an *underwriting* section,[†] which decides which risks to insure, how much premium to charge for them, and which of those risks to *reinsure* with another insurer; an *investment* section, which takes the premiums and invests them to meet future claims; and a claims administration section, which adjusts and processes claims made. Different types of insurance companies take risks in a variety of areas:

- *Life insurers.* On-shore companies providing pensions and life insurance services. Insurance regulation is much more fragmented than bank regulation.

---

* UCITS stands for *Undertakings for the Collective Investment of Transferable Securities.* The UCITS directive defines the regulatory framework for certain classes of unit trust or mutual fund-like investment vehicles within the European Union.

[†] The term *underwriting* in insurance refers to the process of deciding which risks to insure and what premium to charge for them: this is distinct from its use in capital markets where it refers to the agreement to buy a primary security issue if it cannot be sold in the market. In retail banking, the use is similar to the insurance one, referring to the decision whether to make a loan or grant a mortgage and, if granted, what spread to charge. Thus, we see that the spread over a bank's cost of funds on a mortgage is similar to the premium paid on an insurance contract: it compensates the bank for the risk it is taking.

- *Non-life insurers.* Similar to the above but providing property and casualty insurance including fire, theft, negligence, liability, etc.

- *General reinsurers.* Either on-shore or off-shore insurers of insurers, the reinsurer acts as a risk-spreading and aggregating vehicle for the on-shore primary insurers. Since these firms only deal with professionals, the extensive marketing and claims management facilities of a primary insurer are not needed.

- *Specialist reinsurers.* Reinsurers who only take risk in a particular area of activity such as catastrophe or workers' compensation.

- *Financial guarantee companies.* These are very-high-credit-quality insurers writing insurance on the performance of financial obligations. A variety of financial insurance called a *bond wrap* is an undertaking to top up bond cashflows if needed so that investors can have more confidence that a bond will pay coupons in a timely fashion and return ultimate principal.

1.4.3.6 Financial conglomerates

A few years ago, there was a fashion in retail banking for emphasising the virtues of *cross-selling*. The idea was that by marketing a wider range of products to the bank's clients, banks could make more money. For instance, retail clients would be sold not just traditional deposit, loan and mortgage products, but also insurance and investment management services. Partly through historical accident, and partly driven by this idea, a number of groups include both banking and insurance components. Examples at the time of writing include Allianz, an insurance company which owns a bank, DKW; Lloyds TSB, a bank which owns an on-shore insurer, Scottish Widows; and ING, a holding company owning both banks and insurance companies.

Firms which derive significant income from both banking and insurance are called financial *conglomerates*. Issues with these types of firm include the following:

- Is cross-selling really profitable enough to justify the complexity of the group?

- Financial conglomerates naturally have parts of the group which generate cash, such as deposit takers, and other parts which have historically needed cash, such as life insurers. The ability to use one part of the group to fund the other would seem a natural advantage of the conglomerate. However, in practice, this is often made difficult or impossible by the interaction between the multiple regulatory frameworks which must be adhered to.

- Given the different accounting methods and regulatory framework of banks and insurers, is an arbitrage available by moving risk around the group? (Although in theory, massive advantages could be obtained by for instance reinsuring the corporate loan book from an off-shore insurance company, it does not appear as if any firms have attempted to affect this arbitrage thus far. The backlash from regulators would almost certainly destroy any benefit.)

- Do analysts and investors understand and give sufficient credit for the conglomerate model?

A counterexample to the growth of financial conglomerates comes from Citigroup: Citi used to have significant insurance companies within the group, the Travellers companies. These were spun off between 2002 and 2005 leaving Citi as primarily a banking group.

1.4.3.7 Key features

There is increasing overlap between the activities of different kinds of financial institution, but key distinctions remain between how they are treated. The table below summarises these in outline: depending on jurisdiction, it might or might not be possible to construct an entity with any desired characteristics (for instance, an essentially unregulated fair-valued accounting insurance company might be possible in Bermuda but not in the United States).

	Non-Bank Financial Services Co.	Bank	Investment Manager	Insurance Company
Type	U.S. broker/dealer, mortgage company, ILC	Retail bank, commercial bank, investment bank	Hedge fund, mutual fund, UCITS	Life insurer, non–life insurer, reinsurer
Typical legal status	Company	Bank or bank with bank holding company	Company or partnership	Insurance company
Domicile	Usually on-shore	Usually on-shore	Mutual fund or UCITS: often on-shore; hedge fund: often off-shore	Insurer: on-shore; reinsurer: often off-shore
Regulation	Varies	Probably Basel	Varies	Varies
Accounting	Varies. Broker/dealers often prefer fair value when possible	Trading book: fair value; loan book: often historic cost	Usually fair value	Insurance accounting

1.4.4 Some Large Losses in the Wholesale Markets

It is always interesting to see how firms sustained losses in the past. We look at two well-known events: the U.S. bond trading losses sustained by Daiwa Bank; and the interest rate derivatives issues at NatWest Markets.

1.4.4.1 The Daiwa Bank loss: history and context

In 1995, Daiwa Bank was a significant but not globally top-ranking bank: by size, it ranked around 33rd. In the 1980s, Japanese banks were buoyed up by a booming domestic market and abundant liquidity. Their strategy included attempting to increase their market share in the securities business. Since New York is a major centre, that meant setting up U.S. branches. However, many of these banks had few natural clients in the United States, so some became heavily involved in proprietary trading. This process continued through the early 1990s, and by 1995, a number of Japanese banks had significant proprietary trading activities outside their home country.

On 13 July 1995, the senior management in Daiwa were informed of a trading loss of more than $1B in their New York branch. The source of the information was a letter of admission from the man responsible, Toshihide Iguchi. It was not until mid-September that the U.S. regulatory authorities were informed, and meanwhile the bank made a quarterly report to the New York FED for the quarter ending July.

It appears that Iguchi's first loss occurred in 1984 while trading a U.S. government bond. Rather than reveal the loss, he made the first of over 30,000 unauthorised transactions. In part, it is believed that he sold bonds owned by customers, then falsified custody statements. This process apparently went on for 11 years, during which time he is conjectured to have sold over $100M bonds belonging to Daiwa and over $350M which should have been held in custody for clients.

1.4.4.2 Consequences

The reaction of the U.S. authorities was swift: they ordered Daiwa to close all U.S. banking operations and fined them $340M. Part of the reason for the severity of this punishment was probably the delay in reporting the issue and the inaccurate July regulatory filing: to lose a billion dollars due to control issues is unfortunate, but to fail to report it is even worse.

Mr. Iguchi himself was imprisoned for 4 years and fined approximately $2M. The reputational damage to Daiwa was considerable.*

1.4.4.3 Why was the loss not spotted?

The precise details cannot be known after the fact. But we can conjecture some of the factors responsible:

- Lack of supervision of the trader. As a home country national working in a foreign branch, Iguchi probably enjoyed a higher status than would have been afforded to a local, or indeed than he himself would have enjoyed in the home office. Moreover, with over half of NY branch's profits appearing to come from Mr. Iguchi's trading desk, staff may have been less willing to ask questions.
- Mr. Iguchi was in charge of the back office as well as trading. This is the kind of classic failure of segregation of duties that often accompanies large losses.

1.4.4.4 The NatWest Markets interest rate derivatives loss

A rather different example comes from a British bank. From early 1994, market participants noted that NatWest Markets was selling DEM and GBP options in some size. NatWest's prowess in sterling interest rate derivatives was recognised with a Best Bank for GBP swaps award from a well-known trade publication. Certainly, things seemed to be going well for NatWest: in early February 1997, the bank announced what was then a record profit of £1.1B.

---

* The U.S. criminal indictment against Daiwa can be found on the web. A good source of information on this and other large financial loss, including links to news stories and other contexts, is Roy Davies' site at the University of Exeter: www.ex.ac.uk/~RDavies/arian/scandals/.

At the end of February, the bank announced a £90M loss on remarking the GBP and DEM interest rate options books. Five individuals including two risk managers were suspended, and the chief executive offered to forgo two-fifths of his £500,000 performance bonus.*

1.4.4.5 Why did it happen?

It has been suggested that NatWest had two different incompatible risk management systems, one of which was rather old and not very sophisticated. In particular, it seems that the older system was not used in such a way that out-of-the-money options were properly priced [that is, the volatility smile was not used: see section 2.4.2.].

Thus in this case there were no hidden trades: trading activity was transparent to anyone who could access the system. The issue rather was that traders with a strong track record seem to have persuaded the management that their assumptions regarding volatility were correct even though they were increasingly out of line with the market. (This phenomenon of 'trust us, we're right about the vols' has also played a part in other derivatives-related losses.) The bank's control infrastructure seems to have lacked either the knowledge or the authority to challenge traders.†

Exercise. Pick one of the two events above, or another familiar to you such as the Barings losses‡ and identify any of the following you can find: senior management failures; cultural failures; systems or risk-reporting failures; risk management failures (based on correct information); other operational risks.

Now allocate responsibility: who was responsible for the event? Did they act deliberately or just negligently? What do you think of the punishments they suffered?

Finally, suggest three key control improvements.

Several points suggest themselves from these histories:

- Trading activities have often generated significant losses for firms;
- Without good risk management—especially good risk culture—these losses can increase while remaining undiscovered leading to reputational as well as financial damage;
- Segregation of duties and the *four eyes principle*—that all significant activities are independently reviewed by competant, authoritative staff—are a lot easier to discuss than to implement across all of a firm's business.

1.5 SOME EXTERNAL CONSTRAINTS

In this section, we give an overview of two of the key constraints for financial services firms: *regulation* and *accounting*. Both of these can affect what businesses firms do and how they measure the results of their activities, so understanding the relevant frameworks is crucial.

---

* If the reader has an idle moment, he or she might wonder what exquisite calculus of blame led to the fraction two-fifths.

† Details of the regulator's findings in this affair can be found at www.fsa.gov.uk/pubs/additional/sfa008-00.pdf.

‡ Nick Leeson's own *Rogue Trader* is worth reading here.

One important consideration for many firms is *regulatory capital*: the amount of capital that firms are required to keep against the risks that they are running. Capital requirements constrain leverage and hence potential return on equity. Moreover, where regulatory capital requirements differ significantly from economic capital estimates, they can alter a firm's perception of the attractions of a business or product.

A key accounting issue is how unrealised profits and losses are recognised on a position: in mark-to-market accounting, this happens regularly potentially resulting in P/L volatility, whereas in historic cost accounting, unless an asset is *impaired*, profit is amortised over the life of the transaction, resulting in a much smoother P/L.

1.5.1 Regulation

Regulation is a pervasive influence on banks in many jurisdictions, and bankers, at least, have the perception that their industry is tightly regulated. Four main reasons for regulating financial services are often given: the protection of deposits, equalising information asymmetries, the avoidance of social upheaval and the prevention of systemic risk.

1.5.1.1 Protection of deposits

It is difficult for a modern society to function without banking. Governments recognise this and provide *deposit insurance*: retail deposits are wholly or largely protected against bank failure by the state so that individuals can make deposits without having to make a decision on the creditworthiness of a particular bank. However, if a bank then fails, the government (or its deposit insurance agency) has to pay out. Therefore, it is in the government's interests to ensure that this happens rarely. Bank regulation is one way of doing this.

1.5.1.2 Information asymmetries

Information asymmetry occurs when one party to a transaction knows more about it than the other. In retail banking, the bank usually* has access to more information than their clients, and so clients may not be able to make informed decisions in an unregulated market.

Regulation protects a bank's clients by constraining the products that the bank can offer, defining standards of disclosure and constraining the bank's conduct of business.

1.5.1.3 Social upheaval

Governments that allow major bank failures tend not to be popular: this is not a vote winner. Therefore, regardless of the (considerable) other benefits of regulation, governments choose to regulate to safeguard their popularity.

1.5.1.4 Systemic risk

Systemic risk is risk to the functioning of the financial system as a whole. As we saw in the simple bank in the previous section, customers' and interbank deposits are lent to other parties or used to finance a bank's activities. They are used in making loans, buying risky

---

* Although not always. It may be insightful if you have a free hour to see if your retail bank can justify in detail the interest charges on your mortgage. The branch staff usually know less about day count conventions, the precise form of compounding chosen and when interest is charged to and debited from the account than you do.

assets and lending to other banks. The financial system is thus linked together by the credit granted between firms. If one bank defaults, it might cause a domino effect, bringing down other banks and threatening the whole system. Regulation makes this less likely in (at least) three ways:

- Regulatory capital requirements force banks to keep sufficient capital to absorb all but the most extreme losses. This capital is in form of equity, offer loss-absorbing liabilities such as subordinated debt, and possibly mandated deposits at the central bank.

- We saw that our old-fashioned bank had a leverage of about 9:1. That means it can sustain losses of one-ninth of its risky assets before being insolvent. Clearly, the higher capital requirements are—the more capital it has to hold—the safer it is.

- Extra regulatory capital requirements are imposed if a bank has a *concentrated exposure* to the failure of a single counterparty or linked group of counterparties. This helps to reduce systemic risk.

Finally, regulation requires banks to adhere to *minimum standards* of conduct of business, risk management and so on. Some of these regulations aim to ensure that the management has reliable information with which to manage the bank and that the bank is organised in such a way that it can be well managed.

1.5.1.5 Regulatory capital and the BCBS

In Michael Power's delightful phrase,* banking regulation has 'always grown up in the shadow of crisis'. International concerns about systemic risk came to the fore with the failure of Bank Herstatt [discussed in section 1.3.1]. The *Bank of International Settlements* (BIS) had existed since 1930 to facilitate international cooperation, so this was the natural location for a body to promote financial stability and address these concerns. The BIS is located in the Swiss town of Basel, so when the new body was formed, it was called the *Basel Committee on Banking Supervision* (BCBS), or just the Basel Committee for short.

The BCBS is composed of representatives of the central banks (and, where this is different, the banking regulatory authorities) of the leading economies. It provides a meeting place for international cooperation on bank supervision with the aim of improving the quality of banking supervision worldwide. In particular, it has developed a range of standards, principles of bank supervision and statements of best practice which have gained international acceptance.

1.5.1.6 Basel I

The focus of the BCBS has always been the reduction of systemic risk within the international banking system. As banking became more internationalised in the 1980s, it became clear that there was a danger of *capital arbitrage*: banks might simply re-domicile to countries with lower regulatory capital requirements, or banks from those countries might be able to

---

* See Michael Power's *The Invention of Operational Risk*.

undercut competitors based in higher capital jurisdictions. It is worth seeing this in detail as it demonstrates how capital requirements affect profitability. Consider the same $100M loan made by two banks. Bank A has a 4% capital requirement for the loan, whereas Bank B is forced to put aside 8%. Suppose this capital is in the form of equity in each case. This equity has to earn the return demanded by the banks' shareholders. Suppose both banks fund at Libor flat and they both have a target (pre-tax) return on equity (ROE) of 40%. Then:

- Bank A must obtain a return of 40% of 4% of $100M or $1.6M to meet its target ROE;
- Whereas Bank B must obtain a return of 40% of 8% of $100M or $3.2M to meet its target ROE.

Therefore, Bank B must charge a spread of over 3.2% to Libor on the loan for the trade to add any economic value, whereas Bank A can charge, say, 2% and still add value to shareholders.

The only solution to this problem is to standardise capital requirements internationally. However, the problem is difficult because the variety of ways that an institution can take risk is so large: via lending, via the capital markets and so on. All of these should have consistent capital requirements. The BCBS needed to find a compromise which protected the financial system but allowed individual institutions to prosper. Also pragmatically when capital requirements were first introduced, it was important that most institutions could meet them: destroying a country's banking system by regulating it would not have been helpful.

It is in this context that the first capital requirements should be understood. The BCBS agreed to a simple and deliberately crude compromise. This is known as the 1988 Basel Accord, or *Basel I*. The basic idea was that banks had to have sufficient capital to cover basic operations and against the potential decline in the value of loan assets. Each bank therefore had to divide its lending into a number of buckets based on their risk.* Capital was assigned as a percentage of the exposure in each bucket as shown in the following table.

Obligator	Capital Requirement as a Percentage of the Exposure
OECD Central Governments	0%
OECD banks and regulated firms	1.6%
Loans secured by residential properties	4%
Credit exposure via derivatives	4%
Others including corporate lending	8%

The bank had to add up its total capital requirements on the basis of their total exposure in each bucket and regularly report these, plus the capital it actually had, to its regulator. Provided its capital was greater than its capital requirement, all was well. If not, it was said to be *capitally inadequate*, and the regulator would intervene.

---

* For details of all the Basel Committee capital requirements, best practice guidance, etc., see www.bis.org. We have simplified the discussion here as a more detailed account is given in Part III.

In Basel I, 8% was the standard risk weighting; therefore, an asset attracting this charge was said to be *100% risk weighted*: in contrast with this, assets attracting the 1.6% charge were said to be 20% risk weighted since 20% × 8% = 1.6%. The total bank assets multiplied by the relevant risk weight are known as *risk-weighted assets*, and these are often reported by banks as shorthand for the amount of risk on the balance sheet.

These early requirements were extraordinarily crude and only addressed a limited range of risks, notably credit risk in the banking book, but critically they were simple enough to be agreed upon. *Basel I* served the banking system reasonably well until the advent of securitisation technology [as Chapter 7 demonstrates].

1.5.1.7 Beyond Basel I: the market risk amendment and Basel II

Basel I concentrated on capital requirements for credit risk as banking book risk was the dominant risk class for most banks at that time. Clearly, market risk was an important factor too, and the Basel Committee standardised capital requirements for banks here in the 1996 Market Risk Amendment. Of course, before 1996 there had been capital requirements for market risk: Basel simply produced a global standard menu of approaches. Most advanced banks pick *Value at risk* (VAR) [discussed in Chapter 4] from this menu as their preferred means of calculating capital for market risk.

A large programme began in the last years of the twentieth century to produce capital rules which were more sensitive to credit risk than Basel I. The new capital rules, known as Basel II, are much more complex and somewhat more risk sensitive. In addition to the new capital rules, further requirements concern the disclosures required from banks and the supervision of banks' activities.

1.5.1.8 EU regulation

The European Commission in Brussels produces EU directives which aim to promote the EU's internal market, provide a level playing field across the Union and protect the consumer.

Local regulators in each member state work within the framework of these directives, producing local regulations which implement the directives in each member state.*

There have been a number of key directives for financial services firms including the capital adequacy directives (implementing the BCBS requirements on capital), the investment services and banking coordination directives (which make it easier for firms in one EU country to do business in another) and the solvency ratio directive. It would take an entire book in itself to discuss EU financial services regulation in detail, but it is worth observing the following:

- The EU, like all developed countries, has accepted the Basel Committee capital rules for internationally active banks. However, it has gone further: these rules apply within the EU to all investment firms, whether banks or not. This is a much wider class of

---

* This process, of course, offers the potential for different interpretations of what the directive 'really means' in different member states. Even without this process, there is sufficient ambiguity in the internationally agreed rules that one is probably best thinking of Basel as a framework rather than a firm set of rules which are followed identically everywhere. Differences tend to occur in the treatment of more sophisticated instruments, particularly when these have been developed after the publication of the capital rules.

firms, and it gives rise to a potential problem. The Basel Committee is only charged with making rules for large banks. It has no remit to consider the effect of its rules on smaller banks, or non-bank financial services firms. Yet the EU applies these rules much more broadly, often without, it seems, a pause for thought.

- This can give rise to significant asymmetries, notably between the EU and the United States (where the FED will only require the very largest banks to use the Basel II rules, for instance, and the SEC has an approach which it says is 'consistent with the standards adopted by the BCBS' for the largest broker/dealers).

1.5.1.9 Conduct of business

Although capital standards for banks at least are broadly similar internationally, *conduct-of-business* rules are not. These rules determine what you can sell to whom, what information must be provided to different classes of client, local listing requirements and so on. Even within the EU, wide differences exist, although the commission is slowly levelling them through salvos of directives. These differences can pose significant regulatory risk, since though a bank may be regulated for capital purposes in its home country, and may indeed do business physically from there, it is regulated for conduct of business locally, and hence potentially has a different set of requirements to meet depending on the domicile of its client.

1.5.2 Accounting

We talked about daily risk controls earlier in this chapter [see section 1.3.4] in a way that was slightly disingenuous: there was a hidden assumption that the P/L was a *market-to-market* or *fair value* measure. Most trading books are accounted for on that basis: each day, the firm attempts to value each item in the trading book at the price it could be exchanged at in a transaction with a knowledgeable, unrelated willing counterparty. The objective of a fair value measurement is to estimate an exchange price for the asset or liability being measured in the absence of an actual transaction. In fair value accounting, changes in the fair value fall through to the profit and loss account.

Fair value is generally acknowledged as the best accounting measure for liquid products with easily observable prices. If I have 2,000 freely transferable and unencumbered shares of DuPont, why would I choose to value them at anything other than the price I could sell them for? The problem is that financial services firms have assets which are neither liquid nor with a readily established price such as:

- Securities which have never traded since primary issuance some years ago;
- Loans to middle-market corporates which are not transferable and where the corporate has no visible credit spread;
- PE investments in unquoted companies;
- Complex derivatives which may be highly illiquid.

There is a continuum of financial assets ranging from those with the most straightforward price discovery—where fair value is clearly appropriate—to those where it is impossible. Understandably, therefore there is some controversy about the use of fair value for much of a bank's or an insurance company's assets. We need to look at the alternatives to fair value and to understand the framework within which firms are permitted to select one accounting method over another for different assets and liabilities.

1.5.2.1 Establishing fair value

Accounting standards state that in an active market, the best evidence of fair value is a published price quotation. However, where there is not an active market, firms are permitted to use model valuations (for instance, using an options valuation model). The inputs to the model must themselves be based on observable market transactions if possible.

1.5.2.2 Available for sale

Available-for-sale accounting is a variant on fair value. Here the initial valuation is at fair value, but changes in fair value flow through the firm's equity account rather than being recognised as P/L.

1.5.2.3 Hedge accounting

It is sometimes possible to identify an instrument as a hedge to another. That it, its changes in value offset or largely offset those of the hedged item. *Hedge Accounting* permits the two items to be marked together rather than separately marked to market. This treatment recognises the potential accounting asymmetry between a derivative at fair value hedging an item which may not be fair value accounted.

To qualify for hedge accounting, the instrument and its hedge have to be identified and a hedge effectiveness test, known as the 80/125 test,* must be met. Both fair value and cashflow hedges are recognised: for fair value hedges, the value of the joint position is realised in P/L; for cashflow hedges, the value of the residual cashflows after hedging is recognised.

Exercise. Verfänglich AG is a wine merchant based in Koblenz. Its clients are wealthy Americans who appreciate the fine Beerenauslesen and Trockenheeren auslesen it sells. It also sells Swiss wine to the United States (but even Verfänglich's *Direktor* would not call it fine.) Therefore, the company has purchases in EUR and CHF, accounts in EUR and has sales in EUR and USD.

Clearly, its profitability depends on USD/EUR and USD/CHF.

The company's earnings volatility would be reduced with the right FX hedge. Should it hedge using forwards (to lock in profitability) or using options (to give the firm some upside if things go in their favour)?

Should it hedge its expected purchases in CHF versus EUR and then its expected sales in USD versus EUR? That would be clean, but would it be cheaper to hedge from CHF to USD directly? Or should Verfänglich exploit the historically high correlation between USD/EUR and USD/CHF somehow?

Set out its choices, determine which is economically most attractive, and then comment on the accounting required.

---

* The purported hedge must in fact hedge between 80 and 125% of the variability in value of the item it is designated as hedging. This test makes getting hedge accounting for portfolio hedges very difficult.

1.5.2.4 Amortised or historic cost accounting

The amortised cost of a financial asset or financial liability is:

- The amount measured initially;
- Minus principal repayments;
- Plus or minus the cumulative amortisation of any interest;
- Minus any write-down for impairment or uncollectability.

It is typically used in situations where a firm possesses an illiquid interest-bearing asset such as a loan which will be held to maturity. For instance, suppose a bank has a 5-year illiquid security with a stated principal amount of $5M paying interest of 6% semi-annually which it bought at a discount. It would recognise the interest payments (approximately $150,000 every 6 months) as income and slowly amortise the purchase price to par over the 5-year term. In particular, the valuation of the instrument is not changed to reflect changes in the credit spread of the obligator unless the instrument is deemed to be *impaired*, i.e., the amortised asset price does not reflect its economic value, perhaps because those cashflows are unlikely to be received.

1.5.2.5 The effect of different accounting methods

Before we see what method you can apply where, it is worth understanding what is at stake. For corporates, the ability to use hedge accounting can be significant: for financial institutions the real battle is often between historic cost and fair value.

Consider the loan in the situation above. If it was fair value accounting, we would estimate a credit spread for the obligator and PV the principal and interest payments on the loan back along the risky curve thus generated. Based on changes in the obligator *and changes in the overall credit market*, this risky curve would move and hence generate P/L. Over its life, our loan could easily go from a spread of 200 to 300 bps and back to 150 bps just due to changing conditions in the broad credit market. This would change the fair value of the loan during its life by 5% or so and thus introduce volatility into earnings.

Following this argument through to its logical conclusion, if we believe that this volatility has *nothing* to do with the underlying creditworthiness of the loan, then the loan is consuming extra capital just because of its accounting treatment. As before, that capital has to earn a return, and thus, in this simple illustration at least, accounting changes our capital allocation and hence potentially our business decisions.

A sceptic might suggest that the flaw in this argument is the assertion that changes in the credit spread have nothing to do with the obligator's fundamental creditworthiness. However, looking at some of the gyrations of the credit market, it is hard to believe that *all* of them reflect *only* changes in perception about creditworthiness. There are structural trends in the amount of risk capital available that seem to be driven by conditions in the financial market as a whole. For instance, during the Russian and Brazilian crises of the late 1990s, the spread on the U.S. high-yield debt blew out. One cannot reasonably argue that events in Russia or Brazil had much to do with the ability of U.S. corporations to repay their debt. During a crisis, many players withdraw from the market in higher yielding assets of all kinds, causing spreads to go out.

1.5.2.6 IFRS requirements

The required choices of accounting method under International Financial Reporting Standards (IFRS) at the moment* are summarised in the table below.

Broadly, this means that firms will mark-to-market the trading book and historic cost account the banking book.

Category	Instrument	Measured at Fair Value?	Measured at Amortised Cost?
Loans and receivables	Unquoted loan assets with no intent to sell	No	Yes
Held to maturity	Debt assets intended to be held to maturity	No	Yes
Fair value	All derivatives except where qualified for hedge accounting; all actively traded items; any item designated as FV accounted when acquired	Yes, changes in value go to P/L	Only available for items that should be fair value but where this cannot be reliably estimated, e.g., unquoted equity positions such as PE
Available for sale	All assets not in the categories above	Yes, changes in value go to the equity account unless the item is impaired	Only available for items that should be AFS but where this cannot be reliably estimated
Non-trading liabilities	Other liabilities	No	Yes

1.5.2.7 Accounting standards versus regulation

The accounting of financial institutions is in a transitionary situation which may continue for some time. In the 1980s, banks' loan books were all historic cost accounted with fair value restricted to the trading book. During the 1990s and 2000s, credit trading grew dramatically and the distinction between illiquid loans and tradable credit instruments became ever less clear. In response, some accounting standards organisations proposed a wide extension of fair value. Some in the accounting standards community see the use of fair value for all, or substantially all of a bank's balance sheet to be the ultimate goal. However, there is considerable resistance to this view. For instance, after negative comments from the European Central Bank, the accounting standards setters pulled back from their initial proposal and restricted the application of fair value to three situations:

- Where there would otherwise be an accounting mismatch, for instance, if an asset would be held at fair value but a matching liability at historic cost;

---

* Of course, this simplifies a rather complex situation: our aim is to give an overview of the situation in so far as it affects most financial institutions' behaviour rather than to go into the (fairly fast changing, at least at the moment) details. See http://www.iasb.org/ or your auditor for more details.

- Where group of assets managed on a fair value basis in accordance with a documented risk management or investment strategy, for instance in a trading book;
- Where an instrument is or contains one or more derivatives.

The Basel Committee has subsequently issued guidance on the use of the fair value option in banks including seven guiding principles. The concern guiding their comments seems to be that banks might take profits in their loan books by the selective use of fair value when spreads are tight, and then find themselves capitally inadequate later when spreads go out.

Although one can sympathise with the regulators' worries, the problem here seems to be the imposition on a binary distinction on a continuum of liquidity.* For many, perhaps even most, assets in the financial system, it is possible to discern a range of perfectly justifiable fair values depending on the valuation methodology. For some assets, that uncertainty is a material fraction of their mark.

	Advantages	Disadvantages
Fair value	Inherently forward looking	It can potentially be highly subjective and there is a risk of manipulation
	More timely recognition of risks	Use of FV by all market participants may lead to artificial volatility as a small decline is exacerbated by sellers with stop-loss limits
	Provides an incentive to improve risk measurement and management practices	It may not reflect the business model of the firm. Thus it could make the planning horizon shorter and alter behaviour
Historic cost	Inherently 'over the cycle' which may reflect some institutions business plans	Not risk sensitive and so can allow firms to sleep walk into a crisis
	Not susceptible to artificial market volatility caused by short-term panics or bubbles	May fail to give the management insight into the underlying volatility of the business

A final thought: even if as an accounting standards matter you have to or you want to use historic cost accounting for some portfolio, there is nothing to stop you using fair value as an additional piece of risk management information or to inform performance management decisions.

1.5.2.8 Reserves

Some risk managers may take the view that they have never seen a reserve that they did not like. Although one can understand their conservatism, there are two reasons to be more careful about permitting reserves than this:

- First, as an accounting matter, reserves, especially those taken on arbitrary basis, may not be permitted.

---

* Industry participants in the fair value versus historic cost debate can sometimes seem to position themselves on the basis of their firm's asset mix. For instance, some investment banks with rather little lending have suggested that historic cost accounting is 'hopelessly antiquated' for companies primarily engaged in financial services or for companies heavily involved with financial instruments. In contrast, much of the commercial banking and insurance industry has been vociferous in its opposition to fair value.

- Perhaps even more importantly, if a business has a large pot of reserves, there might be a temptation to raid the pot when things go badly, regardless of what the reserves are actually being held for. This kind of reserve manipulation may well be illegal depending on the situation and the jurisdiction, and it certainly deprives the management of important information about the true P/L volatility of the business.

Accounting standards typically instruct that a reserve can only be taken when there is a probable, estimable cause of impairment. Evidence of impairment includes an indication of financial difficulty or delinquency on the part of an obligator; a high probability of bankruptcy or financial reorganisation; or a significant, prolonged decline in fair value. Where there is objective evidence of impairment of a historic cost accounted asset, the carrying amount of an asset should be reduced to the PV of expected future cashflows, discounted at the instrument's original effective interest rate.

Other situations where a sum of money can be held to adjust the value of a position is where, for reasons of convenience, a book is marked using a system but some position marks have to be *adjusted* to properly reflect their true fair value. For instance, a large position might require such a *mark adjustment* to reflect the fact that it could not be liquidated at the market price:* the system would (perhaps automatically) mark the position at the bid price for normal size; an adjustment of, say, 2% of the value would correct this for the likely liquidation price of the position.

In addition to position-specific reserves, there can also be a second, collective assessment of assets. This is designed to allow recognition of losses believed to exist in the portfolio but which are not yet evident. Most banks' provisions for credit risk fall into this category. These can only be made under IAS 39 to the extent that there are adverse changes in the payment status of borrowers, or economic conditions that can be shown to correlate with defaults on the assets held.

1.5.2.9 Risk management under historic cost accounting

As we have noted, risk management rests on a valuation principle: risk is the possibility of change in value. In a historic cost setting, the issue is impairment, as this is the only threat to a change in value. (Of course, a suitably chosen credit loss reserve calculation methodology can bring a nominally historic cost accounted book much closer to fair value and hence to traditional mark-to-market-based risk management techniques.)

The problem with impairment in a pure historic cost setting is that it is a cliff: either an asset is impaired or it is not. Risk mitigants against a spike in impairments include:

- Tough underwriting standards for taking risk on in the first place;
- 'Over the cycle' pricing and reserving which reflects risk over the lifetime of the exposure;

---

* A *reserve* reflects a probable and estimable impairment in the value of a position: a *mark adjustment* reflects a (more or less) certain adjustment to a system valuation which corrects a known deficiency in the model or data being applied to the position concerned.

- Scepticism about the veracity of any particular underwriting model, and ideally the use of multiple models with different assumptions to question the firm's underwriting;
- Competitor benchmarking of underwriting criteria, reserving methodology, and risk appetite;*
- Crude hedges (such as interest rate caps or inflation derivatives) which may help to offset impairment losses in a general economic downturn;
- A liability issuance strategy, perhaps via innovative capital instruments, which passes some of the tail risk of the historic cost books on to other investors.

Some of the difficulty also comes from a *strategic risk* element. Because you often only find out slowly how a historic cost accounted investment is doing, there is a risk that the firm might continue to transact a problematic product for some years before the issue is uncovered. Some insurance companies have failed this way, for example.

1.5.3 Marking to Fair Value

For better or for worse, firms are required to attempt to mark their trading books and some other areas of the firm to fair value.

1.5.3.1 Who is responsible for marking the book?

There are two obvious possibilities shown in the following table.

	Advantages	Disadvantages
Traders mark the book	They have complete responsibility for their P/L and there can be no excuses	Possibility for manipulation or P/L smoothing
	Independent checking can focus on large or problematic positions	In some cases, book marking may be relatively time consuming and it may not be a good use of expensive trading resources
Risk management or Finance mark the book	They (should) have no incentive to mis-mark or smooth the P/L	Traders do not own their own P/L, so they can always blame the marker
	Forces regular oversight of all the positions in the book	Traders should always be the ones closest to the market: getting all relevant information for marking everything might be hard, and the marker may end up relying on the trader anyway

1.5.3.2 Aside: risk marking versus P/L marking

Should the books be marked at the mid market price or at bid/offer? Marking at mid is the typical convention for derivatives books as it gives better greeks: if you mark matching long and short options positions at bid and offer volatilities, respectively, they will show a net risk position.

---

* A cynical observer might suggest that institutions are not punished by equity investors for losses: they are punished for larger losses than their competitors. Certainly, we expect a retail bank to increase bad debt provisions in a consumer downturn. The issue is, how much and how well is the change communicated to the market?

On the contrary, if you mark at bid/offer, the valuation represents the real close-out value of the book, at least for ordinary size positions assuming good liquidity and that nothing else is wrong. Clearly, for cash books this is the best choice.

One solution for derivatives books is to mark at mid for the greeks, but to keep a bid/offer adjustment to reflect the difference between the mid market value and the real close-out value of the book. In addition, an extra adjustment beyond the ordinary bid/offer will be needed for large or illiquid positions.

1.5.3.3 Mark verification

Whoever does the marking, there will be the need for some *verification*. There are typically three elements to this depending on the book under consideration:

- Comprehensive review of all valuations, adjustments and reserves at month end by staff not responsible for the initial marking;
- Some intra-month oversight, often only of a few random or significant positions;
- Regular management reporting of significant variations found and re-marks required.

This activity is sometimes left to relatively junior staff in finance as it is sometimes perceived as unglamourous and hence not worthy work for a risk manager. One problem with this is that it can mean that the risk manager does not have an intimate knowledge of what is in the book. At least if they have to attest to senior management that the marks in the book are all materially correct, then there is some chance that they might look inside it. Also, of course, if the mark is wrong, then the risk information is wrong, so there should be a strong incentive for risk management to be actively involved in the mark verification process.

Moreover, while verifying the mark for 2,000 shares in DuPont might be straightforward, some positions will require delicate judgement especially if they are complex or illiquid. Sometimes this cannot be reasonably expected of the junior finance staff.

1.5.3.4 Structuring premiums

Suppose a team in a bank spends several weeks tailoring a derivative product to meet a client's needs. Clearly, they are going to demand some compensation for that effort, known as a *structuring premium*. In this situation, the client is typically charged more than the offer side of the components to reflect the extra structuring work. Under IAS, though, the client may well be marking the product to market, and they will need the bank to supply marks possibly every month or even more frequently. The problem is

- If you show them the 'real' mark-to-market on day 1, they will know how much structuring premium you have charged and possibly be rather angry.
- If you do not, they will be getting values that do not reflect the close-out value of the position to them, which may pose legal or reputational risk.

The reality is that if the client came back on day 3 and said 'sorry, we don't want the product, we need to close out', many firms would charge them something to do that, but often not the whole structuring premium plus bid/offer. So amortising the structuring premium into the mark over some period of time may not be unreasonable *if* this process reflects the real price you would show them to close out the structure. However, it is obviously important to get legal advice early and often in this situation, and clear disclaimers on what an indicative valuation sent to a client means are vital. Also, of course, marks sent to clients should always be independently produced and properly overseen: there are obvious dangers in marks produced by the sales personnel.

1.5.4 Special Purpose Vehicles and Consolidation

Financial services groups create *subsidiaries* (subs) for a variety of reasons. For instance, to issue warrants on an exchange, one typically needs a vehicle that has a listing on that exchange, and this in turn brings disclosure requirements for that entity. Therefore, we might create a company just to issue warrants, Honest Ron's Warrants (Luxembourg) B.V. say. This sub would issue warrants on the Luxembourg exchange and hedge itself via an inter-company equity derivative with the firm's main equity trading vehicle.

Clearly, our warrant issuance vehicle is ours, in the sense that the group stands behind the vehicle. There may, therefore, be an explicit guarantee (or some other form of credit support) from the group holding company to this sub.

In other cases, companies are created for a single purpose such as the issuance of a particular security, and we do not want them to be part of the group. Such companies are known as *special purpose vehicles* (SPVs). Our desire to stand apart from an SPV may be because the security is explicitly *not* issued with the full faith and guarantee of the parent company or because we do not want the sub's debt to inflate the debt on the group company's balance sheet.

1.5.4.1 Consolidation

A sub is said to *consolidate* if its accounts must be included in the holding company's accounts. In the warrant example, we are perfectly happy for Honest Ron's Warrants to consolidate. Consider though a sub that is created to buy residential mortgages and issue mortgage-backed securities (MBS). We might want to arrange this business and possibly provide liquidity to the secondary market in these securities, but we do not want $1B of MBS appearing as debt on our balance sheet when we do not stand behind its performance. Thus, ideally, we want the MBS issuance vehicle to deconsolidate, otherwise known as *getting it off balance sheet*.

Before Enron, getting an SPV off balance sheet was relatively straightforward under the U.S. FASB rules, and not much harder under IAS. However, one of the suspicions about Enron was that SPVs were used to hide the true amount of Enron's borrowing and hence to disguise its financial condition. As a reaction to the reasonable perception of abuse, the accounting rules on consolidation were changed, and now it is significantly more difficult to get an SPV off balance sheet.

> *Exercise.* SPVs are often domiciled in lightly regulated jurisdictions partly for tax reasons.
>
> Standards of market conduct in these locations can sometimes be different from the ones that pertain to London, New York or Tokyo. Regardless of whether an SPV that was set up as part of one of your trades consolidates or not, if you arranged the sale of securities it issued, there is a risk you may be seen as responsible for its actions. How should the resulting reputational risk be managed?

1.5.4.2 Deconsolidation

Under recent accounting standards, a firm should consolidate a sub that it controls. To avoid consolidation under IFRS, following indicators of control must not be present:*

- Activities are conducted by the SPV on behalf of the parent from which the parent benefits.
- The parent has decision-making powers to control or obtain control of the SPV or its assets.
- The parent has the right to obtain the majority of the benefits of the SPV.
- The parent has the majority of the risk of the activities of the SPV.

Note that these principles are framed in terms of their effect—risk and benefit—rather than their form—equity ownership or voting rights. Also, recent rules have the notion that all SPV activities must be conducted on somebody's behalf and so as to meet somebody's business needs, hence even if you do not consolidate it, your client might have to, and this may not serve their purposes.

---

* Again, the details in SIC-12, IAS 27, 32 and 39, etc. are a lot more complex than we have the space to discuss.

CHAPTER 2

Derivatives and Quantitative Market Risk Management

INTRODUCTION

This chapter reviews some of the basic tools used in market risk management. We start with single risk factors, and look at how they effect both cash and derivatives positions. Then we move onto multiple risk factors and various techniques for risk reporting whole portfolios and businesses.

The theory of options is presented leading to a discussion of what an option pricing model is really telling us. Hedging and implied volatility are discussed, giving some insight into the assumptions of pricing models and their effectiveness in practice.

2.1 RETURNS, OPTIONS AND SENSITIVITIES

We begin with a review of risk measurement for positions involving a single risk factor. First asset returns are examined and we find that although any particular future return cannot usually be known with certainty, the collection of returns often has predictable statistical properties.

Once we have a model of the probability distribution of returns, we can report risk on the basis of those probabilities, and hence produce a simple risk model.

A quick and dirty refresher on options is given as a prelude to a discussion of the reporting of risk for options portfolios. Finally, we return to P/L explanation as a method for gaining some comfort that we really do understand all the risk sensitivities in a derivatives portfolio.

2.1.1 Asset Returns and Risk Factors

The level of many financial variables is uncertain: we do not know what USD/JPY, the closing price of the S&P 500 or 3-month GBP Libor will be in a month's time. However if we examine how many financial variables change, we often find a degree of *statistical predictability*. Specifically if a risky asset S is considered for an extended period, perhaps a few years, and we examine the daily log returns of S, then we often find that large positive and negative returns happen infrequently while smaller returns are more common. A rise or fall of 4% on the S&P 500, for instance, is unusual: $\pm1\%$ or less is much more common. Therefore if we plot the number of times $\log(S_i/S_{i-1})$ is in a given bucket versus that bucket, we often find the following picture:

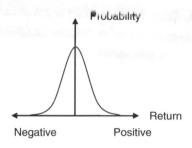

This observation is helpful is two ways:

- Firstly if we have a position whose value depends on a risky asset level S then we can study the risk of the position—its expected P/L distribution—by studying the distribution of risk factor returns. In particular we are often able to translate statistical facts about the risk factor returns into facts about our position, so for instance we might conclude that a fact like 'on average DD will fall by more than 3.7% no more than once a year' means that 'on average 2,000 shares of DD will lose more than $3,150 no more than once a year'.

- Secondly if our position is a derivative whose underlying is the risky asset S, we may be able to use a model of the returns of S to value the position and from there to reason about its risk.

2.1.2 Risk Factor Selection

For a practical risk manager, market moves are only of interest to the extent that they can effect the P/L. Therefore, we need to study how we pick risk factors to represent the sensitivity of a position. A risk factor for a position should be:

- *Effective*—Changes in the risk factor should have a stable and predictable impact on the value of the position.
- *Comprehensible*—It should be clear what the risk factor is.
- *Hedgeable*—If a risk is too big, it should be clear what we do about it.
- *Measurable*—There should be sufficient history and liquidity in the risk factor so that we can understand its behaviour and feel reasonably confident in that analysis.

The selection of risk factors for a position can be very simple:

- If we are long $5M of S&P futures and the index goes up 5%, we make $250K. The risk factor is clearly the level of the S&P index or the price of the future concerned.

But it can be a lot more complex than that:

- If we are long a government bond, that bond's value is susceptible to changes in many parts of the government yield curve corresponding to each cashflow the bond provides. Of course, we often approximate that sensitivity by simple measures like duration and convexity, but they *are* approximations.

It can be extremely difficult to capture the risk sensitivity of some instruments:

- The behaviour of the interest-only piece of the senior tranche of a portfolio of subprime residential mortgages as interest rates and house prices move might well be something that we can only guess at until such a move happens and the security reveals its true behaviour on this occasion.

In practice there are some obvious choices of risk factors when derivatives are not involved:

- For foreign exchange spot and forward positions, the obvious primary risk factors are the spot rates. In addition, we may want to look at interest rate risk if we have long-dated forward exposure.
- For government bonds or Libor-based interest rate positions without optionality, *maturity buckets* are an obvious answer. We look at our exposure in the 0–3 months maturity (or better duration) bucket, the 3–6 months bucket and so on.
- For equities we look at either index equivalents or individual stock positions.

Even here the choice of risk factor is not always entirely obvious:

- Should we use the Dow or the S&P 500 as a risk factor for the broad U.S. equity market?
- Is Royal Dutch/Shell a FTSE 100 or an AEX stock? Should we use the London or the Hong Kong close for HSBC?
- When exchange controls lift on an emerging market currency, is our data still valid?

Clearly, pragmatic choices will have to be made. It is important not just to think about these issues but to document a policy for dealing with them or you run the risk of having an unsuitable risk factor foisted upon you when much turns on the decision. One thing to bear in mind is that it is sometimes better to have a liquid series with a lot of data that is slightly less representative than an illiquid short one that is more representative.

2.1.3 Risk Reporting with a Single Risk Factor

Consider the 2,000 shares of DuPont position mentioned earlier. One reasonable measure of the risk we run in this position is the amount of money we could lose on it. Thus, our simplest possible risk report would be to list the notional size of the position [as we did in section 1.3.1]. The same approach is useful in many markets.

2.1.3.1 Spot FX reporting

For a spot FX book, a list of our exposures in our accounting currency could give one picture of the risk

	Currency				
	ARS	AUD	...	USD	ZAR
Exposure (€M)	1.92	−8.32	...	−39.24	11.35

And we could extend this to a spot and forward book by giving maturity buckets:

	Currency				
	ARS	AUD	...	USD	ZAR
Spot to 1 week	1.92	−9.32	...	−152.3	0
1 week to 1 month	0	1.0	...	14.7	0
⋮					
Over 1 year	0	0	...	24.1	0

2.1.3.2 Bond risk reporting

Where there are too many underlyings to list, we could classify them using some risk-related criteria. For a corporate bond book for instance we might start with:

	Rating				
	AAA	AA	...	BB	B or Below
Exposure (€M)	203.9	131.3	...	42.2	3.87
Weighted average spread	35	72		397	501

A country analysis gives another view of diversification:

	Country				
	Australia	Belgium	...	United States	Global
Exposure (€M)	5.00	5.23	...	100.2	36.41

Some summary characteristics of the portfolio:

Total Size	Weighted Average Spread	Duration	Number of Issuers	Percentage of Non-IG Countries	Number of Defaults This Year
700.3	153.2 bps	5.17 years	81	14.1	1

Interest rate risk reporting:

	Bucket				
	0–3 Months	3–6 Months	...	15–25 Years	Over 25 Years
Exposure (€M)	0	10.1	...	147.3	55.38

Concentration reporting is given as below:

	Rating			
	AAA	AA	...	BB or Below
Largest three issuers	EBRD, NIB and World Bank	HSBC, Swiss Re and Japan	...	Ford, Visteon and Athena

2.1.3.3 Risk factor mapping

We can see in the example above that *proxy* risk factors are often used instead of a more accurate version. The interest rate risk of a real bond with all its sensitivities is approximated by a position in a bucket, for instance. There are several things to consider in this representation process:

- *Bucket boundaries.* Consider a large position in a bond with residual maturity 3 months and 3 days, hedged with a matching short position of maturity 3 months and 1 day. We see no risk in the 3–6 months bucket as the two positions net. But then 2 days pass and the long is still in the 3–6 months bucket but the short is now in the 0–3 months bucket. The risk report shows large positions in both buckets. This is clearly undesirable. One way round this problem is to project a position into two buckets rather than one, so instead of putting *all* the risk of any bond of residual maturity from 3 to 6 months in the 3–6 months bucket, we instead interpret 3 months as a *pivot*, so a bond with exactly 3 months maturity would go half into the earlier bucket and half into the later.

- *Beta factors.* Instead of giving our position in terms of stock, we could express the risk of our DuPont position in index equivalents. Crudely, we could do this just by assuming $1 of any stock was equivalent to $1 of the index, or we could instead use a beta factor. Suppose the beta of DuPont versus the S&P 500 is 1.62. Then 1 CME S&P 500 index future (representing $250 times the value of the S&P 500, with the S&P at 1,276 and DD at $42.58) is hedged by a short position of $250 × 1276/(1.62 × $42.58) or 4,624 shares, so 2,000 shares of DD is roughly 43% of a future.

- *Industry, sector and credit quality proxies.* Sometimes it can be helpful to approximate an exposure by one in the same industry, the same sector or one of the same credit quality. This could be because we want to reduce the number of risk factors considered, because we have better data quality for the proxy risk factor, or because we want to measure a net risk position in the proxy as this will be our hedge instrument.

2.1.4 Forwards and Arbitrage

Arbitrage is a basic idea in finance: it essentially says that there should be no free lunches, so asset prices often arrange themselves so this is the case.

A more sophisticated version of arbitrage relationships arises from the idea that when there is uncertainty *on average* there should be no free lunches: this gives rise to the idea of a *statistical arbitrage*.

2.1.4.1 The risk-free return

One common and reasonable assumption made in financial markets is that there is a risk-free asset associated with a currency: typically, the assumption is that we can invest in this asset and without doubt recover our money with interest later. Perhaps the asset is a zero coupon government bond. In any event, we typically chose to measure relative to this asset to capture the time value of money.

Sometimes it turns out to be convenient mathematically if we assume that this asset grows continuously in value, that is we assume that risk-free assets grow over a period t at $\exp(rt)$ for some continuously compounded risk-free rate r.

2.1.4.2 Futures and forwards

A forward transaction is a *bilateral* agreement between two parties, one to buy an asset in the future and the other to sell it, at a price determined now. A futures contract is an exchange-traded forward transaction. Most futures are *margined*: if the contract value changes against us, we have to pledge margin with the exchange in the form of cash or other *acceptable collateral*.

2.1.4.3 Static arbitrage

A *static arbitrage* is the simultaneous sale and/or purchase of two or more assets to generate a profit. Think of saying 'yes' on two telephone calls to other market participants and, a few days later, a suitcase of money arriving at your door.

Fix an asset and suppose this does not pay coupons or dividends. Suppose we enter into a forward to sell the asset at some point in the future t, and we buy the asset now. The forward price should logically reflect the price of buying the asset now and holding it.[*]

Therefore, if S is the spot price of the asset, our position involves borrowing a cash amount S to fund the purchase, buying the asset with the borrowed money, holding it, delivering it into the forward at expiry and repaying $S\exp(rt)$ on our borrowing, where r is the instantaneously compounded risk-free rate.[†] The strategy of borrowing S, buying the

---

[*] Dividends introduce complications for two reasons: they are discrete payments and future dividends are uncertain, at least until they are declared. The first produces discrete drops in the forward when dividends are paid and the second introduces another source of risk.

[†] Of course this assumes all sorts of things including the ability to borrow at the risk-free rate, zero storage costs for the underlying and no capital allocation against the position.

asset, holding it until maturity, disposing of it and repaying the borrowing clearly *replicates* the forward position.

- If the forward price F is smaller than $S\exp(rt)$, then we buy the forward, sell the spot short,* invest the proceeds at r and close our short position in the underlying with what we are delivered on the forward.

- If the forward F is more expensive than the cost of replication, $S\exp(rt)$, then we make money by selling the forward, buying the spot and borrowing.

Either strategy is known as *spot/forward arbitrage*: this is a simple example of a static arbitrage. Apart from this example, static arbitrages are relatively rare. However, the concept of *replicating* a position by constructing a portfolio of instruments with equal and opposite payoff is important.

In practice, of course, transaction costs and other frictions mean that there is an *arbitrage channel*: if F is more than $S\exp(rt)$ plus the cost of doing the trade, then we should sell it and buy the cash; if it is less than $S\exp(rt)$ minus the trading cost, then we should buy it and sell the cash. Between these two, however, there is not enough juice in the trade to pay our costs, and there is nothing to do.

The principle of pricing two assets to avoid an arbitrage between them is known as *no arbitrage*.

2.1.4.4 No arbitrage on forward rates

If we know the return from 3-month Libor, $L(0,3)$, and the return from 6-month Libor, $L(0,6)$, then the Libor curve predicts what the return from 3-month will be in 3 months time, $L(3,3)$: by no arbitrage we should not be able to make a riskless profit by investing for 3 months and then another 3 versus borrowing for 6 months. Therefore the 3-month rate 3 months forward rate is given by

$$1 + L(0,6) = (1 + L(0,3))(1 + L(3,3))$$

This forward rate is the market's prediction of what 3-month Libor will be in 3 months time. In fact the forward Libors have been a pretty bad predictor of what the actual spot Libors will be in the future, but that is neither here nor there: the key point is that the curve and no arbitrage allow us to derive the forward Libors.

> *Exercise.* Pick a yield curve and calculate the forward 3-month Libors for a few years.

---

* In practice we cannot assume the ability to short an asset such as an equity at zero cost: we need to subtract any borrowing cost to get the forward.

2.1.4.5 Forward rate parity

Another application of no arbitrage comes in FX forward rates. There should be no absolute advantage to borrowing or lending in one currency over another, for then we would always pick the preferred alternative, changing from our home currency into that one. This trading in turn will move spot rates to eliminate the advantage if all other things are equal.

This idea establishes a relation between interest rates, spot exchange rates and forward exchange rates. Suppose you will need $100K in 1 year and you borrow in EUR. Currently €1 buys $1.2017.

Clearly, we can generate our $100K either by borrowing in EUR, turning those EUR into USD at once and investing at USD Libor or by borrowing in EUR, investing those EUR at Euribor and undertaking to convert them into USD in a year's time in the forward market.

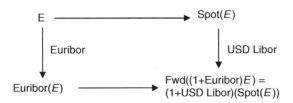

Since there should be no difference between these two approaches, we know that the 1 year forward is related to the spot rate and the two Libors, so the number of EUR *E* we need in either case should be the same.

This relationship is known as *forward rate parity*.

Exercise. Part 1: If USD Libor is higher than Euribor, is the forward rate above or below spot? Calculate the forward rate for USD Libor = 3.4%, Euribor = 3% and check your answer.

Part 2: Is it possible for the yen carry trade to make money and forward rate parity to hold? If so, explain the risks of the trade. If not, explain the misconception.

2.1.4.6 Statistical arbitrage

In a static arbitrage, we know we will make money. A more common situation is a *statistical arbitrage.* Here we *expect* to make money, in the sense that the probability-weighted average over all outcomes is positive, but there may well be situations where we can lose money.

A simple example of this is a weighted dice. Suppose we have a loaded dice that rolls 1, 2, 3, 4 or 5 exactly one-eighth of the time, and a 6, three-eighths of the time. You enter a betting game with this dice where you wager a pound and, if the dice comes up 6, you win £5; otherwise you lose your pound.

The expected payoff of a single round of the game is 15/8th of a pound and the entry fee is 8/8th, so the expected profit is 7/8th of a pound or 87.5p. But it is possible for the game to carry on arbitrarily long without the dice coming up 6. The probability of going 10 rounds, for instance, without seeing a 6 is $(5/8)^{10}$ or roughly 1%.

Notice that a player who did not know that the dice was loaded would calculate very different probabilities for the game. Their estimate of the expected payout is only five-sixths of a pound so they would never pay £1 to play.

Many arbitrages in the financial markets are statistical arbitrages: there is no guarantee we will win on any particular occasion, or on any particular sequence of bets, but if we have enough capital to continue playing and we have estimated the probabilities involved accurately then we should on average make money.

2.1.4.7 The no arbitrage principle

In the spot/forward argument above we argued that the forward price F *must* be $S\exp(rt)$, because otherwise we could make money for no risk. *No arbitrage* arguments like this are fundamental in quantitative risk management. Note though that no arbitrage does not always hold in practice—one comes across examples of arbitrages in the markets with some regularity—it is a reasonable theoretical principle.

There are many markets where a number of participants are aware of the principal potential arbitrages and when they appear these players act to close them: this is some justification for the no arbitrage principle. Thus, on a liquid equity index, the future is almost always within the arbitrage channel discussed above. But in other situations, either the arbitrage is not monitored closely by most market participants or it is too complicated to be easily traded. The practical utility of a no arbitrage argument therefore depends on their actually being market participants willing and able to close an arbitrage should one develop.

2.1.4.8 Not an arbitrage

A position is only an arbitrage if there are *no* unhedged risk factors. One sometimes finds 'arbitrage trades' which are not perfectly hedged, so it is important before taking the term arbitrage at face value to be sure that the position is not, in fact, risky. A particular culprit here is *contingent risk*. For instance, I buy a fixed rate corporate bond, hedge the interest rate risk with an interest rate swap and hedge the default risk with a credit default swap. At first sight this looks completely hedged. However, if the bond defaults and we deliver it into the credit, derivative, we are left with the IRS *which may be off-market and cost money to terminate*. The risk is not usually large, but it is there. Even more blatant, of course, is uncovered interest rate, volatility, FX or other risks: these are not unknown in some purported 'arbitrage' positions either.

2.1.5 Models of Market Returns

Once we have selected the risk factors that influence a position, we need to build a model to determine what P/L change in our portfolio is produced by a move in the risk factors.

This may seem utterly obvious, but it is important to keep in mind that we are engaged in a fundamentally predictive activity in risk management: we *conjecture* a possible future world, and *model* its properties. The kinds of models we build are not like theories in the physical sciences. There the model is disconnected from the thing being modelled: if we come up with a new theory of quantum gravity and it is generally accepted as correct, then the behaviour of cosmic gamma ray bursts does not change as a result. If you come up with a new inflation curve model that becomes widely used, people will trade off it, and the market behaviour may change as a result.* Therefore, our job as risk manager is *not* to find a model which somehow captures deep features of the behaviour of financial markets. Rather it is to find the *simplest* model that captures the features of the market we are interested in for the moment: for the moment, because the market dynamics may change. We then *calibrate* this model by selecting values for its parameters which make the model broadly match the features of the market we are trying to capture.

With this in mind, the key things we know are:

- Market moves are *uncertain*: we do not know what will happen next.
- But over medium timescales, roughly speaking, many markets have some predictable *statistical* behaviour. That suggests we look for a model based on random returns governed by some underlying statistical process.
- Moreover, if we look at returns over periods from months to years, we often recover a roughly bell-shaped distribution, so our desired model should produce a roughly bell-shaped return distribution too.

If we can find a simple random process which can be calibrated to fit our historical return data, we can then use this hypothesised distribution to predict how probable a given return might be in the future and hence, if we have a knowledge of the sensitivity of our portfolio to changes in this risk factor, to quantify risk. Of course, the accuracy of our predictions will depend on how well the distribution we conjecture matches the future returns.

There are (at least) two things to beware of here:

- The future may not be like the past.
- The future is like the past, but we nevertheless make bad predictions because we have fitted a distribution that is not a good reflection of the market's behaviour for the class of returns we are interested in.

---

* To pick one example from many, Riccardo Rebonato in *The Modern Pricing of Interest Rate Derivatives* mentions the appearance of the smile in interest rate markets as participants moved away from a naïve log-normal assumption to more sophisticated models. The fact that the new theories can change trading strategies and hence market dynamics is also made by amongst others Donald Mackenzie who pointed it out in his excellent article "The Big Bad Wolf and the Rational Market: Portfolio Insurance, the 1987 Crash and the Performativity of Economics."

2.1.5.1 Basic distributions

A *probability density* or *distribution* of returns x is a function $f(x)$ valued in $[0,1]$ such that the total probability of getting some return is 1:*

$$\int_{-\infty}^{\infty} f(x)\, dx = 1$$

The expected return $E(x)$ is just the average of this distribution:

$$E(x) = \int_{-\infty}^{\infty} x f(x)\, dx$$

The variance of returns (which, depending on the distribution, may or may not be well defined) is

$$\mathrm{Var}(x) = \int_{-\infty}^{\infty} (x - E(x))^2\, f(x)\, dr$$

We can base a forecast of price behaviour from a set of return paths, which on average have a similar behaviour to the market. What do we mean by similar?

One simple choice would be to match the average and the variance, or in market speak the *forward* and the *volatility* (standard deviation of the log returns). Of course we can think of other properties of return distributions which we might want to model too, such as its higher moments or its autocorrelation: these would suggest different mathematic choices for the fitted distribution.

2.1.5.2 Normal distribution

One of the best-known distributions is the *normal*. This has probability distribution function

$$f_{\mu,\sigma}(x) = \frac{1}{\sqrt{2\pi}\sigma} \exp\left(\frac{-(x-\mu)^2}{2\sigma^2}\right)$$

where μ is the mean and σ the standard deviation of the variable. The mean of the distribution $E(x)$ is just μ and the variance is σ^2. The probability of seeing a return smaller than some constant k is

$$\int_{-\infty}^{k} f_{\mu,\sigma}(x)\, dx$$

The normal distribution with mean 0 and S.D. 1 is known as the *standard normal*, and its cumulative distribution is denoted by Φ:

$$\Phi(k) = \int_{-\infty}^{k} f_{0,1}(x)\, dx = \frac{1}{\sqrt{2\pi}} \int_{-\infty}^{k} \exp\left(\frac{-x^2}{2}\right) dx$$

---

* The rigorous definition is glossed over here in the interests of cutting to the risk management applications fairly quickly. See Geoffrey Grimmett's *Probability and Random Processes* for a relatively gentle introduction to the mathematical framework of distributions and random variables, or David Williams' *Probability with Martingales* for a somewhat faster paced and more advanced version.

2.1.5.3 Log-normal distribution

One simple distribution that gives us the desired bell-shaped distribution of returns is the *log-normal* one. If r is the instantaneously compounded risk-free rate, the risk-free return is $\exp(rt)$, so our baseline increases exponentially. Therefore, we might hope that the logarithm of the risky return obeys some simple law, and the normal distribution is the simplest one we can pick with the right shape.

The log-normal distribution has probability density function given by

$$f_{\mu,\sigma}(x) = \frac{1}{\sqrt{2\pi}\,x\sigma} \exp\left(\frac{-(\ln x - \mu)^2}{2\sigma^2}\right)$$

where μ is the mean and σ the standard deviation of the variable's logarithm.

2.1.5.4 Fitting the log-normal distribution

Suppose we have some series of returns. We can simply take logs then choose the parameters μ and σ so that we get best-fit for the log return distribution. For instance, an estimate of the *volatility* of the return series S_i for $i = 1, ..., n$ is

$$\frac{1}{n-1} \sum_{i=1}^{n} \ln x_i^2 - \frac{1}{n(n-1)} \left(\sum_{i=1}^{n} \ln x_i\right)^2$$

This is usually annualised by multiplying by the square root of n and dividing by the number of years in the period.

2.1.5.5 Normal versus log-normal

When should we use the normal and when the log-normal distribution?

- A random variable might be modelled as log-normal if it can be thought of as the multiplicative product of many small independent factors, such as a stock return going up or down some percentage per day.

- A log-normal variable can never go to zero as it changes multiplicatively (so if we model equity returns as log-normal and identify default of a corporation with zero stock price, default can never happen in a log-normal model).

- Whereas for a normal distribution, the variable can go negative.

Depending on our application, the last of these can be definitive: a model which permits negative interest rates may be undesirable for instance.[*] Therefore, often despite its slight extra complication we prefer to use the log-normal distribution. Where the forward is only a small amount above spot, however, and the underlying is far from zero, using the normal instead of the log-normal distribution often does not make a big difference.[†]

---

[*] This was particularly an issue in JPY when the short rate there was very close to zero.
[†] If it did, a lot of VAR models would need to be changed as typically we use normal rather than log-normal distributions there and (as we are about to do) ignore the forward. Typically volatility is the major risk driver.

2.1.6 Risk Reporting in a Return Model

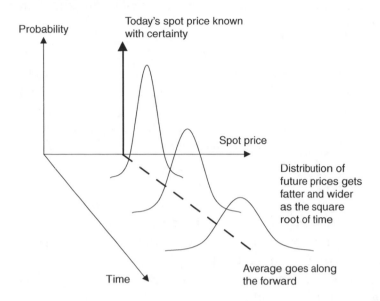

Suppose we have a position in a security which depends on a risk factor with single log-normally distributed returns, the by now tediously familiar 2,000 shares in DuPont for instance.

The distribution of returns tells us how information disperses as we go into the future. If we know an asset price today, it probably will not move far tomorrow. In five years' time, though, it could be much smaller or larger than its current value: there is less uncertainty in the near term than in the distant future.

2.1.6.1 The evolution of uncertainty

The normal distribution disperses information in a rather simple way: uncertainty grows as the square root of time. That is, if the distribution of 1-day returns from some asset has standard deviation 1%, the distribution of 4-day returns will have S.D. 2% since $\sqrt{4} = 2$. The distribution is said to *disperse as the square root of time*.

2.1.6.2 Single factor risk measurement again

It is very unlikely that DuPont equity will be worth nothing tomorrow, so the whole notional or its equivalent in S&P futures is not a very useful risk measure. Instead let us focus on the amount we could lose with, say 1% probability. That is, if we are 99% confident that we will not sustain a 1-day loss holding our position of more than L, then L is a useful measure of the risk of our position, otherwise known as the *value at risk* or VAR.

Our first step in determining L is to discover the annual volatility of DuPont stock. After some analysis (or turning to a market data provider), suppose we discover it is 22%. Assuming 250 business days a year, the daily volatility is 22% $\times$ $\sqrt{250}$ = 1.4% per day, roughly. This, together with the mean, defines our model of the return distribution of DuPont equity.

Looking up the inverse cumulative normal distribution, we find that the 1% return corresponds to a approximately 2.33 S.D. move, i.e., roughly speaking (for reasonable forwards) if f is normal

$$1\% \cong \int_{-\infty}^{-2.33\sigma} f_{\mu,\sigma}(x)\,dx$$

Therefore, $L = 2.33 \times 1.4\%$ or 3.3%: if we hold our position for one hundred days, on only one of them, on average, will we experience a loss of more than 3.3% of notional. This calculation does not tell us how much we could lose: it simply gives a loss, 3.3% of notional, that will not be exceeded more than 1% of days. The 1-day 99% VAR on 2,000 shares of DuPont with DD at $42.41 is approximately $2,800.

2.1.6.3 Linear positions

A position is called *linear* in a risk factor if its P/L varies linearly with the risk factor. For instance our 2000 share position in DuPont is linear in the stock price. Linearity just means that there is no convexity in the position. Thus, equity and spot FX positions are linear, whereas bonds and options are not.

A linear position translates a log-normal (or normal) distribution of returns into a log-normal (or normal) distribution of P/Ls.

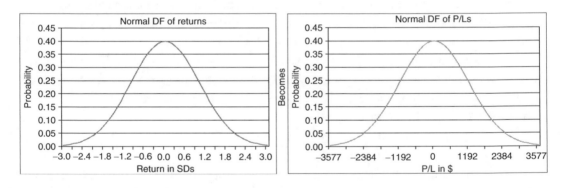

Here we have used the (absolutely trivial) risk factor model of an equity stock position that says that an $x\%$ move in the stock price moves our position value by $x\%$.

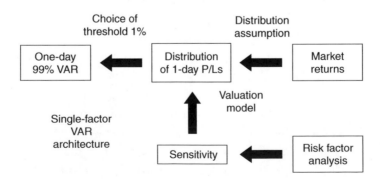

Our valuation model in this situation is very simple. But that should not distract us from the fact that there is one. The illustration above shows how the pieces fit together. Keeping these pieces distinct will become important when we start to look at risk modelling for more complicated positions.

2.1.7 Introducing Options

Linear positions translate a +2% risk factor return into a P/L of 2% of notional. Many positions—including most options—are not linear. In this section and the next one we give a brief recap on options preparatory to examining risk reporting for non-linear instruments. In particular plain vanilla puts and calls on a single underlying are briefly reviewed, some options terminology is discussed, and the well-known Black–Scholes valuation formula is presented.

Let K be the strike of an option with underlying S. There are four simple options positions:

	Nature	Payoff	Option Position Alone Worth More If	Downside
Long a call option	Right but not the obligation to buy	$\max(S - K, 0)$	Underlying goes up	Premium paid
Short a call option	If exercised you have to sell the underlying at the strike price	$-\max(S - K, 0)$	Underlying goes down	Unlimited
Long a put option	Right but not the obligation to sell	$\max(K - S, 0)$	Underlying goes down	Premium paid
Short a put option	If exercised you have to buy the underlying at the strike price	$-\max(K - S, 0)$	Underlying goes up	Strike

Figure 1 shows how the value of a call varies as the time to maturity and level of the underlying vary.

2.1.7.1 Terminology

- *Plain vanilla*. A put or call option with payout given as above is said to have a plain vanilla structure. When the structure of the payout is different in any respect, it is said to be *exotic*.
- *At/out of/in the money*. When the underlying S is at the strike price K we say the option is *at the money*. If an option would not pay out were it to expire at once we say it is *out of the money*. For a call this means $S < K$: for a put, $K < S$. Similarly, an *in the money* call implies $K < S$, whereas $S < K$ for an in the money put.
- *Collar*. A long *zero premium collar* is a long position in a call of some strike K paid for by selling a put of some strike $K' < K$ with the same premium.
- *Call and put spreads*. A long *call spread* is a long position in a call of some strike K and a short position in a higher strike call $K' > K$. A long *put spread* is a long position in a put of some strike, and a short position in a lower strike put.
- *Straddles, strangles and risk reversals*. A *straddle* is a long position in a call and a put at the same strike. A *strangle* is a long position in a call of one strike K and a long position in a put of a lower strike $K' < K$. A *risk reversal* is a short position in a call of

one strike K and a long position in a put of a lower strike $K' < K$. These structures are often traded with spot between K' and K.

- Many options have a single asset or a traded index as an underlying. When we use a composite underlying formed from the price of more than one asset, the option is said to be a *basket* derivative or an *option on a basket*.

When an option can only be exercised at maturity, it is said to be *European* style. If it can be exercised at any time during its life, it is *American* style, and if it can be exercised at fixed dates during its life but not at any point, it is said to be *Bermudan* style.

Exercise. If we think the market is going up and we want a leveraged position, why would we buy a call spread rather than a call? Would your answer differ if I asked about selling a put spread rather than selling a put?

Some people call zero premium collars 'zero cost collars', but this can be dangerous. What is the risk here?

2.1.7.2 Options underlyings

We can in principle use any traded asset as an underlying for an option, although it may cause difficulties in practice if the underlying is not liquid. Even if we cannot *physically settle* the option—deliver the underlying and receive the strike price—it can be *cash settled*—at expiry if the option is in the money, a cash payment is made based on some calculation agent's determination of the value of the underlying. (This calculation agent is often the option seller.)

For the moment we will concentrate on single stock options so our underlyings will be equities. Equity index and FX options can both be treated in a very similar framework. [Interest rate derivatives are discussed in section 2.4.3.]

2.1.7.3 The Black–Scholes formula

Black and Scholes discovered a formula for the value of a call if the underlying asset return distribution is log-normal and certain conditions on trading hold. The formula for a call struck at K with spot at S, risk-free rates of r, volatility σ and maturity t is

$$C(S, K, \sigma, r, t) = S\Phi(d_1) - e^{-rt} K \, \Phi(d_2)$$

where

$$d_1 = \frac{\ln(S/K) + (r + (\sigma^2/2))t}{\sigma \sqrt{t}} \text{ and } d_2 = \frac{\ln(S/K) + (r + (\sigma^2/2))t}{\sigma\sqrt{t}} - \sigma\sqrt{t}$$

Figure 2 shows the behaviour of the Black–Scholes call price as volatility σ and level of the underlying S varies. [We will see how to derive the formula in section 2.4.1.]

2.1.7.4   Exotic options

The literature on exotic options is extensive, and new option payouts are being invented on a regular basis. We just outline here some major classes of exotics:*

- *Average rate options.* Instead of having an option struck at *x*% of the spot at a given time, and payout on the difference between this strike and the value of the underlying at a precise expiry date, we can use averages on either or both of these. Thus, a 1-week *average in* option might be struck at 110% of the average of the closing prices during the first week of the option's life, and an *average out* option might payout on the difference between the average closing price in the last month and some strike. Average rate options—otherwise known as Asian options—are typically rather benign in risk management terms.

- *Currency-protected options.* Here either the strike or the payout (or both) are expressed in a different currency to the underlying.

- *Path-dependent options.* This is a large class of options where the payout depends not just on the final value of the spot but on how it got there. Included here are barrier options (where if spot hits a barrier, an option is either created or destroyed), range accrual structures (where for every day spot is within some range, an amount of payout accrues) and lookback options (where the payout is based on the maximum or minimum value spot attained during the life of the option).

- *Complex basket options.* Here the underlying of the option is not a single asset but a collection of assets. However, once we have two assets, we can base the option payout on the difference between their performance—as in a *spread option*—or select the performance of the best or worst asset. [More assets give us more possibilities, as we shall see in section 11.2.]

- *Compound options.* These are options where the underlying itself has some optionality. A simple example would be a call on a call—where we have the right to buy a fixed call at a fixed strike—but more common examples are situations where we have the right to terminate some derivative structure—such as calling a CB.

- *Volatility swaps.* Rather than using the price of an asset as the underlying for a derivative, we can instead use a derived property of it, such as its volatility to determine the derivative's payout.

2.1.8 The Greeks

The value of a plain vanilla derivative at some point in time depends on the spot price of the underlying, the volatility, the risk-free rate and the time to maturity. Clearly, that gives us four potential risk factors, known as the *greeks*:

- How the option value varies with the underlying spot price;
- How the option value varies with volatility;

---

* There is (much) more on exotic options in the books by Hull and Wilmott (*op. cit.*) and the collection of papers *Over the Rainbow* edited by Robert Jarrow.

- How the option value varies with interest rates;
- How the option value varies as time passes.

2.1.8.1 Delta/gamma

Let us examine the value of a typical call option as the spot price and time to maturity vary [see Figure 1]. There is not a straight-line relationship between the underlying price and spot: the call displays *convexity*. Therefore, we will need at least two sensitivities to capture the behaviour. *Delta* is the first-order sensitivity which gives us the slope of the price/spot function at some point. *Gamma* is the second derivative of the price with respect to spot, and gives us the curvature of the function.

Therefore for a call C

$$\Delta = \frac{\partial C}{\partial S}; \quad \Gamma = \frac{\partial^2 C}{\partial S^2}$$

These two greeks allow us to estimate the P/L for a move in the underlying. If spot moves from S_1 to S_2, the change in P/L due to this (all other things being equal)* is

$$\frac{\partial C}{\partial S}(S_2 - S_1) + \frac{1}{2}\frac{\partial^2 C}{\partial S^2}(S_2 - S_1)^2$$

This is just a Taylor series expansion of the call value $C(S)$ around S_1 truncated at the second term.

2.1.8.2 Vega

If volatility increases, a plain vanilla option is worth more: the underlying can go further in the time to maturity, so the price goes up with increasing volatility [see Figure 2].

The sensitivity of an option price to volatility is called *vega*. For a long call position the $V = \partial C/\partial \sigma$ is positive. If an option is far out of the money, a little extra volatility will not give us that much bigger a chance of getting in the money, so the vega here is low. Similarly for a far in the money option, extra volatility might not significantly increase the chance of getting to the money, so the vega is again typically low. For a (plain vanilla) option, then, the vega is highest around the money.

2.1.8.3 Rho and theta

The sensitivity of the option price to a change in the assumed risk-free rate is called *rho*, written as ρ. For a call it is $\partial C/\partial r$. If rates go up, the forward price of the underlying $S\exp(rt)$ goes up, so the expected value of spot in the future increases. This makes a call more valuable so rho is positive for a long call position.

---

* It is interesting to wonder for a moment what we mean by that. In particular, does "all things being equal" mean that an infinitesimal move dS is accompanied by the passage of some time dt and hence the centre of the return distribution moving $(\mu + \sigma^2/2)$dt up the forward? The usual answer is no, but there may be some intellectual case that $(\partial C/\partial S)$ should be defined this way. Those deltas would be different from the usual ones, though, and that might be confusing.

The sensitivity with respect to the passage of time is *theta* (written θ). For a call $\theta = \partial C/\partial t$. A sold plain vanilla option position is long time decay: if nothing happens, it becomes less valuable every day.

> *Exercise.* Some traders like to be long theta going into the summer holidays. Why is that? Compare and contrast a long gamma, short theta position with a short gamma, long theta one. Is it possible to be both long gamma and long theta?

2.1.8.4 Aside: rho for bonds and duration

We can easily define the sensitivity of the price of a bond to an infinitesimal parallel move in the yield curve $\partial B/\partial r$.

Initially, the bond price is given by the PV of bond cashflows under the old curve. We move the curve up in parallel by a tiny amount, 1 bp say and reprice the bond. This gives a measure of the interest rate risk of a bond that is directly comparable with how we measure it for derivatives: the expression $1/PV(B) \times \partial B/\partial r$ is known as the *Fisher–Weil duration* of B, $D(B)$.[*]

We can use duration to calculate the amount of one bond B_1 needed to hedge the interest rate risk of another, B_2. Suppose the hedged portfolio Π has an amount w_i of each bond. If the position is hedged then it is insensitive to a small move in interest rates. Therefore, $\partial\Pi/\partial r = 0$ and so

$$w_1\frac{\partial B_1}{\partial r} + w_2\frac{\partial B_2}{\partial r} = 0 \ \text{ or } \ w_2 - w_1\left(\frac{\partial B_1/\partial r}{\partial B_2/\partial r}\right) = -w_1\left(\frac{PV(B_1) \times D(B_1)}{PV(B_2) \times D(B_2)}\right)$$

This just tell us how much of B_1 is needed for our portfolio $\Pi = w_1 B_1 + w_2 B_2$ to be insensitive to a small parallel move in interest rates.

2.1.9 Options Risk Reporting

At the overview level, one obvious possibility for risk reporting of an options book is to present the greeks. In practice many traders find *finite* sensitivities like Δ(portfolio price)/Δ(spot) and so on more useful than the pure partial derivatives, so it is common to these finite greeks:

	Sensitivity				
	1% Delta	1% Gamma	1% Vega	1% Rho	1-Day Theta
Exposure ($M)	72.3	6.41	−11.23	−0.66	2.17

---

[*] Historically, a slightly different measure of duration was popular: this was obtained by measuring the bond's sensitivity to a small change in yield to maturity. Such a measure is slightly easier to calculate than the Fisher–Weil duration and is equivalent to it if the yield curve is flat. However, it is not consistent with rho for a non-flat curve, so we prefer the Fisher–Weil measure.

It is also interesting to see where the option expiries are by *time bucketing* the greeks:

	Sensitivity		
	1% Delta	1% Gamma	1% Vega
0–1 week	223.5	0	0
1 week to 1 month	114.3	14.70	1.08
⋮			
4–5 years	−192.4	−7.12	−10.10
Beyond 5 years	−20.2	−3.45	−8.22

Exercise. What is the broad position described above? Why might it have been commonplace an equity derivatives index book in the late 1990s?

If this kind of reporting is used, the bucket scheme can then be used to set limits such as

Risk factors	S&P 500 derivatives risk
Sensitivity measure	Greeks as per bucket report
Delta limit	±$250M per bucket, ±$100M total
Gamma limit	−$20M per bucket, −$25M total
Vega limit	±$12M per bucket, ±$15M total
Application	All U.S. equity derivatives books (R. Long, Desk Head)

Exercise. Why is there no positive gamma limit?

An option is a non-linear instrument so it translates a log-normal (or normal) distribution of returns into some other shape. For instance, for a long call position:

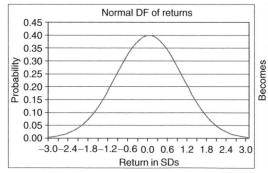 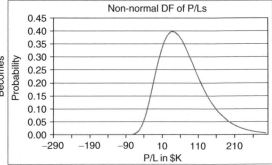

This means that we cannot use the same methodology for obtaining the 1-day 99% VAR as we used for linear instruments of just taking $2.33 \times \sigma \times$ notional. Instead we have to *revalue* the option for a range of returns, obtain the distribution of P/Ls and take the first percentile loss of this distribution. For instance, we could take 1,000 historic 1-day returns (roughly 4 years worth), revalue the option for each return, form the distribution of P/Ls and the 10th worst is an estimate of the 1-day 99% VAR.

2.1.10 P/L Explanation for Options

We have discussed a few sensitivities: delta, gamma, vega, rho and theta. But clearly there are more we could examine. For instance, there are the *vega convexities**

$$\text{vanna} = \frac{\partial^2 C}{\partial \sigma \partial S}; \quad \text{volga} = \frac{\partial^2 C}{\partial \sigma^2}$$

Another example is the effect of rates on delta $\partial^2 C / \partial S \partial r$.

Could any of these be significant? How would we know? One answer is to use the pragmatic test of whether there is an effect on the P/L. We have all the relevant sensitivities if we can successfully explain the P/L [as in section 1.3.4]. The market risk portion of the actual P/L (after funding and other non-risk P/L elements have been stripped out) for each position is compared with a theoretical P/L calculated using the actual market moves and our sensitivities delta, gamma, vega, rho and theta:

$$\frac{\partial C}{\partial S} \Delta S + \frac{1}{2} \frac{\partial^2 C}{\partial S^2} \Delta S^2 + \frac{\partial C}{\partial \sigma} \Delta \sigma + \frac{\partial C}{\partial r} \Delta r + \frac{\partial C}{\partial t} \Delta t$$

If every day actual is close to theoretical, we have enough sensitivities. If not, we need to find the missing information.

2.1.10.1 Dividend sensitivity

A good example of an often overlooked greek is *dividend sensitivity*. Typically, for an equity options book this is small, and moreover diversified, since one stock's dividend increase may be partially offset by another's dividend cut. However, there can be structural changes in dividend yield which can affect an index forward materially and hence produce a significant P/L on an equity derivatives book. This happened for instance during the high-tech boom of the late 1990s: high-tech stocks entered the index, forcing out traditional companies. These new stocks tended to have lower dividend yields than the stocks they replaced, so the overall index average dividend yield fell. The forward therefore went up, and anyone who was short long-dated calls would have lost money unless they had hedged the dividend risk exposure.

2.2 PORTFOLIOS AND RISK AGGREGATION

This section is about trading portfolios. First we look at some common types of portfolios found in financial institutions, then we begin to discuss the tools needed to aggregate their risk factors and give some consolidated measures of risk.

---

* These can, in fact, be important, especially for some volatility trading situations [we return to them in section 2.4.2].

2.2.1 The Varieties of Trading Portfolio and Their Management

This section begins with a review of the major types of risk-taking business. We then focus separately on mark-to-market books and non-mark-to-market risk taking.

2.2.1.1 Types of risk-taking books

Four different types of trading activity can be identified:

- *Flow.* Here the firm's aim is to make money from providing liquidity (also known as earning the bid/offer spread) rather than by taking risk. In these books, risks both can be and often actually are reasonably well hedged. Examples could include equity cash client facilitation books in broker/dealers, mortgage books in some banks where most of the risk is securitised and auto insurance books in some insurers where most of the risk is reinsured.

- *Large client-originated risks.* In this situation risk arises as a result of facilitating clients, but it is too big, complex or specific to hedge at once, or perhaps at all. Examples could include large lending not susceptible to securitisation, concentrated insurance risks such as some catastrophe risks, and some underwriting risk, on primary securities issuance or block trading.

- *Proprietary risk taking.* These are risks the institution has proactively sought out for its own benefit. Examples include pairs trading, credit, yield curve and volatility arbitrage, and some financial reinsurance activity.

- *Risk taking under a client mandate.* This is an investment management situation where a mandate constrains permitted risk taking. Mutual funds are included here, as are hedge funds.

The appropriate risk controls will depend on the type of book.

- Flow books need low limits, speedy risk reporting and tight controls to ensure that trading is confined to risks that can be moved on.

- Large client books can generate risks which dwarf other trading activities. Moreover once you have a risk like this, it might take a while to hedge. Therefore, it is vital to ensure that a hedge strategy is agreed before the position is taken on, and that positions generate sufficient return to meet the capital supporting them given a realistic holding period assumption. Positions here will probably need individual risk management attention.

- Proprietary books will inevitably have higher limits and wider mandates. Here measures to ensure diversification can be important, and for fair value books stop losses are a valuable control. Group think is a particular danger for a prop group, so management should regularly question not just positions but also the assumptions behind them.

- For an investment manager, there will be a focus on *risk versus a benchmark* and on *performance attribution*.

2.2.1.2 Mark-to-market risk taking

The following table indicates some types of trading books found in investment bank trading businesses.

	Cash (i.e., Funded Risk)	Derivatives
Flow	Cash equity Bond market marking (govy, corporate, ABS, etc.) FX spot and forward	Swaps and options market making books Warrant books CDS market making
Proprietary		Risk arbitrage Bond and CB arbitrage Credit arbitrage and repack Volatility arbitrage Proprietary commodity, interest rate, credit and equity books
Structured/large/illiquid	Block trading Primary	Structured finance Derivatives for corporate finance Retail product hedging

2.2.1.3 Documenting proprietary risk taking

A well-disciplined proprietary risk-taking group often finds it useful to document key expectations and controls around a position *before* it is taken on. These might include:

- The market, macroeconomic or underwriting assumptions behind the position;
- The rationale for taking the position using the instruments chosen;
- The expected holding period and expected profit;
- A stop-loss.

2.2.1.4 Client and proprietary books

Some banks prefer to separate *client facilitation* and *risk taking*:

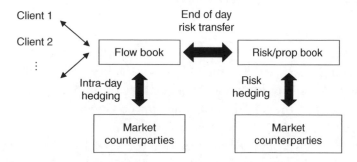

- The P/L of the flow book is money made from clients.
- The P/L of the risk book is money made from taking risk. This book may also be used for the firm's proprietary trading.

- The flow book is automatically risk flattened at the end of each day with a transfer to the risk book. The risk book manager decides how and what to hedge.
- The flow book manager concentrates on making good spreads to clients and only puts on simple intra-day delta hedges where necessary.

The advantage of this arrangement is that as the P/L from pure flow activities can be separated from that due to risk taking, management can see the relative rewards of each activity.

2.2.1.5 Risk taking in the historic cost books

Non-mark-to-market risk taking can also be organised in a similar pattern:

	Banks	**Insurance Companies**
Retail	Mortgage lending,[a] unsecured personal lending[a]	Life insurance,[b] pensions, personal P&C insurance[b] including property, health and auto,[b] investment contracts in the form of insurance
Corporate	Corporate lending,[a] project finance, leveraged finance	Workers' compensation,[b] product and environment liability,[b] marine,[b] corporate P&C,[b] directors and officers[b]

[a] Often securitised.
[b] Often reinsured.

2.2.2 Diversification and Correlation

The average trading book has lots of different risk sensitivities. For instance, a single stock equity derivatives book focussing on the Euro STOXX 50 will have sensitivities including:

- 50 stock prices;
- 50 dividend yields;
- 50 volatilities;
- plus the index level, its volatility, interest rates, etc.

There are too many things to monitor unless we spend a lot of time with the book. For management purposes we need to *aggregate* risks.

2.2.2.1 Two-asset example

Consider two positions: $10M in Google shares and $8M in Boeing Corp. shares. Suppose the volatility of Google is 2% per day and of Boeing, 1%. Proceeding as before, we can calculate the 1-day 99% VAR for each stock. For the sake of simplicity we will use a normal return assumption. Then the S.D. of the change in the Google position in 1 day is 2% of $10M or $200K, and the 99% 1-day VAR is roughly $466K.

Similarly, the 1-day VAR of the Boeing position is 2.33 × 1% × $8M or $186.4K.

But the total risk of the portfolio is not the sum of these due to *correlation*. Google and Boeing do not always move up and down together: negative returns in one asset are typically to a certain extent offset by positive returns in another, so there is a *diversification benefit* to holding both stocks.

In particular, if two assets X and Y have normally distributed returns and the comovement between their returns is characterised by a fixed correlation ρ, then the S.D. of the portfolio of both assets is given by

$$\sigma_{X+Y} = \sqrt{\sigma_X^2 + \sigma_Y^2 + 2\rho\sigma_X\sigma_Y}$$

The correlation between Boeing and Google is estimated as .25 on the basis of historical data, so with $\sigma_X = \$200K$, $\sigma_Y = \$80K$ and $\rho = .25$ the portfolio S.D. estimate is

$$\sqrt{(\$200K)^2 + (\$80K)^2 + 2 \times 0.25 \times \$200K \times \$80K} = \$233K$$

The 1-day 99% VAR for the portfolio is just 2.33 times this, or \$543K. The *diversification benefit* is the difference between the sum of the individual VARs \$466K + \$186K = \$652K and the portfolio VAR \$543K, i.e., \$109K. Our risk in this framework is \$109K smaller than the sum of the risks of the individual assets due to diversification.

2.2.2.2 Correlation and covariance

The correlation coefficient is one measure of the extent to which two linearly related random variables are related. First we define *covariance*: this measures how much two variables vary together. For two assets, i and j, it is defined as

$$Cov_{ij} = E([X_i - E(X_i)][X_j - E(X_j)])$$

The correlation coefficient between assets i and j is then defined as

$$\rho_{ij} = \frac{Cov_{ij}}{\sigma_i\sigma_j}$$

where σ_i is the S.D. of the asset i. If we have a portfolio of assets with weights w_i, the portfolio S.D. is given by

$$\sigma_{port} = \sqrt{\sum_{i=1}^{n} w_i^2\sigma_i^2 + \sum_{j=1}^{n}\sum_{j=1, i \neq j}^{n} w_i w_j Cov_{ij}}$$

The portfolio S.D. is therefore a function of

- The variances (squares of the S.D.s) of the individual assets that make up the portfolio
- The covariances between all of the assets in the portfolio

The larger the portfolio the more the impact of covariance and the lower the impact of the individual security variances—diversification works in the sense that even for assets that are positively correlated, the portfolio S.D. usually increases rather slowly as more assets are added to a big portfolio.

2.2.2.3 Using correlation

The idea of correlation gives us a simple way to *aggregate* risk under the assumption of joint normality. Unfortunately, for it to be completely sound, correlations should be stable: the correlation we measure between two return series should not greatly depend on when we look. This does not hold in practice: for instance, the correlation between global equity markets is unstable and its long-term average has tended to rise slowly over the last ten years. Given this it is worth reviewing a return correlation as a function of time to question the assumption of a single stable correlation. Consider for instance the correlation between the CAC and the DAX: as you can see from the illustration, we could justify pretty much any choice of ρ between .95 and $-.2$ on the basis of this data.

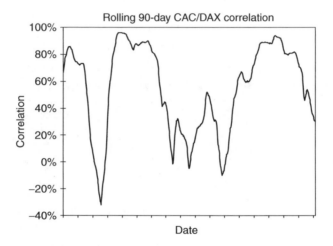

2.2.2.4 Spread options

As an example of a situation where the correlation estimate is particularly important, consider a spread option. This is an exotic option whose payout is based on the difference between two underlyings. For instance, a European CAC/DAX spread call with strike K has payout at maturity

$$\max(\text{CAC level} - \text{DAX level} - K, 0)$$

The pricing of this option depends heavily on the *comovement* of the CAC and the DAX: obviously if they tend to move together, the option is less valuable than if they do not. The standard approach to pricing spread options is to assume that correlation is well defined and to use standard Black–Scholes technology. This gives an option price which depends rather heavily on correlation, as we can see from the illustration. Moreover, the greeks depend on it too. In particular, the option has significant *cross-gamma*: the CAC delta depends on the level of the DAX and *vice versa*.

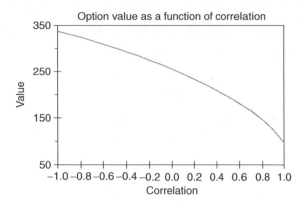

There are two key issues in this situation, a smaller and a larger:

- In the small, what is the right correlation to use for marking or risk managing this position? This is a situation where we might consider marking to our best guess of correlation and taking a reserve for uncertainty. Here if we mark at, say, .3 (assuming we have sold the spread) and take a mark adjustment to a correlation of −.1, our adjustment is roughly 20% of the value of the position. [See section 2.5 for more details on this process.]

- In the large, does the instability of correlation mean that any approach based on a constant correlation is necessarily flawed? In other words, should we be using a model like this at all?

2.2.3 Reporting and Risk in a Return Model: Multiple Factors

Once we have a model of the comovement of returns, we can use it to aggregate risk and hence get a composition picture of the risk in the portfolio. VAR is an example of this technique: it gives us a probabilistic estimate of risk on the basis of a model calibrated to some market data.

Any aggregation technique necessarily loses information—it can never be a substitute for knowing the details of what is in the portfolio—but it can be useful to summarise information about diverse risks.

2.2.3.1 VAR for linear instruments revisited

We have already seen how to calculate the VAR for a position with two risk linear factors via a calculation of the portfolio variance. For linear positions involving more factors, the calculation is similar: we simply determine the portfolio variance using the correlation between risk factors, then the VAR for a linear portfolio is a fixed multiple of that. [In Chapter 4 matrix notation is used to recast this calculation: some alternative means of calculating VAR are also presented there.] The issue, as we have seen above, is that the right correlation between risk factors to use can sometimes not be obvious.

2.2.3.2 Aside: comments

Risk reporting should be as useful as possible given the resources available. Usually this means that a timely one-page report with the most important data on it is better than a 100-page report with more information produced some hours later. Risk managers should not be afraid to add comment to the numbers:

- 'General market risk VAR at $9.5M was well below limits, but we are running a large JPY swap spread position with sensitivity $4.5M/bp' is a lot more useful than knowing just the VAR.

- Also do not be afraid to comment on issues you see as significant. For instance: 'We are also concerned about general widening in emerging market credit spreads. A 100 bps parallel move up in all East Asian and Latin American curves would cost us $14.8M.'

Given the amount of risk data available in most institutions, it is easy to produce a large risk report: the challenge is to produce a useful one.

The following properties are important in a risk report:

- It should be *timely*.
- It should provide an *accurate* assessment of all significant risks and which positions are generating them.
- It should highlight risk *concentrations* and major *changes in risk profile*.
- It should discuss unusual and significant *market events* which may have a bearing on the firm's business.

2.2.3.3 General market risk aggregation

One distinction made by regulators is between *general market risk* (GMR) and *specific risk* (SR). The concept became commonplace in the mid-1990s: the basic idea is that GMR is risk caused by moves in broad market factors such as:

- FX rates;
- Government or Libor curves;
- Equity indices.

SR reflects the behaviour of a particular issuer or obligator that is a specific stock or bond issuer. It can be thought of as *name risk*, and captures phenomena like

- The Libor curve does not move but the credit spread of the corporate bond you own increases.
- The FTSE 100 goes up, but the specific component of it you are long goes down.

Risk aggregation is often done to the level of GMR: as we have already seen, one simple way of understanding a U.S. equity portfolio is to measure its risk in equivalent S&P futures due to the non-normality of an option's return distribution.

2.2.3.4 Multivariate VAR for options

Just as we could not use the simple approach for calculating univariate VAR for an option, so we cannot use the approach discussed above for calculating VAR for portfolios with more than one rate factor due to the fact that the distribution of P/Ls is not multivariate normal.

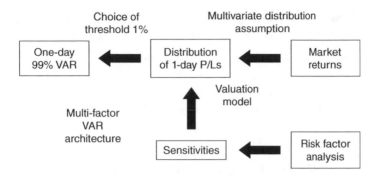

Instead we have to take an approach much like we did for single factor VAR for non-linear instruments: we take 1,000 returns of all risk factors, revalue the portfolio for each day's changes, obtain a distribution of P/Ls and take the 10th worse P/L. This is known as the *historical simulation* approach to VAR [and we discuss it in more detail in Chapter 4].

> *Exercise.* In this approach did we include theta risk? If not, what could we do about it?

2.2.4 Scenario Analysis

Once we have aggregated our risk to the level of general market rule, *scenario matrices* can be used. A scenario matrix shows the change in value of our portfolio as the underlying GMR factor moves.

For instance if we are looking at a $/¥ portfolio in thousands of $, we might find:

Move in $/¥	−8%	−6%	−4%	−2%	0%	2%	4%	6%	8%
P/L	−464	−366	−256	−134	0	146	304	474	556

> *Exercise.* Looking at the table, what can we tell about the position?

For a portfolio including derivatives, we may want to look at our exposure to moves in the GMR factor, moves in volatility and both combined.

Move in $/¥	−8%	−6%	−4%	−2%	0%	2%	4%	6%	8%
Volatility up 10%	230	338	443	515	585	645	737	971	1311
Volatility flat	−464	−366	−256	−134	0	146	304	474	556
Volatility down 10%	−888	−810	−730	−649	−570	−492	−348	−282	−221

This can be displayed either in a table as above or (perhaps less usefully but more colour-fully) as a surface plot [see Figure 3].

2.2.4.1 Scenario limits

One measure of the risk of a position is the largest loss that appears in any box of the scenario: this is called a *scenario loss*.

Scenarios are based on full revaluation of the portfolio so they include all delta, gamma and vega information (plus higher-order convexities in the scenario variables). Therefore, putting a limit on the scenario loss allows us to simultaneously capture delta/gamma and vega constraints.

> *Exercise.* Guess the underlying position from the scenario plot in Figure 3.

2.2.4.2 Scenario versus VAR

The VAR measure introduced earlier using correlation gives us an aggregate risk meas-ure at the chosen confidence interval: it is a *composite risk measure*. The scenario loss is a lower-level risk measure: it looks at the possible loss for a fixed set of moves (so if we make the moves in the scenario bigger, the losses will often increase too). Both kinds of meas-ure have their place: scenarios are helpful for looking at the details whereas VAR gives a higher-level perspective. The problem with scenario losses is that they cannot easily be aggregated since the worst events are highly unlikely to happen in multiple markets simul-taneously. The problem with a single VAR number is that it abstracts away from the detail of the individual positions and risks generating it.

2.2.5 Stress Testing

Stress testing is the process of identifying particular situations which could cause large losses. This is required by many regulators, often as an adjunct to VAR. A typical stress test involves the revaluation of the portfolio after various market factors have been moved. These moves can be generated either by

- History, in which case we have a *historical stress test*.
- Or by a hypothetical market event where likely market moves are estimated. This is a *hypothetical stress test*.
- Finally, we might use a kind of scenario where we examine the impact on a portfolio's value of one or more predefined moves in a particular market risk factor or a small number of closely linked market risk factors without any notion of the event which would generate these moves. This is a *sensitivity stress test*.

2.2.5.1 Historical stress testing

Here large market moves in the past are often used to define historical stress tests. Popular choices include:

- The 'Black Monday' 1987 equity market crash;
- The bond market crash of 1994 [discussed in section 10.2.1];

- The Russian (1998) or South East Asian (1997) crises;*
- Various banking crises, for instance the Nordic one of the early 1990s.

2.2.5.2 Hypothetical stress testing

Possible future events currently considered by some risk managers as the basis for their hypothetic stress tests include:

- Euro break-up;
- U.S. balance of payments, housing market crisis or both;
- Taiwan/China or Middle Eastern conflict;
- Natural disaster (such as a hurricane hitting the East Coast of the United States between Boston and Washington or a large earthquake in California or Japan);
- The default of the firm's largest counterparty;
- The default of a large liquidity provider in a major market (e.g., JPMorganChase in the swaps market);
- A major terrorism event.

2.2.5.3 Sensitivity stress testing

The Derivatives Policy Group of U.S. broker/dealers recommendations[†] includes the following stresses so these can be seen as one benchmark for sensitivity tests:

- Parallel yield curve stress ±100 bps;
- Yield curve twist ±25 bps;
- Each of the four combinations of twist and shift;
- Implied volatilities change by ±20% relative;
- Equity index values change by ±10% relative;
- Currencies move by ±6% relative for major currencies and ±20% relative for others;
- Swap spreads change by ±20 bps (which would take them negative in some currencies …).

---

* The South East Asian crisis of 1997 was a rolling series of market falls. It started in Thailand in July, with some investors betting that the conditions of a pegged currency, growing current account deficit and rising stock market could not continue. Thailand was forced to devalue the Baht, and the local stock market fell rapidly. Speculators moved on to other currencies and stock markets in the region, and there were substantial falls in regional equity markets. Not all the targets were forced to devalue their currencies though: Malaysia opted for currency controls rather than devaluation [as discussed in section 5.5.1].

† See Derivatives Policy Group's "Framework for Voluntary Oversight of the OTC Derivatives Activities of Securities Firm Affiliates to Promote Confidence and Stability in Financial Markets."

Some banks prefer to use moves generated from current market volatility. One approach for instance is to stress test using ±10 S.D. moves in all risk factors.

2.2.5.4 General remarks on stress testing

Whichever sort of stress testing you use it is important to consider stresses which really challenge trader's assumptions and reveal vulnerabilities. For instance, if you are long lots of straddles, nothing happening may be the worst stress.

Stress tests should be designed to avoid group think: regardless of whether the risk manager thinks that the firm's risk taking is well positioned or not, it is important for the stress test to act as a contrarian.

Finally, knock-on effects should be captured in a stress test. Thus, a test which captures a 20% fall in equity markets might be interesting, but such a test should also model the likely illiquidity after such a fall, the impact of a large market fall on implied volatilities, on margin calls in the prime brokerage portfolio and so on.

Exercise. Design a stress test. You should state:

— The aim of the stress test and what parts of the firm's portfolio are to be stressed;
— Why the stress is plausible;
— What assumptions it challenges;
— What risk factors to move;
— How you determined the size of the moves;
— How you determined any correlations you used.

2.2.5.5 Results of stress testing

Once you have a final stress loss figure, how should it be interpreted? In particular, how much is too much? Typically, a stress limit would be set for each stress test expressing the firm's risk tolerance to that event. The difficulty, as with VAR limits, is knowing what to do if the limit is broken. Often more than one desk contributes towards the excess and, since desks hedge each other, a single culprit cannot be identified. Tough (if occasionally arbitrary) action from trading management is typically required at this point.

2.2.5.6 A hierarchy of risk reporting

We have seen a range of market risk reporting techniques now: notional exposures and greeks, scenario reports, VAR and concentration reports and stress tests. Typically, these techniques are used at different stages in the hierarchy: VAR and stress tests can be useful at the aggregate level where many desks are contributing to the risk; scenarios, greeks and more detailed reporting are more useful lower down where we are interested in the details of individual positions or small groups of related exposures.

2.2.5.7 Example: outline VAR report

Consider the following excerpt from an imaginary aggregate risk report:

Desk	Position	GMR VAR							
		IR	FX	Equity	Commodity	Volatility	Total	Limit	Stress
FX	102.1[a]	0.2	8.1	0	0	1.1	8.4	10	18.2[b]
U.S. IG trading	25.2[c]	3.1	0.2[d]	0	0	0	3.0[e]	3.5	7.1
U.S. high yield	14.2[c]	1.2	0	0.2	0	0.2[f]	1.5	2	25.1[g]
Rates	42.2[c]	8.2	0.6	0	0	9.3	16.1[h]	15	21.1
Emerging local currency	8.6[c]	2.7	7.5	0	0	1.2	8.5	10	20.5
...									
Total New York trading	N/A	16.2	15.1	11.2	0.8	19.1	38.4	40	63.4

[a] As with many risk reports, this one is useless without details of what is being measured. 102.1 what? If it is million dollars spot equivalent exposure, that is a lot. If it is thousands of euros, it is trivial.

[b] It would be helpful to know here what the stress test is. Is this the worst loss from a number of stresses? Are they hypothetical, historical or sensitivity? What is the stress limit? Or is not there one?

[c] Do the bond and rates desks use a common risk measure, 10-year equivalents perhaps? Ten-year equivalent whats?

[d] Why has the investment grade trading desk got an FX VAR? Is it a data feed problem or a real exposure? If it is real, should they be taking FX exposure? What is causing it exactly? Why are not they hedged?

[e] Even worse, it appears they are getting some correlation benefit from that FX position. What is happening?

[f] Why does the high yield desk have a volatility exposure? Have they bought caps as a broad hedge against declining rates?

[g] And why is the stress loss so much bigger than the VAR?

[h] That looks like a limit violation. Or is it a data quality problem?

2.2.5.8 Concentration reporting

It is often useful in risk reports to separately identify large positions. The reason for this is that it is important for management to have (and to be able to demonstrate they have) information on all *significant* risks. 'Significant' should be defined, for instance by possible P/L impact. You do not have to report all risks—for instance, the risk of comet hitting the head office building is typically not on the average bank's risk report. But concentrations can dominate the risk of a portfolio, especially in the credit markets, so even if a big position does not appear risky using standard measures, it may nevertheless need to be highlighted.

> *Exercise.* With large risk positions there can be a temptation to ignore the elephant at the table. Typically, these are positions that have been around for a while and they are broadly known by management. Due to the size of the position, the ability to trade around it or hedge it is usually limited. The danger here is sleep walking into a crisis: if you are down 2% every day after a month you have lost nearly 50%. What steps can you take to ensure that this does not happen? How do you keep your eye on the elephant?

2.3 UNDERSTANDING THE BEHAVIOUR OF DERIVATIVES

In this section the pricing and hedging of options is discussed. We start with an overview of the derivation of the Black–Scholes pricing formulae for plain vanilla options on a single underlying, and then see how this result is used in practice. In particular, we look at implied volatility—the fudge the market uses to get the right result from Black–Scholes—and how this phenomenon leads to more sophisticated models of the behaviour of asset returns.

2.3.1 Overview of the Theory of Options Pricing: Black–Scholes and the Replicating Portfolio

This section gives a quick and dirty explanation of the derivation of the Black–Scholes pricing formulae for options.*

---

* Our aim here is not to give an alternative to the standard texts: you can find a more comprehensive account in John Hull's *Options, Futures and Other Derivatives* or *Paul Wilmott on Quantitative Finance*. Rather our treatment of Black–Scholes (properly Black–Scholes Merton) is intended to provide the context for a discussion of the assumptions involved, the nature of implied volatility and the implications of these considerations for risk management.

Suppose we want to model a risky underlying S that grows at some rate given by the forward and which has a random return:

- To get the average growth right we need in a small time dt for S to grow by $\mu S\,dt$ so we get exponential growth along the forward at rate μ, $dS = \mu S\,dt$.*
- The random part of the motion is modelled by simply introducing a Brownian motion (or *diffusion*) W and having S walk around the forward at a speed given by the volatility σ, i.e., $dS = \sigma S\,dW$.
- This random walk increases the variance of the return by $\sigma^2 S^2\,dt$ in a small interval dt, and hence we have the expected square root of time evolution of uncertainty.

Putting these together the Black–Scholes model of risky underlying S is given by

$$dS = \mu S\,dt + \sigma S\,dW.$$

2.3.1.1 The Ito lemma

It is now time to pull a rabbit out of a hat. I apologise for this but the detour to explain the Ito lemma properly is extensive, and there is a long if not entirely honourable tradition of simply stating the result than using it.

The Ito lemma states for a suitable function f of a random walk S driven by a diffusion

$$d(f(S)) = \frac{\partial f}{\partial t}\,dt + \frac{\partial f}{\partial S}\,dS + \frac{1}{2}\frac{\partial^2 f}{\partial S^2}\,(dS)^2$$

And if we take f as log we get

$$d(\log(S)) = \frac{1}{S}\,dS + \frac{1}{2S^2}(dS)^2 = \left(\mu + \frac{1}{2}\sigma^2\right)dt + \sigma\,dW$$

The expectation of a random walk is zero, so $E(S) = \exp(\mu + \sigma^2/2)$ and we are on course to model log-normal returns. Notice the term $\sigma^2/2$: the expectation of the growth of a risky asset S has an additional term depending on the volatility.

Suppose we introduce some derivative whose price V depends on one risk factor, S. Using Ito's lemma on V, $dS = \mu S\,dt + \sigma S\,dW$, and $(dS)^2 = \sigma^2 S^2\,dt$ we get

$$dV = \frac{\partial V}{\partial t}\,dt + \frac{\partial V}{\partial S}\,dS + \frac{1}{2}\frac{\partial^2 V}{\partial S^2}\,(dS)^2$$

$$= \frac{\partial V}{\partial t}\,dt + \mu S\frac{\partial V}{\partial S}\,dt + \sigma S\frac{\partial V}{\partial S}\,dW + \frac{1}{2}\sigma^2 S^2\frac{\partial^2 V}{\partial S^2}\,dt$$

This equation governs the behaviour of *any* derivative on a risky asset following a random walk $dS = \mu S\,dt + \sigma S\,dW$.

---

* For a discussion of why assets should grow at the risk-free rate, or more precisely what the risk neutral framework where this is true is, see Hull or Wilmott (*op. cit.*).

2.3.1.2 Hedging

The only source of risk to V is S. Since this is V's only risk factor for V to be hedged we need some collection of hedges* that react to changes in S in an equal and opposite way to V. Specifically, we need a hedge H such that $\partial H/\partial S = -\partial V/\partial S$.

- The only thing that we need in the portfolio H is units in the underlying, since all we are trying to do is match a first-order sensitivity $\partial V/\partial S$.
- So suppose we have $-\delta$ units of S, where $\delta = \partial V/\partial S$.
- Then our complete hedged portfolio Π is long the derivative V, short δ units of S and we have constructed this so that $\partial \Pi/\partial S = 0$, i.e., the portfolio is risk-free for an infinitesimal move in S.

If we can keep the portfolio obeying $\partial \Pi/\partial S = 0$, then it will continue to be insensitive to risk. Of course as the underlying moves, $\partial V/\partial S$ will change, and we will have to *rebalance* our hedge, buying or selling more underlying. But provided we can move fast enough to do this, our portfolio remains risk free and so, by no arbitrage, it must grow at the risk-free rate.

2.3.1.3 The Black–Scholes differential equation

From here we sketch the derivation of the *Black–Scholes differential equation*. This is the equation that governs the behaviour of the price of *any* derivative V that depends only on a random asset price S under the Black–Scholes assumptions.

- Suppose in time dt spot S moves to $S + dS$.
- Then our portfolio changes by $d\Pi = dV - \delta\, dS$. Substituting we have $d\Pi$

$$= \frac{\partial V}{\partial t}\, dt + \mu S \frac{\partial V}{\partial S}\, dt + \sigma S \frac{\partial V}{\partial S}\, dW + \frac{1}{2}\sigma^2 S^2 \frac{\partial^2 V}{\partial S^2}\, dt - \delta\, dS$$

$$= \frac{\partial V}{\partial t}\, dt + \mu S \frac{\partial V}{\partial S}\, dt + \sigma S \frac{\partial V}{\partial S}\, dW + \frac{1}{2}\sigma^2 S^2 \frac{\partial^2 V}{\partial S^2}\, dt - \delta(\mu S\, dt + \sigma S\, dW)$$

- Using $\delta = \partial V/\partial S$ we get

$$d\Pi = \frac{\partial V}{\partial t}\, dt + \frac{1}{2}\sigma^2 S^2 \frac{\partial^2 V}{\partial S^2}\, dt$$

- But we know the portfolio must grow at the risk-free rate since it is risk free and anything else would violate no arbitrage. Therefore $d\Pi = r\Pi dt$ and so

$$r\Pi dt = r(V - \delta S)dt = r\left(V - \frac{\partial V}{\partial S}S\right)dt = \frac{\partial V}{\partial t}\, dt + \frac{1}{2}\sigma^2 S^2 \frac{\partial^2 V}{\partial S^2}\, dt$$

- Simplifying we get the *Black–Scholes PDE*

$$\frac{\partial V}{\partial t} + rS\frac{\partial V}{\partial S} + \frac{1}{2}\sigma^2 S^2 \frac{\partial^2 V}{\partial S^2} = rV$$

---

* Nassim Taleb's *Dynamic Hedging* is a good reference for all of this section.

This partial differential equation was identified by Black and Scholes as governing the behaviour of derivatives. To price V under their assumptions, we just need to solve this equation subject to the boundary condition given by the option's payout. So, for instance, if we solve it subject to $V(T) = \max(S(T) - K, 0)$, we get the Black–Scholes pricing formula for a call struck at K with maturity T.

It is worth rewriting the Black–Scholes PDE to see how the hedging argument is replayed in the equation: $\delta = \partial V/\partial S$, $\theta = \partial V/\partial t$, and $\Gamma = \partial^2 V/\partial S^2$ so it is

$$\theta - r(V - \delta S) + \frac{1}{2}\sigma^2 S^2 \Gamma = 0$$

In other words, time decay minus the interest on the value of the hedge portfolio (i.e., the funding we are playing) plus gamma hedging P/L is zero.

2.3.1.4 Hedging, again

As we have seen, the Black–Scholes pricing formula is derived through an argument about *hedging*. That is, the Black–Scholes price of an option is derived by considering the behaviour of a portfolio which risk neutralises or *replicates* the option. If we keep adjusting the amount of hedge δ so that $\delta = \partial V/\partial S$, the hedge portfolio H will have equal and opposite payoff to the derivative V (provided our assumptions hold). This makes it clear that an option pricing model is only correct to the extent that the hedge ratios it produces allow us to replicate the payoff of the option by hedging.

The Black–Scholes derivation is a *statistical* argument: the option price is just the cost of hedging, and we derive that cost by integrating over all possible *paths* the underlying could take—that is what we mean by solving the Black–Scholes PDE.

2.3.1.5 Assumptions

It is worth highlighting the assumptions used in this argument:

- *Markets are efficient and complete.* We needed this for the no arbitrage argument that Π should grow at the risk-free rate.
- *Trading is always possible at mid market*, i.e., there are no commissions or bid/offer spreads. We assumed we could instantaneously adjust our hedge at the current risky asset price S.
- *Short selling is permitted*, because δ might be positive or negative depending on V.
- *There is only one risk*: that of the underlying S. In particular, interest rates remain constant and known.
- Returns of the underlying are log-normally distributed with constant volatility σ.

It is possible to generalise most of these assumptions: for instance, adding trading costs or introducing extra sources of risk is not too difficult. The big problem, though, is the log-normal assumption. To see some of the issues volatility must be considered in more detail.

2.3.2 Implied Volatility and Hedging

If asset returns were really log-normal, we would find that options trade as follows:

- There is one volatility that we would need to put into Black–Scholes to recover observed option price on a given inderlying no matter when we looked;
- What strike we looked at;
- Or what maturity option we looked at.

None of these properties is true. The volatility we need to put into Black–Scholes to recover observed option prices is called *implied volatility* or implied vol (and in general any parameter chosen to recover an option price is called an implied parameter). If we study implied vols, we typically find:

- A *strike dependence* of implied volatility. This shape is usually upwards sloping on the downside, and either kicks up above the money or carries on gently down. The former is known as a volatility *smile* and the latter as a *smirk*.

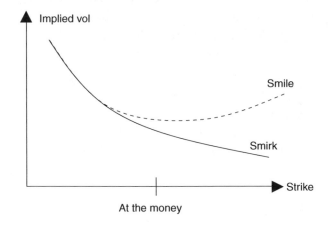

- Often implied volatilities decline slowly with increasing option maturity.

The strike and *maturity-dependent* implied volatility function is sometimes called the *implied volatility surface*. The fact that underlyings have traded options with observable prices means that use usually do not use Black–Scholes to price options in a vaccum: rather we *calibrate* it to observed market prices.

2.3.2.1 Skewness and kurtosis

The presence of the implied volatility surface alerts us to the fact that the option market prices options as if asset returns deviate from log-normal. Higher vols on the downside, for instance, indicate that highly negative returns are somewhat more likely than log normality would predict.

This phenomena is known as *fat tails* (or occasionally *heteroskedasticity*) because there is more weight in the downside of the distribution than we would expect from normality.

Typically, the upside of the vol smile is less steep than the downside, so the implied distribution is fatter tailed on the downside than the upside. There are two standard measures of the fatness of the tails of a distribution and the difference between the upside and the downside: these are respectively *kurtosis* and *skew*.

- Kurtosis is the fourth moment of the return distribution. For a standard normal distribution it is 3 (so Kurt − 3 is sometimes called the *excess kurtosis*):

$$\text{Kurt}(x) = \left\{ \frac{1}{N} \sum_{i=1}^{N} \left(\frac{x_i - E(x)}{\sigma} \right)^4 \right\}$$

> *Exercise.* A 25-delta strangle is one where both the call and the put have a delta of 0.25 in the Black–Scholes setting. What does the price of this combination imply about kurtosis and the curvature of the smile?

- Skew or skewness is the third moment, and for a normal distribution it is zero:

$$\text{Skew}(x) = \left\{ \frac{1}{N} \sum_{i=1}^{N} \left(\frac{x_i - E(x)}{\sigma} \right)^3 \right\}$$

> *Exercise.* A 25-delta risk reversal is one where both the call and the put have a delta of 0.25 in the Black–Scholes setting. What does the price of this combination imply about skewness and the steepness of the smile?

The implied volatility smile, as a phenomenon, indicates that the return distributions implied by option prices have positive excess kurtosis and non-zero skew.

2.3.2.2 Implied volatility as a convention
An implied parameter is just a convention for quoting a price: we might talk about an option 'trading at 30 vol', but all we mean by that is that if we put 30% into Black–Scholes, together with all the other relevant parameters, we obtain the right price for the option.

2.3.2.3 Volatility and hedging
Before we see how to recover the implied return distribution from option prices, it is worth going into a lot more detail on how hedging works. Let us start by assuming we are long a call on a single stock:

- The hedge is to short stock as we are long delta.
- The cash obtained from the short is invested at the risk-free rate.
- If the stock moves up, delta increases on the call, and we have to sell more stock short to rebalance the hedge;

- If the stock goes down, delta decreases, and we buy back some of our short;
- The option gets longer as the stock goes up and shorter as it goes down, so stock moves in any direction make money;
- On the other hand, theta is negative: the option loses money every day due to time decay.
- Finally note that any net cash—in this case the difference between the premium paid for the call and the cash obtained from selling the stock short—is invested at the risk-free rate.

Suppose we rehedge our position every day. The P/L for the day will depend on the stock move, theta, and the interest received or paid on the net cash position. In general it will depend on the stock price moves as follows:

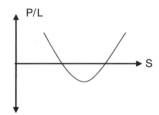

If the stock does not move on a given day we lose a day's theta and make a day's interest on the cash position, giving a net loss. Any stock move produces a gamma gain to offset against this. In the Black–Scholes universe the option price is an estimate of the average outcome of this process over the whole life of the option. Under their assumptions, if the delivered vol equals the option implied vol, then the average gain from hedging will equal the premium paid for the option if we are long and the average cost of hedging will equal the premium received if we are short.

If the delivered vol is larger than the implied vol we pay for an option then moves will be bigger than we have paid for an on average we make money hedging. If the delivered vol is smaller than the implied vol we pay for an option then moves will be smaller than we need and on average we lose money hedging.

	Buy an Option	**Sell an Option**
We think implied vol is	Cheap, so we buy vol	Expensive, so we sell vol
Premium	Paid	Received
Time value	Negative: decreases value of what we have bought	Positive: decreases value of what we have sold
Gamma	Long gamma: hedgers want big moves	Short gamma: hedgers want small moves
Fair value	Gamma profits = theta losses + carry	Gamma losses = theta profits + carry

The option price is just the expectation of the amount of money we will make from hedging. Therefore, we need to understand where we get gamma since if we want to make money hedging a long option position, we do not just need the underlying to deliver volatility: we need it to deliver volatility *where we have gamma*.

Figure 4 shows the variation of gamma by maturity and spot price for a call. Far from maturity the gamma is spread out over a range of underlying values, but closer to expiry most of the gamma is around the strike of option.

To see how this makes a difference, consider two paths of the underlying with the same delivered volatility.*

In one, the spot price goes up and down in a curve, wiggling around with volatility 2% per day (= 31.6% annualised). In the other, the spot price goes nowhere, again with 2% vol. In both cases, the spot price ends up at the same level but there is a huge difference in hedging each path.

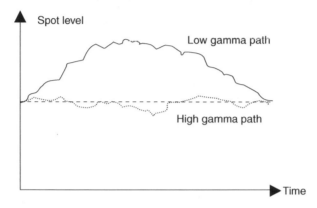

Now consider a long position in an at the money option. If spot follows the first path, it quickly leaves the area where we have gamma and only re-enters it very close to expiry. Even if we bought the option at an implied volatility considerably lower than 31%, we will probably lose money hedging here. On the other path, though, we have gamma all the way through the option's life since spot stays close to the strike. Therefore, we can pay more than the delivered vol and still make money from hedging the option.

This argument makes it clear that the option price for a given volatility is not just the price of hedging at that volatility: it is the average cost over all paths with that volatility of hedging. There may well be paths where we make or lose money within that. Therefore, just because we buy an option at cheap implied vol, we would not necessarily make money, nor would we necessarily lose it if we sell at too cheap a vol.

> *Exercise.* Why do option traders prefer to sell Asian rather than European calls if the notional is large? (*Hint*: think about gamma.)

2.3.2.4 Which vol should we hedge at?

Clearly, the right vol to mark an option is the implied vol: that is the price we could sell it at. But what vol should we hedge at? At first that might look like a trick question: should not

---

* I am grateful to Jim Gatheral for this instructive example, and for (a subtler version of) the following discussion on hedge vol. Note that this discussion is based on real world hedging—with price jumps and periodic rebalances—rather than pure diffusions and instantaneous re-hedging.

we use the market's best guess on delivered vol, i.e., the implied vol? On average, it is easy to see that if we buy an option at implied and hedge at delivered vol, on average we make

$$\gamma \times \frac{1}{2}S^2 \times (\text{realised vol}^2 - \text{implied vol}^2)$$

But thinking about the two paths above, we see that that is not true path-by-path, just on average:

- A high vol spreads our gamma over a wider range of spot moves, so if we hedge a long option position at too high a vol in a range-bound market like the high gamma path above, we do not make as much as we could have done as our delta does not change much within the range.

- A lower hedge vol, on the other hand, concentrates the gamma at the strike and so hedging at this vol makes us more money.

This shows that there is a legitimate place for trader judgement in deciding a hedge vol (but not, of course, a mark vol).

> *Exercise.* If a position is marked at one vol but hedged at another, which vol should a risk manager use for generating greeks or other risk reporting?

2.3.2.5 P/L profile by mandate

If we are long plain vanilla options and hedging, big moves are good and time decay hurts; the opposite is true if we are short. This sometimes allows us to see a trader's strategy just by looking at their P/L history.

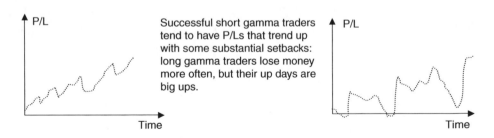

Successful short gamma traders tend to have P/Ls that trend up with some substantial setbacks: long gamma traders lose money more often, but their up days are big ups.

We might expect a book that was naturally long options, such as some proprietary view taking books, to have a profile more like the latter, whereas a short options book that was mostly delta hedged might well look more like the former.

2.3.2.6 Volatility regimes

A volatility smile or smirk is observed in options on most underlyings at the moment. How does it move as spot moves?

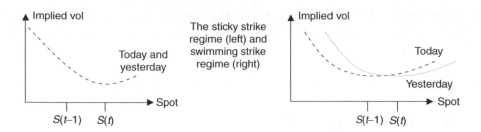

There are two obvious possible answers to that question:

- The shape of the smile is unchanged as the underlying moves, and the at the money vol stays constant. This is called the *swimming strike* regime.

- The absolute position of the smile is unchanged, so as the underlying falls, at the money vol rises. This is called the *sticky strike* regime.

These two regimes have been characterised as 'fear' and 'greed',* the sticky strike regime being typical of falling or uncertain markets, the swimming strike being more bullish.†

2.3.3 Retail Equity-Linked Products: Some Simple Structures and Their Problems

Periods of rapidly rising equity markets tend to attract media attention. Articles appear; investment advisors suggest higher allocations in equity; investors profit. This leads retail investors to contemplate equity investments. However, buying stock with small amounts of money is difficult: it is hard to build a diverse portfolio, commissions and bid/offer spreads take a bigger fraction of the investment and researching stocks is time consuming for a novice investor. Moreover, equity investment can be risky: some players lost 75% or more of their portfolio value in the NASDAQ falls of 2000. Investment providers have addressed this by providing convenient products linked to the performance of broad equity markets, many with features to reduce risk.

2.3.3.1 Guaranteed equity bonds

One of the most popular equity retail products is the guaranteed equity bond (GEB). This is a long-term product, often 5 or 7 years, which offers a guaranteed return of some fraction of principal plus participation in one or more equity markets. The simplest version is just composed of a zero coupon bond—providing a return of 100% of the initial principal—plus a call option on an equity index.

This product therefore offers a fixed-term structure with participation in the rise of the equity markets, diversification and limited downside risk thanks to the guarantee. These features make it easy to market to retail investors.

---

* This terminology for volatility is usually attributed to Emmanuel Derman and colleagues at Goldman Sachs. The idea of local volatility, and much else besides, were invented independently by Derman et al. and by Bruno Dupire: see Derman and Kani's "Riding on a smile" or Dupire's "Pricing with a smile" (both in *Risk* magazine, 1994). There is much more on this in Jim Gatheral's *The Volatility Surface*.

† Notice incidentally that in a falling sticky strike market, a short gamma trader loses money on both the spot move (because of gamma) but also the vol moves (because they are often short vega too). It is difficult to take a position on volatility without also taking a position on the volatility regime that pertains.

The 'magic' version of the GEB from a marketing perspective is a 100% principal guarantee plus a call on 100% of the notional. To see when this is possible, consider the components of the structure:

- Fees and costs. Retail structures require marketing campaigns, brochures and account maintenance. In addition, they are sometimes sold by agents who take a commission which might be 4 or 5% of notional.

- The zero coupon bond depends on rates. In a high-rate environment, paying 100 in 5 years' time only requires an investment of 70 or 75% of notional, leaving 25% to pay for the option; in a lower-rate setting, the zero absorbs 80% or more of the initial investment, leaving less for the option.

- The option price depends on implied volatility, the notional and the strike. If vols are low, then a 5-year ATM call option on the FTSE 100 might cost 20%, giving us a breakdown of fees and P/L: 5%, option: 20% and zero: 75%.

- The notional of the embedded option in a GEB is known as its *participation*. If vols are higher, then the ATM call price goes up to 30% of notional, and we cannot structure a profitable GEB with 100% guarantee, 100% participation and an ATM strike.

The illustrations below show the participation possible as a function of implied volatility and of the level of the guarantee provided.

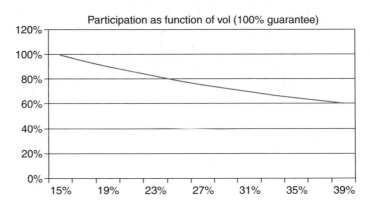

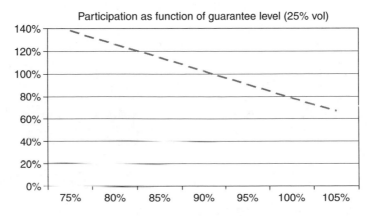

2.3.3.2 The risk profile of a GEB book

Suppose GEBs prove attractive to retail buyers. GEB providers sell a lot of product, and they buy their hedges from investment banks. These banks develop a large short vega position thanks to the sale of long-dated calls. How can they hedge?

- The ideal hedge is to buy back the calls, but there are few natural sellers of long-dated equity index volatility.

- In the early days, some dealers hedged by buying short-dated calls on the exchange to hedge their net vega, then delta hedging the resulting position.

- However, this generated a vega spread position which hurt when the term structure of volatility (the relationship between the implied vols for short- and long-dated options) inverted.

- Some players therefore turned to *volatility swaps* to buy their vega.

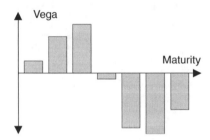

This example shows how risk management needs—the short vega position generated by selling GEBs—generates product innovation. The vol swap was popularised as a way of sourcing long-dated index volatility, allowing GEB hedge providers to continue to support the retail note market without accumulating an untenable risk position.

2.3.4 A Little More about Exotic Options

Exotic options are typically used for several reasons:

- An investor has a particular risk profile that they want to profit from;

- An investment bank has a risk or collection of risks that it wishes to hedge;

- Or an investment bank notices a 'perception arbitrage' between the modelled value of a structure and what investors think it is worth.

A good example of the use of exoticism comes in the development of the retail equity note market.

2.3.4.1 Retail note structuring alternatives

During periods of low volatility relatively simple GEB structures are popular: when volatilities increase, alternative, more complex structures have to be used, as products without a 100% guarantee are hard to sell. Possibilities include:

- *Capping the upside*, so the investor has a call spread or similar structure rather than a call.

- Making the underlying a *basket* of indices rather than a single index to bring the volatility down.

- Using a *downside knock-out*—a form of barrier option—so that if the market falls below 80% of its starting value, the option disappears and the investor cannot profit from any subsequent rises. This makes the option cheaper but can introduce *reputational risk* if retail investors are sold a structure they (can claim they) do not understand.

- Inverting the structure, so instead of buying a call, the investor sells a put, put spread or barrier put. This version, known as a *reverse convertible*, gives a high coupon but loss of principal if the market falls below the put strike.

2.3.4.2 The need for exotic models

If we only want to price plain vanilla options on liquid underlyings, we only need Black–Scholes; all of the information in the market is captured in the Black–Scholes implied volatility. But suppose we have a barrier option: specifically a 5-year call on the FTSE 100 struck at 6,100 that *knocks out* (ceases to exist) if the FTSE 100 ever hits 5,500 during the life of the option. Intuitively we care about the delivered volatility around two levels here: 6,100, because that is the strike, and we know that delivered vol around the strike determines the cost of hedging a call, and 5,500, because we go from having an option to not having one at that level, so there is gamma there too as the delta disappears below 5,500.

We could price this option using a standard extension of the Black–Scholes framework* but that would only give us one volatility input. The FTSE 100, on the other hand, has a smile, and there is a higher Black–Scholes implied vol for 5,500 than 6,100. Which one should we put in our Black–Scholes barrier model?

There is no right answer to this question. A Black–Scholes model cannot capture the expectation of a different delivered vol at 6,100 versus 5,500 as it assumes delivered vol is everywhere constant. We need a new model: an exotic model which captures the information available in the entire volatility surface rather than a single implied volatility.

2.3.4.3 Exotic model desiderata

Typically, banks begin exotic model development with a concrete problem in mind. There is a product or a range of products they wish to price and hedge. The business typically wants the simplest, easiest to use model that deals with the product and its hedges properly.

Therefore, we want a model that is (in no particular order):

- *Technically correct.* It should fit into the established framework of option pricing, (usually) be arbitrage free and so on.

- *P/L variance minimising.* The whole point about hedging is that it should enable us to lock in the premium we have bought or sold the option for. Therefore, the model must produce good greeks in the sense that if we hedge using them, the P/L over the life of the option is close to the option price regardless of market conditions.

---

* The details are standard: see for instance *Over the Rainbow* (*op. cit.*).

- *Descriptively rich enough, but not too rich.* We want the model to capture the phenomenon we are interested in, but it does not have to capture lots of other (perhaps perfectly interesting but irrelevant and potentially confusing) phenomena.

- *Comprehensible.* Since the model will be used on the trading desk by people who may not have a lot of time for mathematical niceties, it is good if we can easily explain what the model does and what it does not do.

- *Easy to calibrate.* It should be quick and easy to take market prices and calibrate the model using them. Once calibrated the model should return market prices for liquidly traded instruments to within bid/offer.

- *Stable once calibrated.* The model should remain over a reasonable range of market moves. If every recalibration produces radically different greeks, traders tend to distrust the model.

- *Intuitive.* It should be straightforward to understand how the valuations produced by the model change on the basis of how the inputs have moved. A model that produces a volatile P/L without the source of that volatility being obvious is typically undesirable, even if it behaves well on a time average.

2.3.4.4 Hedge analysis

Does a model do what it says on the tin? One way to find out is to run it with actual or simulated market data and see how the hedges perform. The process is as follows:

- We begin with current market levels and a sample position.
- The model is calibrated and greeks are produced from the model.
- The greeks are used to construct a hedge portfolio.
- Then a move in market levels is simulated, perhaps a 1 day move.
- The net P/L for the derivative and its hedges on the basis of this move are calculated.
- The model is re-run producing new greeks.
- The portfolio is re-hedged based on those greeks, then time is stepped forward another interval.
- This continues to the maturity of the derivative, giving a net P/L for the hedge strategy over this particular path of market factors.
- The process is then iterated for a number of different paths, giving a distribution of P/Ls.

If this distribution is centred on the model price—so that the model is estimating the real cost of hedging—and it has a tolerable S.D.—so that paths that produce radically different hedge costs are highly improbable—then we can have some confidence in the model. If not, the analysis should pinpoint the circumstances under which the model P/L does not accurately reflect what can be captured by hedging.

Extended exercise. Select a typical long-dated sold option position, and gather a long market data series for each parameter needed to mark the option.

Implement the hedge strategy analysis suggested above, including hedging all parameters (so that for an equity call, for instance, you include the interest rate hedges and the variable funding and stock borrow on the stock hedge). The fair value of the hedge strategy will become spread to the implied vol due to these extra risks and costs: how big a spread, roughly? How does it compare with the bid/offer spread in volatility?

Figure 5 summarises the results of the hedge analysis of a particular model for one path of the underlying S. For the example shown:

- The option has positive delta: as the underlying goes up, so does the option. The hedge is therefore to be short delta, and the hedge amount from the model is shown. Selling the hedge gives us a net long cash position, also shown.

- As the underlying moves each day, we re-adjust the delta hedge according to the model.

- The net P/L position is the combination of the mark-to-market on the option, the P/L on the hedge and interest on the cash position. We assume that the option is originally bought at the model value, so the net P/L starts out at zero.

- Note how a small amount of delivered volatility late in the life of the option close to the strike generates considerable volatility in the delta, and symmetrically in the cash position.

- There is a clear cause for concern: despite diligently following the model hedge ratios, the net P/L has a small but persistent downward trend. It may be that we are short gamma on some unmodelled risk factor: certainly further investigation will be needed to understand why hedging is not capturing all the value the model suggests is there.

2.3.5 Local and Stochastic Volatility

There are two common approaches to deal with the phenomenon of the smile:

- The first approach leads to *local volatility models*. These models allow volatility to depend on strike and time, so we work with a process such as

$$dS = \mu S\, dt + \sigma_L(S,t) S\, dW$$

This simultaneously has the advantage of dealing with the entire phenomenon—by making volatility strike- and time-dependent by definition, the entire smile is captured—and the disadvantage that we have explained nothing: all the information is in the calibration.

- The second approach is to let volatility itself be variable. For instance in a *stochastic volatility* model, it could be driven by a (second) random variable Z. We could, for instance, set $dS = \mu S\,dt + \sigma_S S\,dW$ but with the variance of vol (the square of σ_S) itself as a stochastic variable:

$$d\sigma_S^2 = \alpha(S, t, \sigma_S)\,dt + \beta(S, t, \sigma_S)\,dZ$$

Here we have perhaps the more intuitively compelling idea that volatility itself varies randomly, with drift given by α and with vol of vol β. The random process Z driving the evolution of σ_S can take various different forms, and typically we define some constraint on the comovement of W and Z.

2.3.5.1 Local volatility models

The key insight in local volatility models is that if we have the prices of calls for all strikes and maturities up to T, we can derive a unique consistent implied return distribution.

The local vol can be connected with the familiar Black–Scholes vol σ at strike K in a way that should make sense in the light of the previous discussion about delivered vol around the strike: $\sigma = E(\sigma_L | S(T) = K)$.

2.3.5.2 Example: knock-in reverse convertible

Consider the following structure:

1 year reverse convertible note			
Underlying	ABC Plc. stock	Issue price	€1000
Coupon	9%	Put amount	min(€1000, value 20 shares of ABC)
ABC price at issue	€50	Redemption	Par if ABC stock never falls beneath €40, put amount otherwise

The investor takes the risk that ABC will fall below €40 at some point during the note's life, and will be beneath €50 at maturity. In exchange for this, they receive an enhanced coupon.

These kinds of note are relatively common in some retail markets: for high volatility stocks, especially those with steep skews, they can offer the investor coupons much higher than deposit rates even for short-dated structures.

If ABC stock has a liquid series of options available on it, this kind of structure could be valued using a local volatility model. This would give not just a different price from Black–Scholes—about 5% different for a typical single stock smile—but also different greeks.

> *Exercise.* Build your own local volatility model and calibrate it for a simple knock-in reverse convertible structure. How different are the greeks from those that come from a Black–Scholes barrier model? How sensitive are they to calibration?

2.3.5.3 Issues with local volatility models

There are (at least) two classes of problems with using a local vol model. First, it is only practical to calibrate a local vol model if you have lots of liquid option prices, so that constrains the range of underlyings you can use them for.

Second, in practice the calibration often tends to be rather unstable: the shape of the implied return distribution seems to change rather quickly over time. This would not matter too much in the abstract, but it results in both P/L volatility and unstable hedge ratios. To see the P/L effect, consider the straight Black–Scholes P/L on delta hedging a long call position

$$\gamma \times \frac{S^2}{2} \times (\text{realised vol}^2 - \text{implied vol}^2)$$

In a local vol model, the rough analogue of this (taking liberties with discounting and underlying drift) is

$$\gamma \times \frac{S^2}{2} \times (\text{realised vol}^2 - \text{implied vol}^2)$$

$$+ \frac{\partial V}{\partial \sigma} \times \text{vol drift} + \frac{\partial^2 V}{\partial \sigma^2} \times \frac{\text{vol of vol}}{2}$$

$$+ \frac{\partial^2 V}{\partial \sigma \partial S} \times S \times \sigma \times \text{corr}(S, \sigma)$$

The last three terms are the P/L from the vol process: a vol drift term, a volga term and a vanna term. Unfortunately, these terms tend to be rather volatile under recalibration, which in turn produces P/L volatility.[*] This makes local vol models useful for gaining extra insight into the initial *pricing* of some exotics—particularly those that are strongly path dependent—but sometimes unsuitable for *hedging* them unless we can find a way to neutralise those volga and vanna terms in our hedging too.

2.3.5.4 Stochastic volatility models

Depending on our choices for α, β and the nature of the variation in Z, a range of stochastic volatility models can be created. One of the simplest is due to Heston[†] where we have a mean reverting volatility driven by a random walk Z

$$dS = \mu S \, dt + \sigma_S(t) S \, dW$$

$$d\sigma_S^2 = -\alpha(\sigma_S^2 - v)dt + \beta \sigma_S \, dZ$$

where constant v is average variance; α, speed of reversion to the average; and β, vol of vol. Z and W are related via a correlation $E(dW \, dZ) = \rho \, dt$.

---

[*] This discussion is a simplification of the (much longer and more detailed) account in Martin Forde's "The Real P&L in Black–Scholes and Dupire Delta Hedging" (*International Journal of Theoretical and Applied Finance*).
[†] The Heston model is discussed in Gatheral's and Wilmott's books (*op. cit.*).

In this setting, we can derive an analogue of the Black–Scholes PDE and, using rather heavier machinery, solve it numerically.

What would it mean for a model like this to be good? Ideally we would like a more or less stable calibration, so we should be able to set at least α, v and ρ initially, then need not to change them while still recovering a smile that fits the market at least at the maturity of interest.[*] Given the presence of different volatility regimes, it might not be too worrying if the vol of vol β needed to be remarked occasionally, although that will give rise to volga P/L which the model does not price in.

We are back, if not in the same place, at least with a familiar view. The greeks and in particular the higher-order sensitivities are significantly model dependent for most exotics (and sometimes for the vanillas too). Taste is not a good reason to prefer one model to another here: minimising P/L variance is. This suggests in practice:

- Hedging with similar instruments where possible;
- Trying to create a two-way market so we can hedge like with like;
- And hedging gamma path-by-path, filling in vega holes, and managing volga, vanna and other exotic sensitivities in as model-independent a way as is consistent with tolerable hedge costs.[†]

Exercise. If you can, find the documentation on an exotic model used by your institution. Try to discover:

— What the detailed theory behind the model is;
— How it is implemented;
— How the model is calibrated, and how often it is recalibrated;
— What the model is used for;
— What hedges are used, and how these are risk managed;
— Whether the model gives any exotic sensitivities such as volga or vanna, and if it does, how this information is used;
— How good the P/L explanation is for the model.

Based on your research, what notional or other limits would you put on the model's use?

2.4 INTEREST RATE DERIVATIVES AND YIELD CURVE MODELS

Derivatives whose underlying is one or more interest rates form the largest class of financial instruments, at least by outstanding notional, and one of the most important.

---

[*] If we want to recover the whole surface rather than just the smile at a specific maturity, the problem becomes much harder. Practically though, most of the things we want to price are European and our hedges will typically have the same maturity, so this problem often does not bite.
[†] Ayache, Henrotte, Nassar and Wang's *Can anyone solve the smile problem?* (available from www.ito33.com) is insightful on P/L variance reduction in these kinds of settings.

Interest rate derivatives pose new problems for several reasons:

- Even for a single Libor curve at one point in time t, we are dealing with a collection of underlyings, $L(t, t')$ which give us the return for depositing until t'.
- The forward rate arbitrage argument $(1 + L(t, t'))(1 + L(t', t'')) = (1 + L(t, t''))$ already discussed links spot and forward rates.
- As does the permitted dynamics of the curve.

To see what this means, suppose we start off with a normal shaped yield curve. As time evolves, there are things it is likely to do, and other things that are either very unlikely or impossible by no arbitrage. Some things that may happen to the yield curve as time goes from t to u include

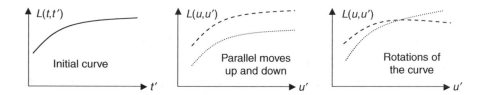

Whereas some of the things that cannot happen to the yield curve—because in both cases we would get a negative forward rate—include

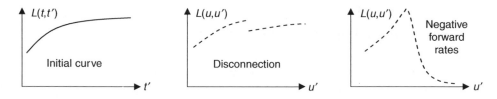

Coming up with a model which captures the observed yield curve dynamics but does not allow 'bad' curves to develop is a hard problem, and a lot is known about it.* We will start with a discussion of interest rate derivatives and their uses, and look at some of the more common models in later sections.

2.4.1 Futures and Forwards on Interest Rates

We have already seen a typical future on interest rates when we discussed the long gilt contract [in section 1.1.5]. Futures contracts on both short-term rates—such as 3-month Libor—are among the most liquid futures contracts in the world. For many currencies, a wide range of maturities is available, known as the *eurodollar strip*.† Alternatively, any (or at least most) desired interest rates can be locked in via an OTC contract.

---

* See for instance Riccardo Rebonato's *Modern Pricing of Interest-Rate Derivatives*, Jessica James' *Interest Rate Modelling* or, at a more advanced level, Rebonato's *Volatility and Correlation*.
† The term "eurodollar" is similar to "eurobond"; it simply indicates an inter-professional wholesale market, rather than something specifically European or EUR-related.

2.4.1.1 The eurodollar strip

The strip offers a range of standardised contracts on 3-month Libor at various points in the future rates, typically with maturities separated by 3 months, and stretching out some years. Most liquidity is concentrated in the front few months, but in G4 currencies even the 2-year future on 3-month rates can be liquid. As usual with futures contracts, the counterparty to the contract is the exchange, so credit risk is small, and typically the contract is *margined*.

2.4.1.2 Forward rate agreements

A FRA is an OTC contract which allows a counterparty to lock in an interest rate at some point in the future. So this is a:

- Bilateral contractual agreement between two parties;
- Where on trade date a notional principal, an underlying rate, a maturity and a contract rate is agreed;
- Some short period (typically 2 days) before maturity, the actual underlying rate is *fixed*;
- At maturity, a payment is made based on the *difference* between the contract rate and the actual rate on the notional principal.

If the FRA maturity is 2 months and the reference rate is 3-month Libor, we say the FRA is two *into* five (5 months being the maturity plus the rate period): this product locks in 3-month Libor 2 months forward. The timeline is shown in the illustration.

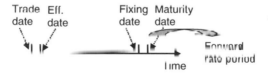

2.4.2 Interest Rate and Asset Swaps

The interest rate derivatives market grew up after the bond market, so some of its features are based on bonds. In particular, if you hold a bond, you might wish to hedge its interest rate risk. As we have already seen, this is not very simple as there are a number of cashflows: the coupons and the return of principal. We could simply use a FRA for each one. Suppose we have a 5% semi-annual pay bond with residual maturity 17 months and $100M notional. This bond is funded by the issuance of a note paying 6-month Libor. Therefore, our risk on the position is that 6-month Libor will go above 5% during the life of the bond. The cashflows are:

- A bond coupon of 2.5% × day count × $100M in 5 months' time;
- A bond coupon of 2.5% × day count × $100M in 11 months' time;
- A bond coupon of 2.5% × day count × $100M plus $100M principal in 17 months' time.

2.4.2.1 Strips of FRAs

Each of these cashflows can be hedged with a FRA. For the first cashflow, suppose we can pay on a 5 × 11 FRA at 4.60%. Then we have

Time	FRA Pay	FRA Receive	Funding Pay	Bond Receive	Net
5 months	4.6%	6 months Libor	6 months Libor	5%	0.4%

A pure spread of 40 bps is guaranteed provided we have matched dates of cashflows, interest conventions, etc. For the next coupon the situation is the same: there consulting the market we find we can pay on a 11 × 17 FRA at 4.82%, netting a spread of 18 bps.

Finally, at maturity of the bond, we have to repay the principal of the funding as well as pay a final coupon. For a 17 × 23 FRA the market rate is 5.11%, so here

Time	FRA Pay	FRA Receive	Funding Pay	Bond Receive	Net
17 months	5.11%	6 months Libor	100 + 6 months Libor	105	−0.11%

We might think that the net profit we have locked in here is 40 + 18 − 11 = 47 bps, but this is not quite true as we get the 40 earlier than we pay out the 11, so there is a PV effect which increases our effective spread. Working all of this out is slightly tedious, and the situation is so common that this structure is packaged up and a single blended rate quoted for it. This is an interest rate swap or IRS.

2.4.2.2 Swaps

A swap is an OTC contractual agreement between counterparties to exchange cashflows at specified future times according to pre-specified conditions:

- Typically with a notional principal;
- And a stated maturity, often close to the date of the last cashflow.

Each set of cashflows might be:

- A fixed percentage of the notional principal;
- A floating percentage depending on some interest or currency rate, such as 3-month Libor or 1-month Libor plus 25 bps;
- Or otherwise determined.

An IRS is a swap in one currency where:

- One set of cashflows is a fixed percentage of the notional principal;
- The other set is determined by a short-term interest rate index such as 3-month Libor in that currency;
- The cashflows on each leg typically occur on matched dates;
- Payment is of net amounts, and initial and final notionals are not exchanged.

This contract is typically transacted under an ISDA master [see section 10.1.1] and documented using a swap confirmation.

2.4.2.3 Swap terminology

- *Trade date.* The date on which the parties commit to the swap and agree to its terms.
- *Effective date.* The date on which interest starts to accrue, often 2 days after the trade date.
- *Payer.* The party who pays the fixed rate.
- *Receiver.* The party who receives the fixed rate.

2.4.2.4 Asset swaps

Some investors, as in the example above, fund at some spread to Libor. Therefore, it makes sense for them to have floating rate assets. Unfortunately, the universe of these assets is rather smaller than that of fixed rate assets. Therefore, investment banks take a fixed rate bond and swap it to produce a synthetic floating rate note. The package is called an *asset swap.* Entering into an asset swap consists of:

- The purchase of a fixed rate bond together with;
- The agreement to swap the scheduled cashflows on the bond for a floating rate plus a spread *x*.

Typically, the bond is pledged as collateral against the swap.

Bonds are often bought in the secondary market to asset swap. There are two different structures common in the market:

- In a *par asset swap*, the investor pays par rather than the market value of the bond (so the bank has credit exposure to the investor if the bond trades at a premium and is long cash if it is at a discount).
- Whereas in a *market asset swap* the market value is paid.

Par asset swaps make it particularly easy to see the all-in spread of the instrument: it is just the asset swap spread. Market asset swap spreads will obviously be lower than par spreads for discount bonds and higher for premium ones.

> *Exercise.* Consider the 17-month 5% bond again and suppose it trades at 101 and asset swaps at a market asset swap spread of 40 bps. What is the par asset swap spread?

2.4.2.5 Building the Libor curve

The market rarely provides instruments which exactly match the dates we have cashflows. For instance in the 5% bond example above we need to know what 6-month Libor will be 5 and 11 months into the future. The easiest way to obtain this information is to build the *Libor curve* then use this curve to estimate any forward Libors we need.*

The basic data for building the Libor curve in most currencies are:

- For spot rates, interbank deposits or other money market instruments;
- The eurodollar strip of interest rate futures prices (and possibly some FRA rates);
- And for the longer maturities, quoted swap rates.

The deposits start us off: they give us the first few Libors. Next the futures: the convention for quoting eurodollar futures is 100 − yield, so if a 2-month future on 3-month rates is quoted at 95.780, the 3-month Libor 2 months forward is 4.220%, is it not?

Not quite. The problem is that the eurodollar future price is not an *unbiased* estimate of the expected spot price in the future due to *convexity.*† To see it in action, think about receiving on a 2 × 5 FRA versus being short the matching eurodollar future:

- Suppose both contracts are traded at 4.220%.
- If rates fall, the net payment on the FRA to us goes down and, because rates have gone down we discount this future payment back at a lower rate.
- If rates rise, the net payment on the FRA to us goes up and, because rates have gone up, we discount this back at a higher rate.
- But for the future the changes in value are monetised immediately via margin.
- The receive position on the FRA is therefore positive convexity and the duration of the future is slightly higher than that of the FRA.
- To get the correct forward curve, we need to adjust the futures price to correct for this phenomena. This is known as a *convexity adjustment.*

Once we get to the end of the maturity of the liquid exchange-traded futures, we typically switch over to swap quotes in building the Libor curve. Then, once the basic data is in

---

* The standard approach to curve building is widely discussed, for instance, in Hull (*op. cit.*).

† Yes, yet another phenomenon called convexity. Convexity appears in so many places because so many relationships in finance are not straight lines, and/or most variables have some volatility. A junior interest rate quant once suggested to me that he believed that whenever his boss had a difference of 1 bps or less between his answer and the "right" answer, he would mutter "just a convexity adjustment" and pass on. This is probably untrue, but it does at least illustrate the ubiquity of wrestling with convexity.

place, we have to decide how to draw a line between the points which will allow us to interpolate a rate at any given date. The difficulty here is keeping a balance between:

- *Smoothness.* Typically, we expect both spot and forward rates to vary rather smoothly with time. Therefore, the yield curve builder should try to construct a smooth curve, perhaps by a technique such as *cubic spline interpolation.*
- *Verisimilitude.* Just occasionally though, the real curve is not smooth. This typically happens when there is sharp rise in the cost of borrowing money for a short period, for instance, at year-end when liquidity is low and many financial institutions are contracting their balance sheets for their quarter ends. Indeed, this phenomenon of a significant rise in costs over the year-end is well enough known to have a name: it is called the *turn*. Clearly, a curve builder should not be so smooth that it flattens real spikes in rates such as those in the turn.

2.4.2.6 Pricing
Conceptually swap pricing is straightforward:

- Once we have the Libor curve, the forward rates at any point follow by no arbitrage.
- This gives us the size of each cashflow in the floating leg.
- The value of the leg is just the sum of the PVs of these cashflows.
- Similarly the value of the fixed leg is just the sum of the PVs of its cashflows.
- And the value of the swap is the difference between these two amounts.

2.4.3 Credit Risk in Swap Structures
Swaps are bilateral agreements: I agree to pay you this series of cashflows; you agree to pay that one. Therefore, you have the risk that I will not perform on the arrangement, and symmetrically I have risk on your non-performance. Typically, the payments are netted, so for any pair of simultaneous cashflows the credit risk is only one way around at a given point in time. Think of a classical upwards pointing Libor curve. The forward Libors are above spot, so the curve predicts that the (fixed rate) payer will have a net cash outflow early in the life of the swap which should be compensated by the higher Libors later in life. In effect this is a kind of loan: the payer is sending out cash now in the hope of getting it back later.

In the abstract, if I am owed a net cashflow by an institution which could default, I should PV that payment along their risky curve. This would be a *credit-adjusted* swap price. However, two factors make practical credit-adjusted pricing more complex than this.

- Most swaps are done under ISDA master agreements. Then, if counterparty defaults, the swap terminates and the net difference between the two legs becomes immediately due and payable. However, ISDAs also net *all* payments between one party and another under the master, so the credit adjustment depends on the entire portfolio of exposures we have with counterparty.

- Second, many swaps are done with some form of credit enhancement such as collateral or downgrade triggers. This can also serve to mitigate the credit exposure on the swap.

[Given this complexity we postpone dealing with credit-adjusted pricing further until section 5.3.7.]

2.4.4 Other Interest Rate Swap Structures

A vast array of different swap structures has been invented: we mention only a few of the more common types.

2.4.4.1 Off-market and prepaid swaps

Most IRSs are transacted *on market*, that is the fixed rate is arranged so that the PV of the fixed rate equals that of the floating rate on trade date. If the swap is *off-market* (that is one of the legs is more valuable than the other) an initial or occasionally final payment can be made.

A related structure is the *prepaid swap*. Here one leg, usually the fixed one, is prepaid at the start of the transaction. This structure obviously gives rise to significantly more credit risk than a standard IRS.

2.4.4.2 Forward starting and amortising swaps

An agreement to enter into a particular swap at some fixed date in the future is known as a *forward starting* swap.

An *amortising* swap has a notional principal which varies according to a predetermined schedule, for instance, to hedge an amortising bond.

All of the structures so far are priced as before as their cashflow timing and sizes are determined by the Libor curve. A more exotic instrument is the *index amortising swap* where the amortisation schedule is not fixed but instead determined by some other financial variable such as an equity index. Clearly, here we need to know not just what the expectation of the index level is, but also what comovement there might be between index levels and interest rates.

2.4.4.3 CMS and Libor-in-arrears swaps

A *constant maturity swap* (CMS) is one where the fixed leg is determined by some constant maturity level, for instance, the 10-year swap rate. Thus, the payer pays whatever the 10-year swap rate is at fixing versus floating.

A *Libor-in-arrears swap* is best remembered as a Libor *set* in arrears swap: the floating rate side of an ordinary IRS is reset on each coupon date and, once set, the payment due is calculated and paid at the end of the period. In a Libor set in arrears swap, the floating leg is set *and* paid at the end of the period.

Both CMS and Libor-in-arrears swaps require a model of the yield curve dynamics to price so they are fundamentally different from FRAs or ordinary IRSs.

> *Exercise.* Why would a corporate client want a Libor-in-arrears swap? Why trade them?

2.4.5 Cross-Currency Swaps

Suppose we work for a European institution which takes fixed-term deposits in EUR at 3-month Libor and we wish to acquire $100M of a fixed rate bond paying coupons 5% in USD to the same maturity. Clearly, we are exposed to a variety of risks if we use the deposits to fund the bond. A hedge might be helpful. What would a complete hedge look like?

- First, we would take sufficient of the EUR deposits and turn them into USD in the spot market to buy the bond.
- Next we need to swap each 2.5% semi-annual coupon in USD for a 3-month Libor coupon in EUR to pay on to our deposit investors.*
- Finally, when the bond expires, we will need a forward FX transaction to turn the $100M back into EUR to pay back our depositors.

The package consisting of this initial exchange of notional, coupons in one currency versus coupons in another, and the final exchange of notional is called a *cross-currency swap*. It comes in four varieties: fixed or floating in one leg versus fixed or floating in the other, and thus allows a fixed or floating asset in one currency to be turned into a fixed or floating asset in another.

Pricing is straightforward: we just need Libor curves in both currencies and the FX spot rate. The forward FX rates follow no arbitrage, and we PV along each currency as usual. Or do we? There is one problem: the existence of the following instrument.

2.4.6 Basis Swaps

Consider a 10-year cross currency swap receive 3-month Libor in USD versus pay 3-month Libor + x in JPY.

If we ignore any slight differences caused by business day mismatches, this should price flat, i.e., the mid market spread x at which this swap will be transacted should be zero since by definition either path around the square shown below should give the same result. But in reality this floating versus floating cross-currency swap often does not trade flat.

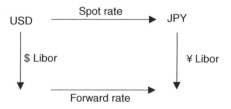

---

* Note that there may be coupon dates in USD that are not business days in EUR and vice versa. Therefore, in practice, depending on the precise business day convention, there may be a slight mismatch of payments.

Any swap of one floating rate for another is referred to as a *basis swap*: a basis swap is determined by a pair of floating rates and a maturity. If the two rates are Libors in different currencies, the swap is a floating versus floating cross-currency swap and the market swap level is known as the (*cross-currency*) *basis swap spread*.

Thus, for instance, we might see a ten year swap for 3-month $ Libor versus 3-month ¥ Libor quoted at 17 bps mid, meaning that the mid market basis swap at 10 years is 3-month $ Libor versus 3-month ¥ Libor + 17 bps.

Basis swap spreads are typically single-digit basis points, but they can be tens or more, especially for USD/JPY.

2.4.6.1 Causes of the basis swap spread

One way of thinking about the presence of the basis swap spread is to consider funding. The Libor curve in our previous framework is both telling us the cost of funding a cashflow in the future, and simultaneously what its PV is. We need to split these two roles. Consider funding first: here we are dealing with a market in *liquidity*.

If yen deposits are cheap and plentiful whereas dollar deposits are more expensive to source then institutions facilitating the conversion of one into the other can charge a premium.

Alternatively, the USD Libor rate can be thought of as the blended rate of funding of large U.S. banks: to the extent that these are significantly better credit quality than large Japanese banks, we could also think of the USD/JPY basis swap spread as some kind of blended credit premium.

Whichever way we think about it, the basis swap spread is telling us that just because we fund in one currency at Libor flat does not mean we can construct funding in other currencies at Libor flat too, at least if we use the basis swap market.

2.4.6.2 Implications of the basis swap spread

The problem with the basis swap spread is that it means we cannot use Libor curves for both calculating forwards and discounting in more than one currency. For a concrete example of this suppose we use the Libor curve to price the following swap S1

pay *f*% fixed USD versus receive 3-month Libor USD

By forward rate parity, the following currency swap S2 will price on market too:

receive *f*% fixed USD versus pay 3-month Libor JPY

But by the definition of the basis swap spread, the following basis swap S3 prices on market (where bss is positive)

pay 3-month Libor USD versus receive 3-month Libor − bss JPY

S1 + S2 + S3 is completely matched, costs nothing to put on, and has positive value, bss, so we have clearly done something wrong. The solution is to construct FX forwards as usual, but apply the basis swap spread in currencies away from our funding currency for discounting. That will ensure we price on market basis swaps to zero PV, and price cross-currency swaps consistently. It will give an arbitrage between long-dated FX forwards and cross-currency swaps, but that arbitrage is real.*

2.4.7 Caps, Floors and Yield Curve Models

An *interest rate cap* is a strip of options on interest rates, usually with a fixed strike. Specifically, we select a rate, such as a 3-month Libor, and a maturity, say 10 years, and agree in exchange for a premium to payout if, on any fixing date, 3-month Libor is above the cap strike.

The fixes are typically separated by the same interval as the rate, so the protection is continuous. Each individual option in the structure is known as a *caplet*.

Similarly, an *interest rate floor* is a contract in which the seller compensates the buyer on each occasion when the observed rate is less than the predetermined strike. Caps and floors of maturities out to 10 years or more are liquid in major currencies.

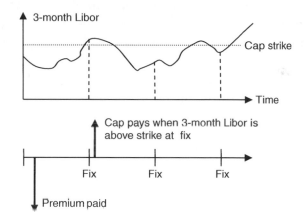

2.4.7.1 Pricing interest rate caps

Fisher Black, of Black–Scholes fame, noticed soon after the original piece of work that commodities did not adapt well to the Black–Scholes framework. The main problem is that the spot price often has rather different dynamics from the forward price, due to issues of deliverability, squeezes in the spot market and so on. So he reworked the framework with the future as the basic variable, and derived a price for plain vanilla options based on the assumption that the future price had log-normal returns with volatility σ.

The resulting formulae, known as the *Black* model, looks rather similar to the original framework. For instance, for a call on a forward F

$$C(F, K, \sigma, r, t) = e^{-rt}(F\Phi(d_1) - K\Phi(d_1 - \sigma\sqrt{t}))$$

---

* Once we have paid bid/offer on both sides of the arb and accounted for the credit risk or the cost of credit risk mitigation, though, this arb may not meet a reasonable ROE target, so it may be more theoretical than real. The exception to this in the past has occasionally been in yen.

where

$$d_1 = \frac{\ln(F/K) + (1/2)\sigma^2 T}{\sigma\sqrt{T}}$$

This model can be used to price caps. For each caplet:

- The underlying is the relevant forward Libor rate.
- The maturity is the fixing date.
- The implied volatility is taken from a broker quote for that maturity and strike cap.
- The 'risk-free' rate is the Libor rate to the caplet maturity.

While it is of course formally inconsistent to assume both a constant risk-free rate to discount our payoff back by and a stochastic underlying rate, the procedure works perfectly well in practice. We simply price each individual caplet and add the prices up.

The market quotes cap prices as a *flat* volatility, that is a single volatility number. The convention is that if we use this vol for each caplet, we get the right price. As before, that does *not* mean that the Black model captures all of the relevant features of interest rates, simply that the market uses flat Black vols as a convenient way of quoting cap prices.

2.4.7.2 Black on the forward dynamics

The Black model is perfectly adequate for plain vanilla caps, but it does have a couple of obvious issues:

- Some practitioners take the view that interest rates are mean reverting, and this model does not capture that property: as time goes on, rates evolve further and further away from the original value.
- There is no relationship between different maturities: clearly a 15-month cap on 3-month Libor is just a 1-year cap with an extra caplet, yet the Black model treats each caplet as an independent variable. There is no way of capturing the intuition that these need to paste these together to create a consistent curve.

2.4.7.3 Mean reversion

Vasicek developed an interest rate model which accounts for the first objection by including mean reversion. In the Vasicek model, the short rate is assumed to follow

$$dr = \theta(\alpha - r)\,dt + \sigma\,dW$$

Here α is the long-term average short rate, and the parameter θ controls the speed of mean reversion.

Exercise.   Examine the history of short rates in a currency that interests you and conduct some statistical tests. Does the data support the idea of mean reversion?

Models like Black's and Vasicek's are called *one-factor short rate models* because there is only one source of uncertainty, *W*, and this affects the short-term interest rate. There are a number of further models of this type.*

2.4.7.4 The whole curve

Any one factor model predicts the whole yield curve: the expected future value of the short rate determines today's long-term rate, so the probability weighted average of the future short rates should lie on the current curve.

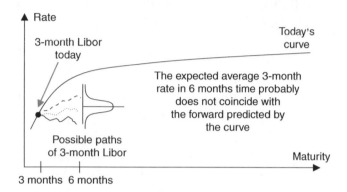

Sadly, this will only happen by the most fortunate accident: a one-factor model does not have enough parameters to model the whole curve.

If we are only dealing with plain vanilla caps and floors as described so far, this is fine. We do not *need* to recover the whole curve, just to have a model that gives us an appropriate P/L volatility minimising delta hedge, and the delta hedge from the Black model is typically broadly acceptable. For something like a CMS, though, we will need a more sophisticated model.

2.4.7.5 The Ho and Lee model

The simplest way to capture the whole curve is just to introduce a time-dependent drift into the Black world using

$$dr = \mu(t)\, dt + \sigma dW$$

This is known as the Ho and Lee model. Again, we have not really explained anything: just increased the number of calibration variables to give us more freedom.

2.4.7.6 Chooser caps

A lovely product for highlighting the challenge of the full curve is the *chooser cap*. This is just like a plain vanilla cap, with the restriction that only a certain number of caplets can

---

* Hull or Wilmott (*op. cit.*) deal with some of them. It might be helpful to keep in mind when reviewing an interest rate model the questions: what future yield curves does it permit, how do different rates comove and what is the calibration procedure for the model.

be exercised. Thus, a 6 out of 12 chooser cap on 3-month Libor would be a 4-year cap on 3-month Libor with the restriction that a maximum of 6 of the 12 caplets could be exercised during the life of the structure. Here both history and comovement matter—if a high 3-month Libor going into a fixing means that future Libors are likely to be high too, then the future caplets are more valuable and perhaps would be better off waiting and not exercising the current caplet—whereas if we have short-term mean reversion, a high current 3-month Libor means future Libors are less likely to be high and so we should exercise the current caplet. Thus to decide which caplets to exercise and value this product properly we need to capture information from the market about the comovement of the different Libors. Before we turn to that, though, let us look at the final important vanilla Libor derivative not discussed so far.*

2.4.8 Swaptions

A swap option or more commonly *swaption* is an option to enter into a swap at some point in the future. Thus, it is a single option on a long-term rate: in contrast, a cap is a series of options on short-term rates.

For example, we might pay a 70 bps premium for the right but not the obligation in 1 year's time to enter into 5-year IRS receiving 5% fixed versus 3-month Libor flat. This is a *European receiver swaption*: European because we can only exercise it and enter into the swap at the option's maturity in a year's time; and receiver because it is the right to receive fixed.

Both cash-settled swaptions—where the option payout is the PV of the underlying swap—and physically settled swaptions—where the swap is actually entered into—are possible. The terminology is 1 *into* 5, the first number being the maturity of the option and the second that of the underlying swap.

At any point in time, the various swaption quotes provide information on the price of optionality across the curve: the 1 into 5 depends on the volatility of the 5-year rate; the 1 into 6, that of the 6-year rate; and the 2 into 5, that of the 5-year rate but for a year longer than the 1 into 5.

2.4.8.1 Hacking swaption valuation

We can cram swaptions into the standard Black–Scholes framework by choosing the par swap rate as an underlying. This is inconsistent with using Black for caps but that does not matter particularly unless we want to hedge swaptions with caps or *vice versa*.

2.4.8.2 Extending the Ho and Lee model

Just as we dealt with the smile in local volatility models by introducing a time- and strike-dependent volatility, here we could introduce a time-dependent drift, or a time-dependent volatility, or both. For instance, we could have a model of interest rates with

$$dr = \mu(t)\,dt + \sigma(t)\,dW$$

---

* We have completely side stepped the problem of derivatives on the government curve such as bond options. The arguments here are broadly similar and can be found in the standard sources. It is important to bear in mind that these markets are now well segmented: Libor instruments are used to hedge positions on the Libor curve; government bonds, on the govy curve; and when the twain meet, there is swap spread exposure.

This is still a one-factor model—there is only one source of uncertainty—but by including a time-dependent drift and time-dependent vols, we can match all the cap volatilities and the initial shape of the yield curve. This gives one harder-to-calibrate model, rather than a family of easier to calibrate Black models.*

2.4.9 Exotic Interest Rate Derivatives

We have already seen several exotic interest rate derivatives—CMS and index amortising swaps, chooser caps—and there are many more. For instance,

- *Yield curve options* pay out on the difference between two points on the curve, such as 6-month Libor and the 5-year par swap rate.
- *Bermudan swaptions* are swaptions exercisable on a specific schedule of dates, such as each fixing date.
- *Knock-out caps* incorporate barrier option features into an interest rate cap.

2.4.9.1 Libor market models

To model a product like a yield curve option, it will be important to have a model with more than one source of uncertainty: the value of a yield curve option depends on twists and inflections of the yield curve. One good approach is the *Libor market model* (LMM). An advantage here is that the model is based on observable market variables, and so is not so difficult to calibrate as some other multivariate models.[†]

We begin with some fixed times at which market-traded caplets fix: suppose for ease of discussion these are (the first business day following) 1st January, 1st April, 1st July and 1st October each year, and we will list them in order $k = 1, 2, \ldots$. The natural underlyings in the model are, therefore, the 3-month Libors between these dates each year, and we will write $L_k(t)$ for the 3-month Libor between k and $k + 1$ observed at time t, and $\sigma_{jk}(t)$ for the volatility of this rate at time t under factor j

The $L_k(t)$ are the variables. how many stochastic factors should drive them? If we pick one, W_1, we could write

$$dL_k(t) = \mu_k L_k(t)\,dt + \sigma_{1k} L_k(t)\,dW_1$$

This is just a family of one-factor models, each driven by the same stochastic variable, but with different drift functions and volatility functions for each Libor: in general, these will have a complex structural form, with μ_k and σ_{1k} depending not just on t but also all the $L_j(t)$s up to the current time as well. Here we can match the initial term structure and

---

* It is worth remembering that using Black means we have a separate Black model for each cap maturity and strike and one for each swaption maturity and strike, each model being used only for that (product, maturity, strike) triplet. There is no connection between any of the (differently calibrated) Black models.
† Our discussion of the LMM model—also known as the BGM model after one of the several groups who independently discovered the model, Brace, Gatarek and Musiela—is very much a sketch which sidesteps a number of important points for the sake of brevity. A more detailed account can be found in Hull, Rebonato or Wilmott (*op. cit.*).

recover all the caplet vols, but the model constrains the types of permitted movements of the curve as all the Libors move under the influence of one variable.

To capture the possibility of parallel moves, twists and inflections of the curve, we need more stochastic variables, W_j. Then we can define $E(dW_i dW_j) = \rho_{ij} dt$ as usual, perhaps with something of the form

$$dL_k(t) = \mu_k L_k(t)\, dt + \sum_j \sigma_{jk} L_k(t)\, dW_j$$

Again, each drift and volatility function at a given point will, in general, depend on the current level and history of all the Libors.

How many stochastic variables do we want? One obvious answer is one for each Libor, but that gives too many functions to calibrate. It is usually better to pick a small number of variables, perhaps representing the first few principal components of movements of the yield curve [see Chapter 4], and to enforce some functional form on the drift and volatility functions.*

2.4.9.2 Using the right model

Suppose:

- On 27 February 2006 we sell a 4-month swaption into a 1-year swap.
- On 1 March 2006 we buy a 4-month swaption into the same 1-year swap.

In each case, the terms of the transaction are the same (although for safety we will use different dealers), the curve has not moved materially in the meantime and the swaps refer to 3-month USD Libor.

> *Exercise.* If the bid/offer spread is the market norm, and the second dealer prices using the same mid market vol as the first, why might we be very happy with the position?

Clearly, we are long volatility for a short period. Why would that vol be particularly useful? The answer is that the period chosen includes an FOMC meeting when a decision on U.S. short-term rates is announced. Delivered vol around FOMC meetings is typically much higher than at ordinary times, so that vol would be significantly more valuable than a typical few days' long vol position. One would need a different kind of model to capture this effect: one that allowed us to calibrate to market vols yet specify a differential delivery of that volatility depending on key dates of interest to the interest rate markets.

---

* Rebonato makes the important point here that the contribution of Libor market models is not just the mathematical form of the models themselves but also their calibration procedures: see Riccardo Rebonato's *The Modern Pricing of Interest Rate Derivatives*.

2.4.9.3 Swap book risk measures

The techniques discussed earlier can be extended easily to swap books. Complexity arises from the three-dimensional nature of the problem:

- The *rate duration* dimension, i.e., we are dealing with an option on 1-year rates or 2-year rates;
- The *option maturity* dimension, i.e., we are dealing with an option that expires in 3 years or in 4 years;
- The *strike* dimension.

Typical risk reports would examine:

- Net interest rate deltas, gammas and vegas by bucket in each dimension;
- Scenarios based on parallel and non-parallel yield curve moves and moves in the volatility surface;
- Sensitivity to movements in the basis.

Notice that the instruments that go into the yield curve builder (the deposits, futures and swap rates) define a natural bucket structure and it makes sense to use this since rates away from these benchmarks are interpolated. Therefore, if we use the 10-year swap rate as a curve building input, we should have corresponding bucket in the risk report—9 to 11 years say—and this bucket should actually be hedged with 10-year swaps.

2.4.9.4 Suitability

The ability to invent, price, trade, risk manage and profit from increasingly exotic derivatives can lead firms into trading them without considering all the implications. A good example of the kinds of problem that can arise is the case of Bankers Trust (known as BT).

During the mid-1990s, BT had become a leader in the interest rate derivatives market and drew much of its profitability from this area, having largely withdrawn from its old business base of corporate lending. However, competition was considerable and as bid/offers spreads contracted in standard products, it was natural to look at more exotic derivatives with higher profit margins. One of the areas BT chose was *power swaps*: structures depending on powers of Libor such as Libor squared.

Some of the bank's clients sustained heavy losses on these products, and two of them, Gibson Greetings and Proctor & Gamble, successfully sued BT, asserting that they had not been informed of or had been unable to understand the risks involved in the product. Moreover, several tape recordings of BT staff supported the contention that the salespeople knew that the client did not understand the product, and valuations given to clients may have been manipulated.

The resulting financial settlement and reputational damage was significant. Indeed some commentators have suggested that (along with losses sustained in the Russian crisis) these events contributed to the sale of BT shortly thereafter.

With 20/20 hindsight at least, one could have seen this coming. How could you prove that a corporate client had a genuine need for a swap like 5.5% fixed versus $Libor^2/6\%$? And if they did not need it, how can you show after losses have been sustained that it was suitable for them and that they understood it? This is always a risk with exotic structures, and one that should have senior management oversight.

2.5 SINGLE-NAME CREDIT DERIVATIVES

One obvious way of taking corporate credit spread risk is to buy a corporate bond. But holding a bond involves taking a variety of other risks including interest rate risk and liquidity risk. Moreover, if we want to buy a bond, the position has to be funded so our profit depends on funding cost. *Credit derivatives* allow us to transfer pure credit spread risk without all of the other complications of dealing with corporate bonds.

In this section we will look at *single-name* credit derivatives: those that pass on the risk of a single obligator. [Chapter 5 has a discussion of portfolio credit derivatives.]

2.5.1 Products

A (plain vanilla) credit default swap or *CDS* is:

- An OTC bilateral contract where in exchange for the payment of premium;
- The protection seller agrees that if any one of a number of *credit events* occurs on a *reference instrument*;
- They will compensate the protection buyer for the difference between the value of the reference instrument after the credit event and par.

Typically, the reference instrument will be some senior debt security or loan, and the allowed credit events will include default, so the protection seller will compensate the protection buyer if there is a default by the reference obligation.

After a credit event, the holder of the reference instrument will get the recovery value, so the protection seller will suffer a loss of par minus recovery. Typically, this can happen in one of two ways:

- In *physical settlement*, the protection buyer can deliver the reference instrument (or one of a number of agreed deliverable instruments from the same obligator ranking *pari passu* with the reference instrument) and receive par.
- In *cash settlement*, recovery is determined by some means, such as a poll of dealers after the credit event, and a cash payment of par minus recovery is made.

Unlike a typical option, the premium in a CDS is usually paid *periodically* and this payment terminates in the event of a default. Quotation by dealers of bid (offer) CDS prices is as basis points per year paid (received) for protection.

In the early days of the CDS market, the reference instrument was a key part of the contract: it alone was deliverable. But most obligators have *cross-acceleration* on all their senior debt, so in the event of default all senior instruments become due and payable, and all of them will receive the same recovery as they rank *pari passu*. Therefore, having a single deliverable simply made credit derivatives less liquid and less useful. Permitting a wider range of deliverable instruments from the same issuer enhanced liquidity in the market. Just as the range of deliverable bonds into the long gilt contract means that the seller has a cheapest to deliver option, so when a CDS has a range of deliverable instruments, the protection buyer has the option to deliver the cheapest one.*

2.5.2 Credit Events and Documentation

As with many other derivatives, ISDA provide standard documentation for CDSs. The CDS market has grown rapidly from its infancy in the middle 1990s, and the documentation has gone through several iterations as the market evolved. In this section we look at some of the risk management issues pertinent to CDS documentation.

2.5.2.1 Credit events

The range of permitted credit events on a CDS is usually chosen from:

- *Bankruptcy.* This is straightforward: the obligator is bankrupt under some definition, often that of its country of incorporation.

- *Failure to pay.* The obligator has failed to pay a material sum under an obligation. There is often a grace period during which this failure can be *cured*. (These materiality and grace clauses are often also included in the other events too.) The failure to pay event assists in situations where local law may not have recognised a bankruptcy, but the obligator is nevertheless not performing.

- *Obligation default.* A default is declared on an appropriate obligation. The obligation may be documented under a different law from the obligator's incorporation, so again this may not correspond to bankruptcy.

- *Obligation acceleration.* Obligations can accelerate—that is become immediately due and payable—under a number of circumstances including default. Since payments on different obligations are due at different times, one obligation may default before the others do. Typically, cross-default clauses then accelerate all the other debt, and this clause captures that.

- *Repudiation/moratorium.* Here the obligator either declares a moratorium on payments or repudiates the obligation as theirs. These clauses are typically more useful for sovereign than corporate CDS.

- *Restructuring.* This covers a change in the terms of the obligation which is materially less favourable to the holders than the original, such as a reduction in interest, a reduction in principal or a lowering of seniority.

---

* There are some credits where it is generally accepted that the CDS market is significantly bigger than the cash market. If these credits ever suffer a credit event, there will then be a scramble to buy the bonds as some market participants will be short, i.e., they will own physically settled CDS protection but not a deliverable instrument. This may inflate the post-default bond price and so effect cash settled CDS too. In this case there may well be significant divergence between the eventual recovery to holders of senior obligations and the recovery obtained by those who had gone short the credit in the CDS market.

2.5.2.2 Restructuring

What does 'materially less favourable' mean? Clearly we can have a range of restructuring events from ones that cause minimal losses—such as an agreement to allow coupons to be paid a few days later—to ones that cause losses of large fractions of the notional. The combination of the former and the cheapest to deliver option came together in a well-known case in the CDS market, Conseco. Here there was a minor restructuring which resulted in a small change to the prices of existing bonds. However, the obligator had some long-dated bonds in issue which were trading well below par before the restructuring. After the restructuring the bond's price did not change much. However, the occurrence of the restructuring event allowed protection buyers to deliver them into the CDS, claiming par. Protection sellers claimed that the credit event—an extension of the term of some of Conseco's loans—was not material. However, restructuring was captured by the documentation used at that time, and many protection buyers prevailed. Thus, the market came to realise the value of the cheapest to deliver option in the presence of restructuring.

Following this, various modifications to the restructuring definition have been proposed, notably restrictions on the residual maturity of the deliverable instruments. North American counterparties now typically prefer to trade without restructuring as a credit event at all: European banks typically need restructuring to be included to have a CDS recognised for regulatory purposes. Hence in many cases the market has split with one price for CDS including restructuring and one without.

2.5.2.3 How many curves?

In Chapter 1 we discussed a corporate obligator's credit curve: this is the curve defining the spread of the obligator's bonds at various maturities. How should CDS spreads relate to this curve?

A first glance, there should be little difference between the excess spread of a corporate bond over the Libor curve, the asset swap spread of that bond and the CDS spread at the same maturity: in each case the holder of the instrument is being compensated for the risk of default of the obligator. But that is not quite the whole story for several reasons:

- An asset swapped bond is a convenient instrument whereas an ordinary corporate bond has fixed coupons and these have to be managed. It might also be illiquid. Therefore, the corporate bond may have a liquidity premium, and inconvenience premium,* or both in its spread. These might not be present in the asset swap spread.

- Depending on the credit events chosen, the CDS can be triggered by a broader range of events than pure default. Indeed, strictly, given that each credit event adds extra risk, there should be a hierarchy of CDS curves, with a lower spread for a CDS only

---

* In my local supermarket, the price of ready rolled puff pastry is roughly double that of the same pastry unrolled. It takes less than 10 min to roll pastry, and a rolling pin, a flat surface and a small amount of flour are all that is needed. If the pastry manufacturer can charge a premium for allowing the cook to avoid the inconvenience of rolling out their tart base, then an investment bank can certainly charge a premium to an investor for not having to hedge the interest rate risk of a fixed rate bond.

triggered by failure to pay and a higher one for all six events. Moreover, the cheapest to deliver option in a CDS is potentially more valuable the wider the range of deliverable obligations. Therefore, a CDS with a single deliverable obligation should be slightly cheaper than one with a wide range of deliverables, especially if potentially highly illiquid obligations can be delivered.

This discussion shows that we need to think not just of a single senior debt credit spread curve for an obligator, but of multiple curves depending on the precise nature of the risks transferred.

Exercise. Suppose you buy a senior corporate bond and purchase a CDS on that corporate with maturity identical to that of the bond. Try to list *all* of the financial risks the position is exposed to.

2.5.2.4 Credit-linked notes

One of the problems with buying CDS protection is that one does not know if one's counterparty is able to pay should a credit event occur. For a bank counterparty this may not be a significant issue, but if we want to allow a broader set of investors to buy credit exposure using CDS, then it would be wise to collateralise the exposure.

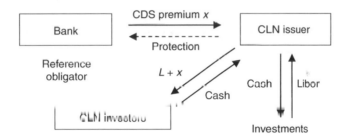

The *credit-linked note* (CLN) does this. We set up an SPV to issue a debt instrument. The cash raised is invested in very high credit quality paper, often AAA FRNs. The SPV sells protection to the sponsoring bank under a CDS, receiving a premium of x bps running for providing protection on a reference obligation. The note holder receives an enhanced coupon in exchange for taking the risk of a credit event.

The SPV's investments are pledged as collateral against the swap, and the note holder receives a coupon of Libor + x (or slightly less) provided the reference obligator does not have a credit event. At maturity the investments are liquidated giving the cash for the SPV to repay the final principal. If there is a credit event, then sufficient of the investments are liquidated to settle the CDS so that the note investors will lose principal. Thus, the performance of the CDS does not depend on that of the CLN buyers as it is collateralised by the SPV's investments.

The credit quality of the CLN—at least to the extent of the spread x—depends on the bank's performance under the CDS. Some banks use this as an argument to use their own floating rate paper as collateral, often giving them access to Libor flat funding.*

2.5.2.5 Total return swaps

An asset swap allows an investor to buy and fund an investment and receive a Libor-based coupon in exchange for taking default risk on it. But some investors want exposure without funding, perhaps because their balance sheets are small or they have relatively high funding costs. They need a different structure:

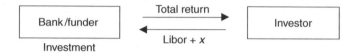

In a *total return swap*, the funding counterparty buys and funds an instrument and enters into a swap where they agree to pay any cashflows on the instrument and any changes in value, positive or negative, to an investor. The investor pays funding at Libor plus a spread, known as the *total return swap spread*. In risk terms this can be thought of as compensation the bank has for the possibility of joint default of the investor and the investment. In market terms, it is rent for use of the bank's balance sheet and the bank's low funding costs.

Exercise. Suppose an investor funds at Libor + 100, and wishes to buy a bond that asset swaps at Libor + 75. How would you determine their breakeven total return swap spread?

The total return swap illustrates an important theme in structured finance: the splitting of *funding* from *risk taking*:

- The best institutions to *take risk* are often investors such as hedge funds, mutual funds or pension funds. While the last of these do not have to leverage themselves, hedge funds and some mutual funds do to meet their target returns.

- Meanwhile there is a range of large, safe, low-cost funds institutions who are happy to use their balance sheets and earn a small spread to Libor, but do not want to take much risk.

---

* The credit exposure of the SPV to the bank could be mitigated if necessary. If the sub consolidates, investors should treat the CLN as a risky corporate issue of the bank (on Libor plus the bank's spread) plus a written CDS on the reference instrument. If it does not consolidate *and* the sub has sufficient credit support to continue paying the CLNs coupon even in the event of the bank's default, then it should treat the CLN as the collateral plus a written CDS. The difference is the bank's credit spread, and hopefully the accounting will follow the economics of the transaction.

A total return swap allows both parties to participate in an asset in different ways: taking risk but not owning or funding it; or owning and funding it but not taking default risk.

Total return swaps are similar to repo in that both structures allow a party to have the benefits and risks of owning a bond without having to finance it. However, the repo market primarily offers short-term funding for high-quality bonds whereas longer-term TRS funding is often available (at a price) even for lower credit quality assets.

2.5.3 Credit Derivatives Valuation

There are two steps in valuing a credit derivative:

- Decide on a framework which expresses the probability of a credit event occurring and calibrate this to the available market data.
- Calculate the value of the credit derivative in this framework.

2.5.3.1 A simple model

Either a credit event happens or it does not. Suppose the probability of a credit event in a given period is PD. Then if we have an exposure to a credit paying a spread s and ignoring rates, with probability $(1 - PD)$ we receive $(1 + s)$, and with probability PD, we receive some recovery R.

Therefore, if s is fair compensation for the probability of default, we should be indifferent to taking no risk and have our notional, or lending it at s and taking credit event risk:

$$1 = (1 + s)(1 - PD) + PD \times R$$

2.5.3.2 Credit event intensity

Let us refine this very simple model a little: suppose the probability of a credit event happening to a given reference instrument in a small time interval dt is $h\,dt$, where h is known as the *hazard rate*. Then the probability of surviving without a credit event* to time T is given by

$$\exp\left(\int_0^T - h(t)dt\right)$$

Consider a risky bond. For each cashflow, either there is a credit event before the cashflow or there is none.

If there is, we will receive some recovery value R, otherwise we will get the next cashflow. If r is the risk-free rate, it is straightforward to derive the risky PV for a cashflow at time T:

$$RPV_T = e^{-\int_0^T (1-R+r(t))h(t)dt}$$

Then we know that a risky bond price must be just the sum of the risky PVs of its cashflows, so we can invert this to derive the term structure of the hazard rate.

This setting is sometimes known as a *default intensity* model.

---

* In reality the recovery will probably be a function of the type of credit event that occurs. Typically, we think of the credit event as default and R as the senior debt recovery. However, the market convention, at least most corporate senior debt, is to set $R = 40$ or 50, and bundle everything into the hazard rate.

2.5.3.3 Credit default swaps

It would be tempting to think that all we have to do once we have a default intensity model is put the protection leg into it and work out the probability-weighted value of it. But this is not enough for most CDS as the *premium* leg is risky too: if we are paying 120 bps premium every 3 months on a 1-year CDS and a credit event happens in month 4, then we only pay two coupons of 30 bps each—one a few days after the CDS is transacted and another 3 months later—before taking advantage of the protection.

Therefore, we have a tree of events through time:

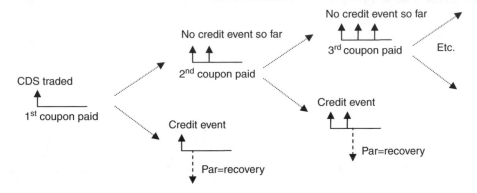

Consider a CDS with N periods, starting at time zero, and let the time interval between CDS coupons be Δt. Assuming h is constant over a coupon period, we get one term in the expression for the value of the CDS for each branch of the tree above. If r is constant, the CDS has value

$$\sum_{k=0}^{N-1} e^{-\sum_{j=0}^{k} h(t_j)\Delta t}(1 - e^{-h(t_k)\Delta t})\left[\sum_{j=0}^{k} e^{-rt_j} c_j - e^{-rt_{j+1}}(1 - R)\right] + e^{-\sum_{j=0}^{j=N-1} h(t)\Delta t}(1 - e^{-h(t_N)\Delta t})\sum_{j=0}^{N} e^{-rt_j} c_j$$

The first term comes from nodes where the credit event happens, and the last one from survival to the end of the swap.

Exercise. A *guaranteed premium* CDS is one where the entire premium leg must be paid regardless of a credit event. The guaranteed premium CDS spread is less than the ordinary CDS spread as it is not risky. How does the difference vary as the CDS spread increases?

2.5.4 Risk Reporting for Credit Derivatives

In general, a credit derivatives book may have:

- Interest *rate risk* from fixed rate bonds and from fixed coupons on default swaps (the CDS premium is paid as a running coupon stream, so if rates go up it becomes less valuable);

- *FX sensitivity* from any instrument not denominated in the bank's home currency;

- *Credit spread sensitivity,* to widening or narrowing CDS, bond or asset swap spreads;

- *Credit event risk,* including possible documentation basis risk if we have hedged a CDS with another CDS with different credit events;

- *Deliverable risk* on potential mismatch of deliverables (if we have sold protection on one swap and bought protection on another with different deliverables, then we may have to sell the instrument we are delivered in an illiquid post-default market and buy something we can deliver into the other swap);

- *Recovery risk* if we are exposed to the actual recovery after a credit event;

- *Equity risk* if we are using equity as a proxy hedge [see section 11.4 for a further discussion of this];

- *Maturity mismatch* where differences in maturity—for instance of a bond versus a CDS referencing and hedging it—create a forward credit position;

- Plus potentially *contingent* versions of some of these whereby if a credit event occurs it affects one instrument, such as a bond, but not its hedge, such as an IRS [we discussed this in section 2.1.3].

2.6 VALUATION, HEDGING AND MODEL RISK

In this chapter so far we have seen how to use:

- Some assumptions about the dynamics of a market (such as log-normality of returns);

- Together with assumptions about market properties (such as its completeness).

To produce a model—often via the solution of a PDE—which can be used for pricing a range of derivatives. This model is calibrated to the market using inputs such as volatility. Now we look at the risk management implications of this process

2.6.1 Mark-to-Market and Mark-to-Model

2.6.1.1 Mark-to-market

In Chapter 1 we discussed some of the issues with the concept of mark-to-market, noting that variable liquidity implies that the ability to discover each day the price at which two informed willing counterparties would agree to exchange an asset is questionable for many, perhaps even most financial assets. Rather than one fair value, it is often more realistic to think of a spread of values which define the likely price of a possible transaction. This uncertainty in value gives rise to *valuation risk* for all but the most liquid positions.

2.6.1.2 Mark-to-model

For derivatives, fair value is often not established by mark to market.* Rather we *calibrate* our model using some market prices—such as interest and FX rates—and some inputs

---

* This is sometimes the case even for exchange-traded options: firms may prefer to mark to a vol curve rather than to exchange prices. This is because these prices can sometimes be stale, especially for away from the money options.

which recover market prices for liquid derivatives—such as implied volatility. The model is then used to value our derivatives positions. This process is referred to as *mark-to-model*, and issues arising here are often termed *model risk*.*

Note that a derivatives model is not just the function that takes market parameters and produces a price: the model should be thought of as the mathematic function *together with its calibration*. The Black–Scholes call formula for DD used with 30% volatility is different from the same formula used with 32% volatility: Black and Scholes assume that volatility is constant, so for them there cannot be a state of the world where DuPont vol is both 30% and 32%. We can use Black–Scholes to mark a portfolio of plain vanilla options on DD with different strikes and maturities, but each different implied vol we use gives a different model.

2.6.2 Marking and Model Risk

There are several obvious places the process of mark-to-model might go wrong:

- The parameters we put into the model might not be right;
- The implementation of the model might not be right;
- The model assumptions might not be right.

We will deal with the first two of these sources of risk in this section, and the last in the next one. Typically, these risks cannot be precisely quantified, but we can mitigate both the frequency and the severity of the errors caused.

2.6.2.1 Marking vanilla parameters

Just because a parameter is not exotic does not mean we do not need to be careful about its value. Part of the trouble here is that there can be multiple inconsistent sources for the same parameter. Should we mark retail warrants to the (high) vol we sell them at, or to the (several points lower) vol of the OTC hedges we buy back in the inter-professional market, for instance?

Another issue is that what was a liquid product can become illiquid as the market moves. ATM volatility becomes OTM volatility, and this can make it harder to mark a product as time goes on since its parameters are no longer the liquidly traded ones. Long-dated FX and equity options can have material vega and still be far from the money, so this can be a real issue. There are a number of price discovery services who amalgamate data from various institutions and sell some form of composite answer back to the market, but because these are not traded volatilities, there is the danger of manipulation if the contributors all have the same position and hence the same incentive.

Similarly in interest rate markets, discovering the right implied vol to mark some positions may be difficult, and care may also be needed in areas such as correlations or some less liquid swap spreads.

---

* The following discussion is based on material originally developed jointly with Tanguy Dehapiot.

The goals of the mark verification process should be *accuracy, consistency* and *auditability*. Accuracy is obvious, and wherever material accuracy is impossible, a mark adjustment to reflect uncertainty should be considered; consistency is important too as a documented marking methodology removes the temptation to smooth P/L; auditability means that we should be able to recover the price chosen by following the chosen methodology.

2.6.2.2 Marking exotic parameters

Any parameter which cannot be observed from liquid-traded products is called *exotic*. (Nearly all) correlations, forward volatilities and mean reversion parameters are exotic parameters, for instance.

Exotic options do sometimes trade, so it might be thought that we can observe some exotic parameters from option prices. A quanto equity option depends on FX/equity correlation amongst other things, and it is in principle possible to infer a bid correlation and offer correlation if we know the bid and offer prices of this option and all the other parameters (volatilities, etc.). However, the spreads on quantos are often so wide that the bid/offer spread in the resulting implied correlation is rather large.

Given the need for a consistent methodology for exotic parameters, one approach is to use historic estimates. However, this is always problematic: not only might the past be a bad reflection of the future, but also there is the danger of inconsistency when we mix a historical estimate for one parameter—such as correlation—with implied estimates of others—such as volatilities. The resulting covariance matrix might well not be well formed. Nevertheless, it is often the best that we can do.

Finally, it is important to be pragmatic. For instance, simple no arbitrage arguments give the forward volatility skew from the quoted implied volatility surface. The theory here is impeccable, but forward starting equity options tend to be sold from the spot skew rather than the theoretical forward skew as the arbitrage is difficult to put on in practice and anyway many traders think that this theoretical forward skew is usually too flat. Here it is more important to follow the market consensus than to rely on a theoretical argument that is not observed in the market.

Exotic parameters are often inherently uncertain. Therefore, typically one would value a position using the best guess as to the parameter, and take a *mark adjustment*—a sum of money held back from the P/L—to reflect the uncertainty.

2.6.2.3 Model validation

Some of the issues that can arise in model implementation include:

- *Theory errors*: the wrong equation was selected.
- *Coding errors*: the implementation of the solution to the equation is wrong. Problems here also include convergence or arithmetic issues where the implementation is sometimes correct, but when used at the desired degree of accuracy or with particular inputs, materially wrong answers can be obtained.
- *Lack of portability or security*: the model does not work when moved to a different environment or is compromised by insecurity.

A useful control here is a *model validation process*: trading should not be permitted until a model has been validated by an independent team. Traders will apply for approval giving details of:

- What product the model is for;
- The estimated deal size and volume for the product;
- What assumptions are made;
- What the theory behind the model is;
- How it has been implemented;
- What the domain of applicability is expected to be;
- Any factors not captured by the model.

All of these topics would anyway need to be discussed in the model's documentation, so the overhead of this part of the process is not high. The model is then reviewed by the model risk team:

- The application is reviewed and mathematics used is verified.
- The model prices and greeks are tested under a wide range of parameters. The model review team will probably produce a quick and dirty independent implementation of the same (or a different but comparable) model.
- Limiting cases are verified.

The group will review the model performance and write a report indicating:

- What testing has been done.
- Any required changes needed before trading.
- Reserves or mark adjustments to be imposed on the model valuations.
- Suggested additional risk limits needed to constrain the firm's exposure to unknown parameters, the performance of the hedge strategy or model errors.

This, together with the model documentation produced by the model developer will be made available to all users of the model.

2.6.2.4 Managing model risk

A mature attitude in model risk is *management* rather than elimination. We cannot hope for a perfect suite of models, perfectly calibrated: rather we are aiming for an adequate set of models, given what we are trading, consistently and reasonably calibrated, together with sufficient reserves and mark adjustments that when an issue arises, we will already have identified it and kept a sum of money aside to cover its likely impact. The visibility of this risk class is important too: the bank should *expect* model risk-related issues to arise in an exotic options trading business.

Of course, we can always spend more money to enhance our models, hire more quants, improve our calibration procedures and so on. The judgement that should be made is where such spending provides a sufficient return, compared for instance with increased mark adjustments, given the inherent uncertainties in calibrating any model.

A simple model used in a sophisticated way—including an understanding of the simplifications it is based upon—is often a lot less dangerous than a sophisticated model used in a simple way.

2.6.3 Hedging, P/L and Mark Adjustments

A model price typically reflects the expectation of the P/L resulting from executing a hedge strategy. Therefore, an accurate market depends on validating our ability to actually capture that P/L. Here we need two tools:

- A good P/L explanation [as discussed in section 2.1.5] so that we can monitor the effectiveness of the greeks produced by the model in predicting the P/L. If the unexplained P/L shows a statistically significant trend in either direction, this is a sign of model risk.
- A practical hedge strategy analysis which estimates how much P/L can actually be captured by hedging [as discussed in section 2.3.2].

2.6.3.1 Example: barrier options

Some barrier options have a large gamma near the barrier. This makes hedging difficult as large hedge trades are needed for small moves in spot. Only a practical hedge analysis will capture this, and it may be necessary either to limit the size of the notionals permitted, or to take a mark adjustment based on the gamma at the barrier, or both.

2.6.3.2 Models in markets

One problem with some model review processes is that once a model is out on the trading floor, it can be difficult to keep track of where the model is being used. While assumptions like no bid/offer spread, arbitrary market depth and hence the ability to instantaneously re-hedge, and freedom from arbitrage, may not be too dangerous for a particular exotic model used with the S&P or USD/JPY as an underlying, one may be less sanguine about their applicability to, say, a less liquid NASDAQ stock or USD/ZAR.

It may therefore be appropriate to identify either a range of underlyings where exotics can be traded, or one or more traders who are sufficiently knowledgeable that they can be relied on to decide on the suitability of a proposed exotic trade. Trader mandates, ideally implemented via systems controls on what can be booked, should then forbid the use of exotics beyond this permitted area.

2.6.3.3 Situations where model risk mitigation may be needed

Some of the situations where a position or book should be considered for a mark adjustment, a reserve or a risk limit include:

- Marking to mid—a bid/offer adjustment is required;
- Sensitivity to uncertain parameters;
- A model assumption of stationary parameters which may be unwarranted;

- Model re-hedge assumptions which may be unwarranted;
- Model is known or suspected to fail to capture some material market features.

If a mark adjustment or reserve is decided up, the process is:

- Define the *calculation rule*. This is generally based on sensitivities, so we might, for instance, decide on a daily adjustment of 10% of the book vanna, or 20% of the exotic option vega. The process is therefore dynamic with a fixed rule.
- Define the conditions under which the adjustment or reserve can be *released* (e.g., deal unwind).

2.6.3.4 Postscript

Lane Hughston once asked an interesting parenthetical question:*

> Some eminent researchers, including B. Mandelbrot, for example, have maintained, in effect, that the Brownian paradigm is fundamentally flawed as a basis for asset price dynamics. I am not so sure that the majority of practitioners would agree with this point of view (does that mean that the better part of derivatives risk management as it is currently practiced by major financial institutions is similarly flawed?), but one should not ignore the fact that such words are being uttered in some quarters.

Lane's point is a good one. So much of quantitative finance is built on normal and log-normal distributions that if they are inappropriate, we might have concern for the utility of the whole framework.

The convenience of the mathematics of Brownian motions may not help matters: diffusions are easy to work with, and there are a lot of tools available if they are used. However, there is a danger that because we have a hammer, everything looks like a nail.

Despite a legitimate concern about the overuse of diffusions, the answer to Lane's question is no: it is both possible for the Brownian motion paradigm to be fundamentally flawed *and* for the better part of risk management as practiced to be effective:

- We have already seen that the Brownian paradigm *is* flawed. The existence of smiles, the term structure of vols, the instability over time of volatilities and correlations [and as we will see in Chapter 4, persistent autocorrelation]: these all point towards a fundamental failing in the sense that the assumption of Brownian motion of a risk factor is not sufficient, alone, to describe the dynamics of financial assets. Instead we must introduce more and more complications to our models: term- and strike-dependent volatility and correlation, volatility regimes, etc.
- But these complications *are* enough to rescue some tools for many practical purposes.

---

* See Lane Hughston's "The Past, Present and Future of Term Structure Modelling" in *Modern Risk Management: A History.*

We do not use option pricing to determine a fair value in the abstract: we use it to interpolate between known prices, or occasionally to extrapolate beyond them. Exotics are hedged with vanillas priced in the same setting, and a sensible exotics trader will do this in such a way as to minimise his sensitivity to model risks including distribution risk. If we:

- Calibrate to all the relevant liquid market prices;
- Minimise our sensitivity to exotic parameters and take mark adjustments for uncertainty in them;
- Hedge with instruments that have similar sensitivities (so that we buy vega, vanna and volga back, for instance, rather than just delta hedging);
- And most importantly to try develop a two-way market so that we can hedge exotics with exotics and so that our practice, however flawed, influences the market practice.

Then it is unlikely (but not impossible) for us to go too far wrong. Financial mathematics is not about discovering some hidden truth about the universe. Rather, at least as it is applied to risk management in most financial institutions, it is about finding a reasonable model that captures enough of the salient features of the market for the range of behaviours we are interested in. Most financial institutions do not care that they are using interest rate models whose handling of caps and swaptions is formally inconsistent. The models are good enough for what they are used for. The danger comes when they are used for something else ...

2.6.3.5 Hitting the heart of model risk

It is natural in a competitive market that firms will try to exploit their advantages as fiercely as possible. This happens in modelling as with any other area: sometimes products are invented either specifically to exploit another firm's model risk position, or with the happy accident that they emphasise it. A good example of a heavily model sensitive product is the *double no touch*.* This is a form of option with two barriers set some distance apart, and initially with the underlying between them. If the underlying price never goes above the upper barrier or below the lower one, the holder is paid a fixed sum; otherwise they receive nothing. Clearly, the delivered volatility close to both barriers will be important in determining the cost of hedging this option and hence its value, and that will be rather difficult to do without some form of modelling of the full volatility surface. Any firm that dipped a toe into trading the more benign sorts of barrier options without that technology might find themselves drawn into something that is much more model sensitive.

Risk managers need to be acutely aware of the dangers of creeping exoticism and its attendant model risks: just occasionally they really are out to get you.

---

* Others include chooser caps, some forms of multi-asset option and most structures with "power reverse", "perpetual convertible" or "callable dual currency" in the name.

Part Two

Economic and Regulatory Capital Models

CHAPTER 3

Capital: Motivation and Provision

INTRODUCTION

This chapter concerns the need for capital in financial institutions and how that capital is provided. We look at both the optimisation of capital, its allocation—allowing 'better' businesses to grow at the expense of 'worse' ones—and its provision through the issuance of *capital instruments*.

3.1 MOTIVATIONS FOR CAPITAL

Equity capital is the bedrock of a firm: it provides both financing and support in times of trouble. There is no requirement to make any payments on it (although there may be a heavy reputational impact if dividend payments are cut). Equity support makes a firm's debt much safer than it would otherwise be, or equivalently allows a firm to take much bigger risks than it otherwise could.

3.1.1 What Is Capital for?

Financial institutions sometimes take positions that make losses. Even worse, the whole institution can lose money on occasion. Capital is required to absorb these possible losses: it allows the firm to continue in the event of an unexpected loss (UL).

Our definition of risk was anything that could cause *earnings volatility*. If we want the firm to have a certain level of safety, such as to default no more than once in a thousand years, then we need enough capital to provide a buffer against all but once-in-a-thousand-year risks.

3.1.2 Earnings Volatility and Capital

Consider two possible earnings histories:

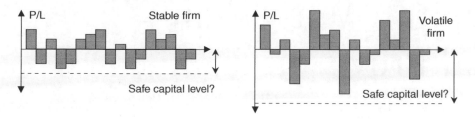

For the stable firm, when losses happen, they are relatively small, so the buffer required is modest. For the volatile firm, on the contrary, losses are more unpredictable, so more capital is required.

A strong earnings history does not guarantee that any losses experienced will be small, partly because for many institutions at least, the loss distribution is *asymmetric*—small positive earnings are rather likely, but there is a small probability of large losses. This is a characteristic feature of credit (and operational) risk loss distributions: since these effects often dominate market risk exposures, the firm's P/L distribution as a whole often has this shape.

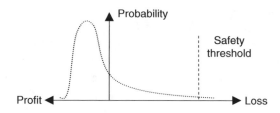

Exercise. What important information would daily P/L smoothing by one business unit deny the management even if the month end P/L was always correct?

In many cases—including writing insurance and making loans—we are paid up front to take the risk of future losses. In this case, it is common to consider the income and the loss distribution separately, moving the whole shape to the left. Here the income on a position must support two elements:

- The *expected loss* (EL) on the position. This is just the average amount we expect to lose on taking this kind of risk.
- The cost of the capital required to support the UL on it.

Typically, we would expect to support the EL on the positions from its return or carry, and the UL by *capital*. The EL is typically subtracted from income via a provision or reserve, leaving a net revenue after costs and provisions. If we have an explicit provision for EL, the capital required to support a position is UL–EL. The UL is determined by:

- The model of the loss distribution;
- The safety threshold under that distribution, that is, the probability of loss we are willing to bear.

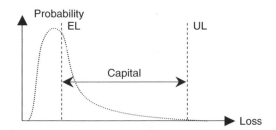

3.1.3 The Optimisation Problem

There are many views on what a financial institution's mission is, depending on its jurisdiction, culture and ownership. Many of them however include an element of *maximising returns to shareholders*.

3.1.3.1 Return on capital
What does that mean?

- First (at least in some asset return models), a shareholder cannot hedge the risk of default, so the firm has a duty to keep its default probability low.
- This is a practical duty too since many markets—including the CP and long-term OTC derivatives markets—are closed to firms of a rating much less than A A– or so.
- Next, there are various other practical constraints such as *regulatory risk*—a firm cannot operate if regulators do not permit it, so they need to be satisfied—and *reputational risk*.
- Within these limits, the firm must maximise return on capital.

The capital problem therefore boils down to deciding how much capital we need to assign against each of the risks being run then to optimise risk taking to maximise return. This capital is called *economic capital*, as it is an estimate of the 'true' capital required for all the risks the firm wishes to measure made using its own *internal model*.

Suppose very roughly that we can raise debt at 4% interest and shareholders demand a pre-tax return on capital of 40%. Then if a position requires 10% of its notional in equity under a certain model and there are no ELs, the position must return at least (90% × 4%) + (10% × 40%) = 7.6% or 360 bps over the bank's marginal cost of funds. This is the minimum *hurdle* that ensures that the position is giving sufficient value to shareholders.

3.1.3.2 Value creation

From the discussion above it is clear that value creation is idiosyncratic. It depends on the institution's:

- Required return on equity capital;
- Cost of debt;
- Capital model;
- Safety threshold.

Another way of looking at things is to consider the minimal return we could possibly tolerate and not destroy value. This happens when the return on a position is greater than the cost of running it. If our cost of equity capital in our example is 12%,* the minimum return would be $(90\% \times 4\%) + (10\% \times 12\%) = 4.8\%$. At this level, we are paying for the cost of running the position but nothing more: no value is being created.

The *risk-adjusted return on capital* or RAROC on a position is defined as

$$\frac{\text{Revenue} - \text{Expenses} - \text{EL} + \text{Risk-free return on capital}}{\text{Capital required}}$$

The risk-free return on capital term is included since capital is real: shareholders have actually paid cash for their shares, so shareholders' funds are (at least in theory) invested at the overnight risk-free rate so that they are always available to support losses.

RAROC was first popularised by BT in the 1980s, and it, or a variant of it, is still used to determine the value contributed by positions in many firms today.

If a firm has a hurdle rate for return on capital such as 40% in the example above, the *shareholder value added* is simply

$$(\text{RAROC} - \text{Hurdle rate}) \times \text{Capital required}$$

> *Exercise.* As a shareholder, would you prefer capital to be returned that could not be profitably used, or for it to be warehoused until a use was found for it?

3.1.3.3 Risk-based economic capital

A risk-based economic capital model estimates the capital required for a position by using measures of the risk being run to deduce the position's P/L distribution and hence its UL. An example of such an internal model is VAR: earlier [in section 2.1.5], we saw that the UL for holding 2,000 shares of DuPont for 1 day at a 99% safety threshold was $2,800. To a good approximation, the EL of this position is zero, so (assuming that the only risk factor is DD's stock price) the equity required to support this position is just the UL, $2,800.

---

* Estimating the cost of equity capital is not straightforward: unlike debt, it is not raised frequently and the trade-off between price appreciation and dividends makes shareholders' expected return for bearing earnings volatility difficult to determine. Approaches include the use of the CAPM combined with a dividend growth model or competitor benchmarking: Chris Matten discusses this further in *Managing Bank Capital: Capital Allocation and Performance Measurement*.

The capital estimates from a risk-based capital model are only as good as the risk meas-ures going into it. If we are missing a source of earnings volatility, then the model will underestimate capital requirements.

3.1.4 Capital, More or Less

Once we have a capital model for all the firm's risks we have decided to model, we can cal-culate the total economic capital required. What if this is more or less than the amount of capital we actually have available?

If the required economic capital is less than the total capital of the bank, we could argue that we are disadvantaging shareholders by decreasing ROE.* If the required capital is more than the capital we have, then we are taking too much risk and the firm's probability of default will be higher than our target threshold.

The capital allocated to various risks, and therefore to various activities within the firm, is useful for more than just determining whether the overall level of risk is correct:

- To align individuals' incentives with those of the firm, performance measurement and compensation should be based on RAROC. Unlike total revenue, this measure takes into account the value to shareholders of an individual or business group's contribution.

- Different business groups' RAROCs indicate where we should be improving returns or reducing capital allocated and where the firm should be allocating more capital. Of course, matters are not quite as simple as that, as we cannot always generate twice as much return by deploying twice as much capital, and the frictional costs of increas-ing or decreasing the size of some businesses, such as retail banking, can be large. Moreover, diversification is important: putting all your eggs in the highest RAROC basket offers no protection against a downturn in that business area.

Exercise.   Does your answer about returning unused capital change if you are a senior manager within the firm?

Over the next few chapters, we will look at how to calculate the capital required for port-folios of various risks: Chapter 4 deals with market risk, Chapter 5 with credit risk and Chapter 6 with operational risk. The pieces come together at the end of Chapter 6 when the total amount of capital required to support the risks in a firm is discussed.

3.2 CAPITAL INSTRUMENT FEATURES

By 'capital', we mean a security—something that has been issued and paid for—whose terms gives the issuer some flexibility over when payments on it are made. Capital can be created using a range of instruments. The most common form of capital is equity, but there

---

* Matten (*op. cit.*) gives the example of two banks with identical risk which differ only in the amount of equity capi-tal they have. The one with more capital will earn more, as it will get at least Libor on its invested excess capital, but it will have a smaller return on equity.

are also other sources: the more equity-like an instrument is, the better the capital it provides is. The term *hybrid security* broadly defines a range of securities which are positioned between the equity and senior debt and which provide some measure of capital benefit. A *capital instrument* is either true common stock or a hybrid security.

There are four broad areas where we require an instrument to have some contribution for it to be a capital security:

- *Loss absorption.* The security must absorb losses on a going concern basis, allowing an issuer to avoid liquidation in times of stress.

- *Permanency.* The security should provide a buffer for an extended period of time.

- *Flexibility and the ability to defer payment.* The issuer must have discretion over the amount and timing of payments made on the instrument.

- *Default performance and freedom of action.* The instrument should not cross-default with senior debt—increasing the size of the sum that becomes due and payable in the event of a senior debt default—and there should be few, if any, covenants restricting the issuer's freedom of action.

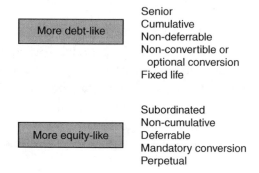

We now look at the structural features that help a security to achieve these aims.

3.2.1 Seniority and Subordination

In the event of the winding up of a company, the orders of creditors to be repaid liquidation in the process would typically be:

- Any truly supersenior creditors, such as the tax authorities or (in some jurisdictions) the pension fund;

- Any effectively supersenior creditors, such as banks where loan covenants may help the bank to shoulder its way towards the front of the queue of seniority;

- Senior bondholders;

- Subordinated bondholders and hybrid security holders depending on their precise subordination;

- Equity holders.

More subordination brings us towards the equity end of the spectrum, and therefore some measure of subordination is to be expected in a capital instrument.

3.2.2 Deferral and Dividends

Making a debt security subordinated does not give it any loss absorption capability: it just means that the holder has a larger loss in the event of default. In addition, we need the ability to stop paying interest on it: something we do not have on senior debt. This is called *the issuer's right to defer*: when it is present, failure to pay a coupon is not a default event. Deferrable securities are either *cumulative*—deferred coupons accrue interest at the same rate as the principal and must be paid in the future before the security is redeemed—or *non-cumulative*—the missed payment is ignored.

Obviously, non-cumulative securities are riskier for the holder and offer more loss absorption capability for the issuer, so a higher spread will be demanded for them.

Equity should be the riskiest class of security, as equity holders get whatever is left after everyone else has been paid. Therefore, we would not expect a hybrid instrument to be deferred *and* a dividend to be paid on common stock: the equity holder should suffer first. This leads to two increasingly common features of hybrid instruments:

- A *dividend stopper*, which prohibits dividend payments on equity if a hybrid security is deferred;
- A *dividend pusher*, which forces payments on the hybrid if dividends are paid on common stock.

3.2.3 Maturity and Replacement

Equity securities are perpetual, so for an instrument to be 'capital-like', it should have a long life at issuance. This can be achieved either through a long stated maturity or by making the security *perpetual* and *callable* after some period. The call permits the issuer to redeem the security, but only if it wishes to do so. To incentivise the issuer to call, a coupon *step-up* is often included, so the cost of servicing the instrument rises if it is not called [as discussed in section 1.2.3]. The callable step-up perpetual structure gives investors some compensation if a maturity extension is forced upon them and issuers the flexibility not to call if they are in severe stress.

In addition, *replacement language* is often also included whereby the issuer undertakes only to redeem the instrument if it is replaced with capital of an equivalent or better quality.

3.2.4 Convertibility and Write-Down

Equity is the best, but the ability to turn a security you have issued into equity is almost as good. Therefore, a *mandatory conversion* feature whereby the issuer will repay the security in equity rather than cash is a great aid in producing an effective capital instrument. [Mandatory convertible securities are discussed further in section 11.3.1.]

A *write-down* feature allows the issuer the discretion to reduce the principal amount of the capital securities on the occurrence of a defined event such as negative net income for more than one quarter. This gives it some flexibility to continue its business without being overwhelmed by having to find cash for a principal redemption.

3.2.5 Example Capital Securities

Consider three examples ranked from the most to the least effective in providing capital:

- A 20-year, mandatory convertible, subordinated non-cumulated fixed rate preference share. This ticks most of the boxes: it has good loss absorption and flexibility due to its non-cumulative deferability, and the combination of long life and mandatory conversion gives permanency.

- A perpetual subordinated bond with the right to pay coupons in equity and the right to convert up to year 10. There is an issuer call every year from years 7 to 10, and the coupon steps up by 150 bps in year 10. This is slightly less effective: although the subordination and coupon mechanism give some flexibility and loss absorption, the coupons are still due immediately. Moreover, conversion into equity cannot be forced, and if the bond has not been converted by year 10, it turns into a simple perpetual subordinated FRN which has little benefit in terms of flexibility. The step-up incentivises the call, so the instrument will behave much like a 10-year subordinated CB, at least until the issuer's credit spread widens significantly.

- A trust-preferred fixed rate security. Here an SPV is set up in a tax-efficient jurisdiction to issue securities. A partnership is set up, with the SPV as one partner. The SPV capitalises its partnership interest with the cash raised from the security issuance. The partnership then makes a subordinated loan to the issuer where the loan terms are designed to provide the desired flexibility and deferral capabilities.

  Since the SPV's only asset is the partnership interest, which in turn is only backed by the subordinated loan (plus a tiny amount of extra cash from the other partnership interests), the securities will carry the risks and rewards of the loan.

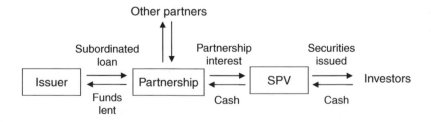

This structure shows one of the key features of capital instrument design: we want something as equity-like as possible, to have all of the flexibility, permanency and so on that characterise equity. But for tax purposes we want it to be debt, since payments on debt are a before-tax expense, whereas dividend payments are after-tax. This structure with an SPV and a partnership achieves a desirable tax and accounting treatment for some issuer domiciles.

Another issue for a capital security is how it is accounted for. In particular, convertible securities might require earnings to be reported on a *diluted* basis, i.e., as if they had been converted. This is undesirable as we have a lower ROE without actually having the benefit of the extra E, so some convertible securities have been structured to avoid this using *contingent conversion* features.*

3.3 REGULATORY CAPITAL PROVISION

The motivations for issuing capital securities are the following:

- *Economic capital benefit.* If the firm has more opportunities available than economic capital, it should issue more capital to allow it to take more risk. The firm's all-in cost of capital may be lowest if this is done via the issuance of a hybrid capital instrument than in a rights issue.

- *The provision of regulatory capital.* Where regulatory capital requirements are in excess of the firm's economic capital estimates, as they often are, a firm should issue the instrument that gets the most regulatory benefit for a given funding cost regardless of its economic capital benefit.

- *The provision of ratings agency capital.* Ratings agencies treat some hybrid securities as providing a measure of *equity credit*, that is, they give the firm benefit for a certain proportion of the issued notional of the security as equity for ratings purposes. Thus, issuing a hybrid capital security can safeguard a rating by reducing the agencies' perception of its leverage. The credit given varies from 75% of notional or more for very equity-like hybrids to 25% or less for more senior debt–like instruments. It may be that the cheapest way of ensuring a desired rating is to provide ratings agency capital via this route.

Given that the average cost of raising funds via a hybrid is usually much less than that for equity, there is a danger that firms will use hybrid instruments over equity wherever they can. Since even the best-structured hybrid is not as effective as equity, regulators have rules that constrain their application.

3.3.1 The Tiers of Basel Capital

Banking and (in some jurisdictions) insurance regulation recognises three tiers of capital.

3.3.1.1 Tier 1

Tier 1 instruments are the most equity-like. Included in tier 1 capital are retained earnings, ordinary paid-in share capital and—as *innovative tier 1*—deferrable irredeemable non-cumulative preference shares.

---

* See Izzy Nelkin's *Handbook of Hybrid Instruments: Convertible Bonds, Preferred Shares, Lyons, Elks, Decs and Other Mandatory Convertible Notes* for more details.

3.3.1.2 Tier 2

This is split in two. Upper tier 2 includes general provisions, revaluation reserves and certain hybrid capital securities including perpetual deferrable cumulative preference shares and perpetual deferrable subordinated debt, both without step-ups.*

Lower tier 2 includes other subordinated long-term debt such as dated cumulative preference shares with at least 5 years' original maturity and perpetual subordinated debt that does not otherwise qualify.

3.3.1.3 Tier 3

This includes subordinated debt of at least 2 years' original maturity.

3.3.1.4 Deductions

A firm is required by regulators to make certain deductions directly from the notional of issued capital securities to determine an effective amount of capital. These deductions include:[†]

- *Holdings in own paper.* Clearly, a bank should not be able to get credit for raising short-term senior debt and using the proceeds to buy its own capital instruments, so any positions in its own securities are deducted.
- *Goodwill.* Goodwill represents the value of an acquisition in addition to its tangible assets and liabilities: it can be thought of as quantifying an acquired company's reputation, intellectual property or expertise. Since the accounting test for determining the value of goodwill on a balance sheet is judgemental, regulators typically prefer to strip this item out.
- *Unpublished losses.*
- *Declared but not yet paid dividends.* Both of these are obvious: if we have identified a loss or agreed to pay a dividend, that capital is gone.
- *Material holdings in credit institutions.* In order not to overstate the amount of capital in the financial system as a whole, holdings of capital instruments issued by *other* banks are deducted from a bank's capital.

The detailed constraints on the required relations between the tiers depend on the type of firm and the jurisdiction, but common requirements include the following:

- Tier 1 > Tier 2 and ½(Tier 1) > Tier 2 sub debt. This keeps 'enough' capital in Tier 1, restricting the use of hybrid instruments.
- Innovative Tier 1 < 15% (or sometimes 25%) total Tier 1. Again, this restricts the amount of capital instruments which are not pure equity.

---

* A step-up incentivises a call, so a step-up security is less equity-like than the same structure without a step-up.
[†] This is an overview of deductions required by EU investment firms: rules for other types of firm may differ.

Moreover, typically no capital instrument can be repaid without regulatory permission: this will not be granted if doing so would breach one of the ratios above.

3.3.1.5 Capital structure example

Suppose a bank has $10B of equity capital. A reasonably efficient and practical capital structure for regulatory purposes would then be as follows:

Tier 1		Upper Tier 2		Lower Tier 2	
Equity	$10B	General provisions	$120M	Dated prefs[c]	$1.2B
Retained earnings	$300M	Perp prefs[b]	$700M	Sub debt[d]	$4B
Innovative tier 1[a]	$1B				

[a] This could be something like deferrable non-cumulative mandatory preference shares.
[b] Perpetual deferrable cumulative fixed rate preference shares.
[c] Ten-year cumulative preference shares.
[d] Step-up callable perpetual debt.

This is not particularly aggressive: banks typically prefer to have some 'headroom' in their capital structure in case they need large amounts of new capital quickly, perhaps to make an acquisition.

3.3.2 Insurance Capital

There is much less international concordance on insurance capital requirements than those for banks, and standards differ very significantly between the United States, the EU and significant offshore insurance company domiciles such as Bermuda. EU standards are slowly becoming more risk based, and capital instrument requirements are similar to those for banks (permitting 15% innovative Tier 1, for instance, and constraining Tier 2 capital to be less than Tier 1).

3.3.3 Consolidated Capital

Another deduction from capital not mentioned above is *investments in unregulated subsidiaries*. Consider a holding company that owns both a regulated bank and an unregulated activity such as a leasing company. Clearly, some of the holding company's capital is being used to support the unregulated activity, and the risks of these activities do not appear on the bank's balance sheet. That capital is therefore not available to support the regulated activity, and it should be deducted in the determination of the amount of holding company capital available to support the bank.

The use of the same unit of capital more than once is called *double leverage*. Here is an example of how it might happen:

- A holding company has $6B of equity and $10B of debt.
- It has two subsidiaries: one with a capital requirement of $4B and a funding need of $2B, the other with capital needs of $3B and funding requirements of $7B.

- Looking at each subsidiary individually, they look perfectly adequate.
- However, on a consolidated basis, the parent has used $6 of equity *and* $1B of debt to capitalise its subs. It is *undercapitalised* or *capitally inadequate* by $1B on a consolidated basis.

Regulators have obvious concerns about the possibility of this situation, especially in the context of financial conglomerates. The problem is that the regulators of the subs need to look up into the holding company to see the full picture and this is not possible if it is unregulated. The solution is a lead regulator for the holding company who can review its consolidated capital adequacy.

3.3.4 Capital Management: Issues and Strategies

Like most features of the financial markets, managing capital requires a certain understanding of psychology as well as technical know-how.

3.3.4.1 New capital issuance

There is nothing worse than having to find new capital in a hurry: the markets will usually detect the whiff of need and charge a large price for the capital raised if they grant it at all. Most firms therefore try to avoid this. Tools here include:

- Limits on the amount of capital instruments that can expire (or where a call will be expected) in any time bucket;
- Diversification of capital instruments across investor bases and currencies;
- The use of optional or mandatory convertibles to access investors who are not interested in more debt-like securities.

3.3.4.2 The capital game

Regulatory capital is a kind of game. Only equity and retained earnings can absorb losses in a completely unconstrained way, and hence offer complete protection up to their full value. Mandatory convertible debt can also allow the firm to continue as a going concern if conversion can be forced at a time when formally tapping the equity market might be difficult. But the true capital benefit of lower quality capital instruments is unclear: it is not that they are 50% effective, as a ratings agency equity credit might indicate. Rather it is that they are 100% effective in some scenarios and 0% in others, perhaps a less helpful situation.*

Part of the problem is that a deferral would signal to the market that the issuer was in trouble, and so it would be likely to trigger all sorts of adverse behaviour such as the closing of the interbank market to the issuer. In that sense, a deferral might actually make default more rather than less likely. Even lowering dividend payments would send a negative signal, which might well trigger a change in ratings agency outlook if not an actual downgrade. This means that firms tend only to waive dividend payments when times are very bad; otherwise, retained earnings or reserves may be raided to provide dividends.

Regulators know this, and some, at least, are candid about their lack of trust in anything other than retained earnings or common stock as capital. This leads to a kind of chase in capital instruments: issuers invent securities with a lower all-in cost of funds and try to persuade regulators and ratings agencies to give them credit for them. Sometimes this works and sometimes not, but when the gamekeepers feel that the erosion of the quality of capital has gone too far, regulatory capital requirements are raised.

3.3.4.3 Signs of trouble

Bank failures over the years suggest some warning signs that may be worth looking for in commercial and retail banks. Before we leave our discussion of capital, we touch on some signs of when more of it might be needed. All of them are fairly obvious, yet various commentators on banking failures, notably the U.S. Savings and Loans crisis of the 1980s, have pointed out that they are often missed.† The indicators are:

- Very rapid asset growth;
- A rapid increase in profitability or market share or both, particularly created by a business segment that is new to the bank;
- Above-market rates offered on deposits;
- A high ratio of secured-to-unsecured funding;
- A large percentage of lending or capital markets activity in high-yield areas;
- A large percentage of assets in complex products given the bank's size and risk management systems;
- A high percentage of off–balance sheet recourse obligations relative to those on B/S;
- The total regulatory capital only just sufficient to cover the regulatory capital requirement or tier 2 capital close to the maximum permitted given the amount of equity.

---

* Perhaps the most poignant indication of this would be if a firm gave no credit for instruments other than retained earnings or common stock in their own economic capital adequacy calculations. If they themselves do not believe in these instruments, how can they expect regulators or ratings agencies to?

† See George Kaufman's *The Failure of Superior Federal Bank, FSB: Implications and Lessons* (Testimony before the U.S. Senate Committee on Banking, Housing, and Urban Affairs).

Any investor in a capital security would be well advised to look for these early warning signs as well as investigating the structure of the proposed investment. In many situations, small or mid-sized losses have no impact on a bank's capital securities due to the high incentives against deferral or failure to call. Once a bank fails, though, its effect on subordinated instruments can be severe.

Exercise. Suppose you own a particular capital instrument and the issuer is in distress. What features of:

— the security you own and
— the regulatory system

would give you some comfort? Which is more important for the protection of your investment? Does your answer change if the issuer is an insurance company rather than a bank?

CHAPTER 4

Market Risk Capital Models

INTRODUCTION

A risk-based capital model uses risk measures to estimate the tail of the loss distribution and hence the capital required to support that risk. The usual process is:

- Choose a small number of representative risk factors as proxies for the many risk sensitivities of the full portfolio.
- Select a model of how those representative risk factors change.
- Select a model of how the portfolio P/L depends on the risk factors.
- Calculate the P/L distribution resulting from the risk factor movements.
- Select some confidence level of the distribution as a representative UL and thus determine the required capital for the portfolio.

We have already seen this process in action earlier when we calculated the 99% 1-day VAR for a portfolio of Google and Boeing [in section 2.2.2]. In this chapter we look at how to extend the approach for that mini portfolio to firmwide market risk models.

4.1 GENERAL MARKET RISK CAPITAL MODELS

General market risk models simplify the thousands of risk factors in a portfolio into a few tens of representative market variables. Typically, these variables are one equity index per country for equity risk, FX spot versus the USD for FX risk and government or Libor rates at a few points on the curve per currency. This gives a tractable set of variables which capture most of the risk not due to issuer-specific factors for a range of portfolios. In this section we look at methods of calculating market risk capital in this setting.

4.1.1 Value at Risk Techniques I: Variance/Covariance

One of the simplest methods for calculating market risk capital is to assume that:

- The returns of each risk factor are normally distributed.

- The joint return distribution is multivariate normal, i.e., we can completely describe it by giving all the variances and covariances.
- And, the portfolio responds linearly to changes in each risk factor.

4.1.1.1 Calculation

Suppose we have m risk factors and the portfolio is fully described by the positions $w = (w_1, w_2, ..., w_m)$ in each risk factor.

Let the S.D. of returns in the ith risk factor be σ_i, and the correlation between the ith and jth risk factor returns be ρ_{ij}. The covariance matrix is then the $m \times m$ matrix

$$\Sigma = \begin{pmatrix} \sigma_1^2 & \rho_{21}\sigma_2\sigma_1 & \cdots & \rho_{m1}\sigma_m\sigma_1 \\ \rho_{12}\sigma_1\sigma_2 & \sigma_2^2 & \cdots & \rho_{m2}\sigma_m\sigma_2 \\ \vdots & \vdots & \ddots & \vdots \\ \rho_{1m}\sigma_1\sigma_m & \rho_{2m}\sigma_2\sigma_m & \cdots & \sigma_m^2 \end{pmatrix}$$

The portfolio S.D. is given by the matrix multiplication $\sqrt{w\Sigma w^T}$, where T gives the matrix transpose. Using linearity, the VAR is simply some multiple of this (e.g., 2.33 $\times$ for 99% VAR or 1.64 $\times$ for 95%) as shown in the illustration.

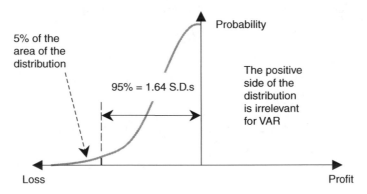

This technique is known as the *variance/covariance approach* to the calculation of VAR.*

4.1.1.2 Example

To calculate VAR in this setting, we just need a matrix of net delta positions in each risk factor w and similar length return series for each risk factor to allow us to calculate the σ_i and ρ_{ij}. As a mini-example, consider the following 26 1-week returns for three risk factors:

1	0.75%	−0.16%	0.32%	−0.95%	0.3%	1.83%	1.67%	0.34%	1.96%	−0.73%	−2.57%	0.08%	0.53%
2	−0.17%	−0.59%	−0.7%	0.48%	0.19%	−1.55%	−1.45%	−0.49%	−0.21%	0.01%	2.48%	−0.85%	−0.4%
3	−0.98%	−0.39%	0.04%	1.84%	1.09%	−0.07%	−0.24%	1.02%	−1.42%	−0.34%	0.3%	−0.33%	1.26%

* For further details on various VAR approaches and related techniques, see Phillipe Jorion's *Value at Risk: The Benchmark for Controlling Market Risk.*

The half year of data above and below is obviously far too little for a real model, and we only have three series rather than the tens or hundreds of a real GMR model: still, it is enough to illustrate the calculation.

1	0.81%	1.44%	−1.21%	0.02%	−1.21%	−0.22%	0.52%	−2.54%	−0.57%	1.67%	−1.14%	−0.96%
2	−1.44%	−1.31%	0.21%	0.67%	1.91%	1.11%	−0.04%	0.42%	−0.14%	−1.12%	0.11%	2.88%
3	−2.66%	−1.69%	−2.13%	2.64%	−0.76%	1.53%	0.49%	0.36%	0.14%	0.50%	−0.88%	1.26%

The correlations are:

- Between series 1 and 2: −72%
- Between 2 and 3: 38%
- And between 1 and 3: −18%

The first two returns come from two FX series, both versus the USD, so their correlation is quite negative, whereas the third is an equity series and so fairly uncorrelated with the FX series. The covariance matrix Σ for these series is

$$\begin{pmatrix} 0.000154 & -0.000103 & -0.000028 \\ -0.000103 & 0.000133 & 0.000055 \\ -0.000028 & 0.000055 & 0.000154 \end{pmatrix}$$

Suppose we have \$10M of net delta in the first risk factor, a short of \$5M in the second and an exposure of \$8M in the third. The exposure matrix $w = (\$10M \quad -\$5M \quad \$8M)$ and the portfolio S.D. is $\sqrt{w \Sigma w^T}$ or

$$\sqrt{(10M \quad -5M \quad 8M) \begin{pmatrix} 0.000154 & -0.000103 & -0.000028 \\ -0.000103 & 0.000133 & 0.000055 \\ -0.000028 & 0.000055 & 0.000154 \end{pmatrix} \begin{pmatrix} 10M \\ -5M \\ 8M \end{pmatrix}}$$

This is approximately \$173K. Since we have so little data, it would be prudent to look only a short distance in the tails, so we will estimate the 80% VAR. This corresponds to 0.84σ since 20% of the normal distribution lies beyond 0.84 S.D.s. Therefore, the 80% 1-week VAR estimated using variance/covariance methods is 0.84 times the portfolio S.D. or \$145K. We would expect to lose more than \$145K holding this portfolio no more than 1 week in five.

4.1.1.3 Errors in variance estimation
The VAR estimate above depends only on the positions and the covariance matrix. Given the central role of the variance and covariances, it is worth examining how accurately

we can estimate them. Suppose we have some length of return series which we assume is normally distributed: how does the *sample variance* estimated from the data relate to the actual variance of the generating process?

It is a standard statistical result that for a large number of data points the variance estimated from n data points is distributed according to a normal distribution with mean σ^2 and S.D. $2\sigma^4/n - 1$. This means the standard error in the S.D. for 1,000 data points at an estimated volatility of 20% is roughly 0.5%: we do not have enough data to estimate more accurately than that.

Often we choose data series with a relatively short weighted average life (perhaps using exponentially weighted moving average [EWMA]) to be responsive to recent market events: this is reasonable, but it gives rise to worse errors in variance and covariance estimation since effectively less data are being used. These effects limit the possible accuracy of VAR estimates.

4.1.1.4 Well-behaved covariance matrices

There are an awful lot of data series required for even a modest whole bank variance/covariance model: usually at least 10 equity indices and possibly as many as 50; key FX and interest rates in 5–70 currencies; and any additional factors such as volatilities or swap spreads. You can quickly end up with a 100×100 or bigger covariance matrix: this contains 4,950 separate pieces of data.

Unfortunately, there are some issues with using matrices of this size. Not least, for a covariance matrix to make sense, it must be *positive semi-definite*. One easy way of understanding this condition for a square matrix Σ is that Σ is positive semi-definite if $w\Sigma w^T$ is non-negative for any w, that is, we never get a negative VAR no matter what portfolio we use. It is relatively easy to ensure that small matrices are well behaved, but for large ones, it becomes more difficult. This is especially the case since matrix manipulation algorithms are often ill-conditioned for larger matrices, so the implementation of variance/covariance VAR and related techniques for large numbers of risk factors can be problematic.

4.1.1.5 Principal component analysis

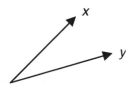

One way around this problem is to observe that many firms' VARs are typically driven by a small number of general market variables, such as the volatility of global equity markets. We can understand the factors driving the VAR as follows: think of the starting risk factors as a non-orthogonal coordinate system like the one shown in the illustration. Here moving along the x-axis involves moving along y too, since they are not at right angles. In portfolio terms this is variation in the FTSE caused by a movement in the S&P. What we need is to separate out the variation of the FTSE that is correlated with the S&P and that which is purely idiosyncratic. This gives us an *orthogonal* coordinate system called the *principal components* of Σ. We will list the components in the order of their contribution to the total variance, so the first few components drive much of the volatility.

To gain another intuition into what is going on, think of yield curve buckets as risk factors. Suppose we have six of them: 3-month, 1-year, 2-year, 3-year, 5-year and 10-year rates. Correlations here are high, particularly for the longer rates: the 5- and 10-year rates tend to move together. If we do a principal component analysis (PCA) on real yield curve return data for these points, typically the first principal component corresponds roughly to a parallel move of the curve up or down: this explains most of the volatility. The second corresponds to steepening/flattening, the third to a bend in the middle of the curve and so on.

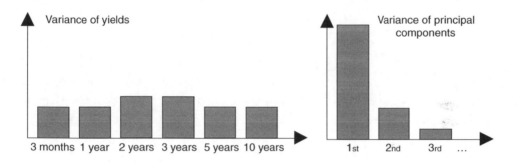

The principal components are derived as follows:

- Let the variances of the principal components be $\lambda_1 > \lambda_2 \ldots > \lambda_m$. Consider the first principal component and suppose there is some matrix β_1 which transforms our deltas in terms of the original risk factors into a new representation in terms of the first principal component.

- The variance due to this principal component is $\beta_1^T \Sigma \beta_1$, so $\beta_1^T \Sigma \beta_1 = \lambda_1$ and thus $\Sigma \beta_1 = \lambda_1 \beta_1$. By itself this does not fully determine β_1 since we can stretch it by an arbitrary amount, so we impose a scale condition $\beta_1^T \beta_1 = 1$.

- The second principal component is determined similarly: we want to find a β_2 and λ_2 such that $\Sigma \beta_2 = \lambda_2 \beta_2$, $\beta_2^T \beta_2 = 1$ but orthogonal to the first, so additionally $\beta_2^T \beta_1 = 0$. This process decomposes Σ into the representation

$$(\beta_1 \beta_2 \ldots \beta_m) \begin{pmatrix} \lambda_1 & 0 & \ldots & 0 \\ 0 & \lambda_2 & \ldots & 0 \\ \vdots & \vdots & \ddots & \vdots \\ 0 & 0 & \ldots & \lambda_m \end{pmatrix} \begin{pmatrix} \beta_1^T \\ \beta_2^T \\ \vdots \\ \beta_m^T \end{pmatrix}$$

The vectors $\beta_1, \beta_2, \ldots, \beta_m$ are called the *eigenvectors* of Σ, and the variances λ_i are called the *eigenvalues*.

4.1.1.6 Example

Consider the short threesome of return series discussed above. The eigenvalues are 0.00027, 0.00013 and 0.0000351, so the first two explain most of the variance. The associated eigenvectors are

$$\begin{pmatrix} 0.640 \\ 0.633 \\ -0.435 \end{pmatrix} \quad \begin{pmatrix} 0.463 \\ -0.134 \\ 0.876 \end{pmatrix} \quad \begin{pmatrix} 0.613 \\ 0.762 \\ -0.207 \end{pmatrix}$$

One of the reasons for doing PCA is that we may well discover that most of the portfolio variance is captured by only a few components, so we can work with much smaller covariance matrices in terms of these new coordinates without much loss of accuracy. Here if we only worked with the first two components, the covariance matrix Σ^* is

$$\begin{pmatrix} 0.00027 & 0 \\ 0 & 0.00013 \end{pmatrix}$$

(Note the zero covariances: the principal components are orthogonal.)

The deltas in terms of the new coordinates are $w^* = (6.09M, 12.3M)$ and so our estimate for the portfolio S.D. using just the first two components is $\sqrt{w^* \Sigma^* w^{*T}}$. This is $173K, demonstrating that the difference between this estimate and the full calculation is minimal in this case.

4.1.1.7 Marginal contribution

Owing to correlation, it is difficult to understand which positions diversify risk and which intensify it. The *marginal VAR contribution* of a position measures the extent to which a small change in it affects VAR. The portfolio variance $\text{Var}(w)$ is given by $\text{Var}(w) = w\Sigma w^T$ and for the jth risk factor we want to know how the VAR changes for a small change in the position $\partial \text{Var}(w)/\partial w^j$.

Differentiating the expression for the portfolio variance, we find that this is given by the covariance between the return on the jth risk factor and the portfolio return x^w

$$\frac{\partial \text{Var}}{\partial w^j} = 2\text{Cov}(x^j, x^w)$$

This gives us some clue about what to do if the VAR is too big: the risk factor with the largest absolute marginal contribution gives the single hedge that reduces the VAR fastest.

4.1.2 Value at Risk Techniques II: Revaluation and Historical Simulation

For portfolios with significant convexity in any risk factor, we cannot use variance/covariance estimates; instead we have to estimate the change in value of the portfolio for each change in risk factor.

4.1.2.1 Portfolio revaluation

There used to be two methods of doing this:

- In *delta/gamma approaches* a correction is made to the net delta position in each risk factor accounting for the net gamma. In the past this was sometimes useful for large portfolios when computing power was less easily available than it is today;

- Now, however, *full revaluation* is commonplace: here each instrument in the portfolio is revalued for each change in the risk factors. A variant of this, which can be used where the VAR system itself cannot revalue all positions, involves each trading system providing a scenario matrix of revaluations for each risk factor. The VAR system then interpolates between these to obtain the change in portfolio value for a risk factor change.

Suppose we have the technology to revalue our portfolio for a change in risk factors. In the *historical simulation* approach we use series of historical returns to calculate VAR.

4.1.2.2 Calculation overview

If there are m risk factors and we are using n days worth of data, we will need $m \times n$ risk factor returns x_i^j. We then create a distribution of portfolio P/Ls as follows:

- For each day in the dataset i, assume that the portfolio suffered that day's returns $x_i^1, x_i^2, ..., x_i^m$;
- Revalue the portfolio for these changes and collect these P/Ls into a distribution;
- The 95% VAR is the 95th percentile of this distribution.

The advantage of this approach is that it makes no distributional assumptions at all: the precise returns of the data series—however fat or thin tailed they are—are used to generate the P/L distribution.

This approach can also be used for estimating marginal VAR contributions, but the VAR system (or the individual trading systems) will have to calculate m shocked revaluations per day corresponding to a small change in the delta in each risk factor, so this can be computationally intensive.

4.1.2.3 Example

Using the same data and positions as before, the returns in order are

−357K	−301K	−278K	−246K	−189K	−139K	...

The fifth worst loss, roughly corresponding to the 80th percentile, is −189K, so the historical simulation VAR estimate is $189K. If we had more points we could fit a distribution to these P/Ls and get a better estimate of the required percentile, but with only 25 P/Ls there is little point. Still, our crude historical simulation VAR estimate is not too far from the variance/covariance calculation of $145K.

4.1.2.4 Data sensitivity

The historical simulation approach only depends on 1 or 2 days' worth of data: the rest are irrelevant. To see this, imagine labelling each P/L as it comes out of the revaluation engine.

Day 1's returns generate P/L 1; day 2, P/L 2 and so on. Thus, in our example the first day's P/L is just given by the position in each risk factor times its return:

$$\$10M \times 0.75\% + -\$5M \times -0.17\% + \$8M \times -0.98\% = \$5.3K$$

If we carry on labelling the P/Ls with the day they came from, we find for our mini-example that the fifth worst loss comes from day 24:

$$\$10M \times -1.14\% + -\$5M \times 0.11\% + \$8M \times -0.88\% = -\$189K$$

It is this day alone that determines the VAR in our example. If we have a more practical calculation with $n = 1,000$ (roughly 4 years) there will be 1,000 P/Ls, but the 95% VAR is just determined by the 50th worst one. This will come from some day (or pair of days if we are interpolating the P/L distribution) from the 1,000 available. So we are still only using at most 2 days' returns from the 1,000.

4.1.2.5 Oversampling

For 1-day VAR there are 1,000 1-day returns in a 1,000-day series, so taking the 50th worst for 95% VAR does not seem too extreme: there are plenty of points in the tail beyond that. For 10-day VAR though, there are only 100 non-overlapping 10-day returns, and we will look at the fifth worst: the situation is even worse for 99% 10-day VAR.

More P/Ls can be generated from the data by *oversampling*: for each sample we pick a random day and take the 10-day return from there. This allows us to get more than 100 10-day returns from our 1,000 days of data.

We do not necessarily get a statistically more accurate estimate by using more and more oversamples, but a limited degree of oversampling can help to improve 10-day VAR estimation in some historical simulation approaches.

One further step instead of taking 10-day periods is to pick 1-day returns at random to construct a composite 10-day return. Thus, a single simulated 10-day return might be composed of the composite return from days 342, 85, 128, 429, ... and 201, and we could use a thousand randomly selected such composite returns to calculate VAR. The problem with this approach is that it does not capture any autocorrelation [as discussed in section 4.2.3] in the data: we are assuming that the returns are independent.

4.1.3 Value at Risk Techniques III: Monte Carlo Approaches

If we do not have enough historical data to produce a good historical simulation VAR, we can make the data up. The idea is to take random samples from a multivariate return distribution that is calibrated to represent the data. For $x\%$ N-day VAR, the modelling proceeds in five steps:

- A multivariate return distribution is assumed: often, but not always, this is multivariate normal with some covariance matrix.
- Historical data are used to calibrate this distribution. In the normal case, this involves estimating Σ.

- A set of random numbers is generated somehow and scaled so that they have the correct distribution: this represents a possible N-day return on each risk factor.
- The portfolio is revalued using this return.
- The previous two steps are repeated until we have sufficiently many possible P/Ls to estimate the $x\%$ of the portfolio distribution. This is the VAR.

4.1.3.1 Using the Cholesky decomposition

Suppose we have m risk factors which are assumed to be distributed with covariance matrix Σ. For one risk factor we can generate random numbers that are normally distributed with variance 1 by taking uniformly distributed random numbers between 0 and 1, and applying the inverse of the cumulative normal distribution to them, as in the figure.* If we do this with m risk factors, we will recover all the correct variances but none of the covariances: the correlations will be zero (assuming our random number generator is working correctly).

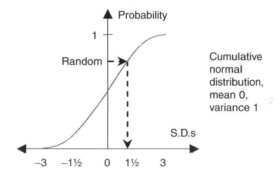

A common technique of producing *correlated* random numbers is to use the *Cholesky decomposition*. For a square matrix A, this produces a matrix B with the property that $BB^T = A$, so that it can be thought of as (one definition of) the square root of a matrix. It is useful because if we have a matrix rans of m random numbers normally distributed with mean 0 and variance 1 and Σ is a covariance matrix with Cholesky decomposition C, then C^T rans is jointly normally distributed with covariance matrix Σ.

4.1.3.2 Example

The Cholesky decomposition C of the previous covariance matrix is

$$\begin{pmatrix} 0.01241 & 0.00000 & 0.00000 \\ -0.00832 & 0.00799 & 0.00000 \\ -0.00226 & 0.00451 & 0.01134 \end{pmatrix}$$

If we generate three streams of random numbers n days long, and multiply C^T by this $3 \times n$ matrix, then the resulting three streams of random numbers will have a covariance matrix

---

* Techniques also exist for other distributions: see for instance Umberto Cherubini's *Copula Methods in Finance*.

close to Σ (and it will be closer the bigger the *n* is). These can then be used in a Monte Carlo simulation to calculate VAR.

4.1.4 Relative VAR

VAR is not just used as an absolute risk measure: it is also useful as a relative risk measure. Suppose, for instance, that an investment portfolio *P* has its returns measured against a *benchmark B*. Then we might look at the VAR of *P* − *B*, that is, long the portfolio and short the benchmark. This would give a measure of the risk of the investments versus the benchmark at the chosen confidence interval.

4.1.5 Backtesting, VAR Exceptions and the VAR Hypothesis

How could we gain some confidence that a VAR model was accurate? Clearly, if we are predicting a 99% 1-day VAR, we would expect to see a loss bigger than the VAR about 1 day in a hundred. Backtesting is the process of examining the number of *P/L exceptions*—losses bigger than the VAR—versus the predicted number.

4.1.5.1 P/L cleaning

One of the issues with this process is that if the total P/L is used a fair comparison is not being made. The VAR is based on holding the portfolio over the holding period without change. The real P/L in contrast contains the results of intra-day trading including new deals, changes in mark adjustments and reserves, commissions and fees paid and received and other non-trading P/L. Therefore, our first step is to use the P/L explanation [discussed in section 1.3.4] to produce a *clean* P/L which only contains the contributions of holding yesterday's portfolio:

$$\text{Ordinary (dirty) P/L} = \text{Clean P/L} + \text{New deal P/L} + \text{Changes in mark adjustments} + \text{Non-trading P/L}$$

4.1.5.2 The backtesting process

The next obvious step is to plot the clean P/L versus the VAR and examine the number of exceptions. This typically gives a view as illustrated with P/L mostly within the interval defined by the VAR and its negation.

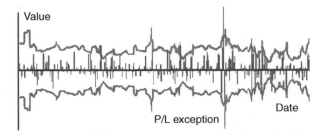

Backtest summary: VAR and P/L

The first reaction of a risk manager to a VAR versus clean P/L picture is often relief: there are usually fewer backtest exceptions than the model suggests. Although having fewer exceptions is better from a purely financial viewpoint than having too many, both are equally wrong statistically:

- If the VAR is systematically too low, the model is underestimating the risk and you tend to have too many occasions where the loss in the portfolio exceeds the VAR. This can lead to an increase in regulatory capital [discussed in Chapter 7], so it has serious consequences and most firms tend to prefer any errors to be in the other directions.

- If the VAR is systematically too high, the model is overestimating the risk (and the regulatory capital charge will be too high).

But to see what we can conclude from, say, having three exceptions in 2 years when the model suggests we should have five, the statistical properties of VAR have to be examined.

4.1.5.3 Statistical limits on backtesting

The trouble with exceptions from a statistical perspective is that for the popular VAR confidence intervals of 95 and 99% they do not occur very often, so it is difficult to get enough data to be sure that our model is accurate:[*] even at 95% multi-year backtests are needed to get much confidence in the model. This is problematic: once a firm has developed a model, it typically wants to implement it quickly.

The process can be improved by using the model to calculate VAR each day at different confidence intervals using the same model, say 80, 90, 95 and 99%, and testing the number of exceptions to each VAR.

Exercise. Comment on the advantages, in terms of the opportunity to improve its risk measurement, and the disadvantages, in terms of extra costs and regulatory scrutiny, if a bank discovers that its VAR model is not accurate. On the basis of your discussion is it worth doing more backtesting than the regulatory permitted minimum?

4.1.5.4 Excess loss over VAR

Backtest exceptions at various levels of confidence are interesting statistics in reviewing the usefulness of a VAR model but in a sense they throw away too much information: we look at the fact that there has been a loss bigger than the VAR, but not at how big it was. Therefore, some commentators focus attention on the *excess losses over VAR*[†] (sometimes known as the expected shortfall):

$$\{P/L | P/L < VAR\}$$

These are best analysed in a framework that looks at risk in the tails such as extreme value theory (EVT) [discussed in section 6.2.1].

---

[*] Jorion (*op. cit.*) gives a more extended discussion of the statistical properties of backtesting.
[†] See Alexander McNeil, Rüdiger Frey and Paul Embrechts' *Quantitative Risk Management: Concepts, Techniques, and Tools.*

4.1.5.5 The VAR hypothesis

The VAR hypothesis is simply that the clean P/L is distributed as predicted by the VAR. Typically, if we look in detail at backtest exceptions, we discover that there are three main areas of concern:

- Exceptions tend to be clustered together in times of market volatility.
- The VAR hypothesis, to the extent that we have enough data to test it at least, becomes less easy to accept the greater confidence interval.
- Exceptions can be caused by SR rather than GMR factors.

Therefore, in the next section we look at each of these issues.

4.2 SOME LIMITATIONS OF VALUE AT RISK MODELS

VAR is in many ways a crude measure of market risk. Its limitations include:

- The simplification of the complexities of portfolio risk into positions in a relatively small number of risk factors.
- The assumption that market data is available for all risk factors that accurately captures the risk of holding the position going forward.
- The assumption that the portfolio is fixed over some time horizon, often 1 or 10 days. For some assets this may be reasonable, but the risk horizon for others might be much longer than that due to their illiquidity. Moreover, if a simple $\sqrt{t}$ rule is used to scale from 1- to 10-day VAR, another distributional assumption is introduced: 10-day VAR is in general only $\sqrt{10}$ times 1-day VAR for normal returns.

In this section these limitations are examined in more detail and some extensions to VAR are considered which help to alleviate their effect.

4.2.1 Specific Risk

GMR VAR involves a problematic simplification: corporate bonds are represented as if they were risk-free instruments, individual stocks look like positions in the index and so on. Why not just have a data series for every underlying and capture both GMR and SR at once?

4.2.1.1 Data requirements

For firms dealing principally in liquid securities and liquid derivatives on them, this approach may be rational: instead of having the tens of data series for a GMR VAR model, we keep a return history for each security and use those directly. The data requirements of this approach are considerable: there will probably be thousands of series in use at any one time, and the data cleaning group will be rather larger than for a GMR model.

4.2.1.2 Illiquid assets

Illiquidity impacts risk measurement in two ways: by making it difficult to estimate the return distribution, and hence to calculate risk measures; and by making the assumption that the position can be liquidated on a short horizon more problematic.

The first problem for a few illiquid or newly issued securities can be dealt with by using proxies provided that the sensitivity of the VAR to their variance and covariances is not large. If we have a good history of an issuer credit curve, for instance, a return history for a newly issued bond from that issuer can be estimated. In extremis we might even use a generic instrument if a reasonable correlation assumption can be made.

The second issue is problematic as it implies that the holding period may be position dependent. [This is discussed further in section 9.2.1.]

4.2.1.3 Mark-to-model in VAR

One problem with specific risk VAR comes with mark-to-model instruments. Consider 7-year 140% strike FTSE implied volatility. We might well need to estimate this to mark a large U.K. retail equity product we have issued. A good product control group would combine broker estimates, data vendor data and perhaps the prices of traded products to obtain a reasonable estimate of this parameter as part of their price testing process. This value would probably be augmented with a mark adjustment if the firm's exposure was large enough to capture the inherent uncertainty in this vol. Given the time needed for the mark review process and the illiquidity of the implied volatility concerned, however, it is highly unlikely in practice that the position would be remarked every day: we would not then have a daily price history of this market variable. Moreover, vega could well be the biggest risk in the derivatives book: delta would typically be hedged, and vega would often dominate gamma for a long-dated product. If our VAR just captures the FTSE as a risk factor, it will be missing an important sensitivity and one whose volatility is difficult to estimate.

In a GMR VAR model we can deal with this problem by having a small number of implied volatility series: there is no pretence that we are capturing *all* risk factors, so proxying FTSE 7-year implied vol by the VIX is no worse than proxying Boeing return variance by that of the S&P. For SR models, though, the situation is more troubling: particular illiquid implied volatilities—or implied correlations—are sometimes likely to be much more important risk drivers than particular corporate bond returns. In this situation there is little value added by diligently using the return series of every small bond position and ignoring difficult-to-find mark-to-model risk parameters such as long-dated vol.

4.2.2 Volatility and Correlation Instabilities

At least three problems bedevil us in the calculation of any normal-distribution-based risk measure: large positive and negative returns happen more often than the normal distribution predicts, return correlations are not constant and return volatilities are not constant.

4.2.2.1 Large returns

The fatness of the tails of financial return distributions has already been mentioned. The following table shows the effect: for a long data series of USD/JPY, we record the actual number of returns larger than 2, 3, 4, 5 and 6 S.D.s observed versus the number predicted by the best normal distribution.

The further out in the tails we go, the worse things get. At 2 S.D.s, the situation might be acceptable; at 4, the error is two orders of magnitude.

Number of Standard Deviations	Number of Events Predicted from Best Normal	Number of Events Observed
2	91	86
3	5	25
4	0.12	10
5	0.0011	6
6	0.000004	3

4.2.2.2 Correlation instabilities

The figure in Chapter 2 of rolling CAC/DAX correlation should also give a warning. For this pair over the time horizon explored, at least, it is hard to gain much confidence that correlation is a meaningful statistic. Certainly, things would appear better if we used a longer window to observe correlation, but doubts would remain that only a non-parametric test for the existence of a process covariance would assuage.*

> *Exercise.* If your firm has a VAR model that depends on a correlation assumption, find the correlation it is most sensitive to. Determine a historical range for that correlation using a reasonable range of data. Calculate the sensitivity of the VAR to a tiny move in correlation, and hence find the approximate potential error in VAR due to correlation instability.

4.2.2.3 Volatility instabilities

For volatility a related phenomenon is observed: volatilities are stable and low for some periods of time, but then they tend to jump when bad news hits the market as prices seesaw more wildly. We examine this next.

4.2.3 The Holding Period Assumption

One property of the random walks we often use for modelling returns is that they are history-free or *Markovian*: the probability of a large change today does not depend on whether there was one yesterday or not. This behaviour gives us the convenient $\sqrt{t}$ evolution of uncertainty that allows us to get the 10-day VAR from the 1 day just by multiplying by $\sqrt{10}$.

If real asset returns are examined, this assumption is often found not to hold. Instead, *volatility clustering* is observed: if there was a large return yesterday—positive or negative—a large one today is more likely. If nothing much happened yesterday, a quiet day today is somewhat more likely.

---

* For a further discussion see Beniot Mandelbrot and Richard Hudson's *The Misbehaviour of Markets* or Edgar Peters' *Chaos and Order in the Capital Markets*.

This phenomenon is known as *positive autocorrelation*: the return series depends on its own history.* This means that using random days returns historical simulation will typically give lower 10-day VAR estimates than one which uses actual experienced ten day returns as the former does not capture the clustering of multiple days of big moves.

4.2.3.1 ARCH and GARCH

One class of models which incorporate autocorrelation are the autoregressive conditional heteroskedasticity (ARCH) and generalised ARCH (GARCH) ones. In the simplest ARCH model, we have a long-term variance ω and today's variance σ_t^2 depends on this together with how large a move we had yesterday x_{t-1}:

$$\sigma_t^2 = \omega + \alpha(x_{t-1} - E(x))$$

where α measures how important yesterday's return is versus the long-term average: more autocorrelated series have larger values. Once this model is calibrated it can be used to predict tomorrow's volatility on the basis of today's return, and hence to give a more accurate VAR estimate given current market conditions.

Generalized ARCH or GARCH models generalise ARCH models by allowing the variance to depend not just on yesterday's return, but also on yesterday's variance:

$$\sigma_t^2 = \omega + \alpha(x_{t-1} - E(x)) + \beta\sigma_{t-1}^2$$

where we need $\alpha + \beta < 1$ in order for volatility to be bounded. Generalisations where the volatility of one series depends on the returns or volatilities of another are clearly possible: the more coefficients we have, though, the harder the calibration problem becomes.

ARCH/GARCH models with positive autocorrelation naturally produce fat-tailed return distributions but the evidence about whether they produce the right sort of fat tails is mixed: the volatility smiles implied by ARCH-based models often do not fit the options market very well.

4.2.3.2 The effect of autocorrelation

Positive autocorrelation is nasty: it reduces VAR in quiet markets, encouraging us to take more risk, and then ramps it up in volatile markets, pushing the same position over the VAR limit and hence forcing the position to be cut in the worst market conditions. Given this behaviour it is rather unclear why a firm would want a VAR model for risk management purposes which incorporated autocorrelation even if it could easily develop one. Some firms certainly take the view that it is better to use conservative volatilities in the

---

* Autocorrelation is measured using a statistic parameter known as the Hurst exponent. For a random walk, the Hurst exponent is 0.5, whereas for the DJIA, it is roughly 0.63. See Carol Alexander's *Market Models: A Guide to Financial Data Analysis* for more details on the measurement of autocorrelation, GARCH, and the implication of these approaches for volatility modelling.

first place than to find vols suddenly increasing just when the firm most needs to hold its nerve.

4.2.4 What Is the VAR Good for?

After this catalogue of issues, the reader may be forgiven for thinking that VAR is a risk measure of limited usefulness. This is not true—a VAR calculation is worth having—but it must be used with care.

4.2.4.1 VAR requires good infrastructure

One of the biggest advantages of VAR is that it forces a firm to build a firmwide market risk infrastructure and to address the data quality and systems reliability issues within it. Even if each system is 98% reliable, if 30 independent trading systems feed the VAR calculation—not unusual in a large bank—then the consolidated number will only be good slightly more than half the time. Since the entire risk infrastructure, including stress and scenario testing, compliance with limits and regulatory capital calculation, is likely to feed off much the same information, getting good VAR numbers most days gives some confidence in the rest of the infrastructure.

4.2.4.2 Less is more

Some of the issues with VAR relate to the tails, so a pragmatic approach to dealing with them is simply not to try to calculate the 99%—or even worse 99.9%—VAR. Distributional issues are much less pressing at 95%, and backtests have more statistical accuracy. Therefore, using 1-day 95% VAR as the internal standard makes a lot of sense: if 99% 10-day is needed for regulatory purposes, then the 95% VAR can simply be scaled.

Another cultural issue concerns how VAR is communicated. Given all of the potential issues stating VAR to the nearest dollar is meaningless and, worse, potentially confusing in that it gives an air of accuracy that the number does not deserve. Two significant figures are more than enough:

Firmwide 95% 1-day VAR	$24M
Limit	$30M

The use of a 95% VAR also focuses on losses larger than the VAR: even with intra-day trading, they will probably happen. Thus, there is little possibility of getting false comfort from the fact that a loss bigger than the VAR has not happened (yet). The following characterisation captures what the VAR is telling you.

The VAR is the amount you could lose providing that you don't lose a lot.

4.2.4.3 Uncaptured risk

If a trader wished to game the VAR calculation taking risk in ways that are not measured, then they almost certainly can. We could even imagine a game of cat and mouse:

VAR Risk Factors	Risk Position Not Captured
GMR	Corporate bond with IR risk hedged by treasuries
Security level SR	Volatility risk in OTC derivatives
Implied vols included	Term and strike structure of volatility
Vol surface included	Exotic sensitivities, correlation positions
...	...

The fact that these games are possible does not invalidate the use of VAR: it just means it must be used cautiously, in conjunction with other controls and reporting. Market risk has too many dimensions for a single measure to be able to tell the whole story for every portfolio.

> *Exercise.*   Design a one-page report which gives a reasonably accurate view on your firm's market risk and is flexible enough to incorporate changes in circumstances. You may not use a point size smaller than 11 for the fonts you use.

4.3 RISK SYSTEMS AND RISK DATA

In this section we look at the architecture of a risk aggregation system such as firmwide VAR and suggest some features that might make it useful beyond the production of a single aggregate risk number. This leads to a discussion on how to select and clean market data for use in such a system.

4.3.1 Effective Risk Reporting

Earlier in the book we outlined the architecture of VAR calculation (repeated below), and in this chapter we have seen various implementations of that design.

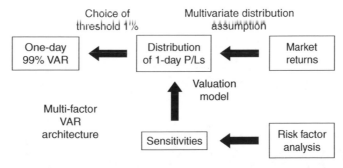

Within this simple box diagram for most banks there is a lot of data and processing: large amounts of market data, lots of positions and many valuation models. This seems an awful lot of material to produce one number. Once we have built a VAR system, is there more we can do with it than simply calculate a firmwide VAR?

4.3.1.1 Breakdown by desk

The first obvious cut of an aggregate risk measure is to see how the overall number is produced from the various desks.

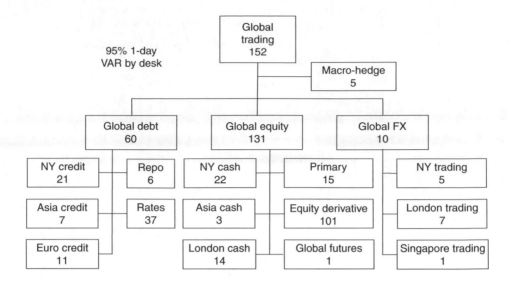

For instance, the first three levels in the hierarchy might look something like the illustration above.

4.3.1.2 Breakdown by risk factor

Next it is helpful to be able to pick any point in the hierarchy and see which risk factors are contributing to the overall VAR. For instance, if we have a small number of risk factors, we might split out a high-level node like the equity derivatives business as shown below.

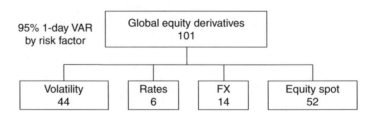

This quickly allows us to see desks taking inappropriate risks: should not equity derivatives, for instance, be hedging that FX risk?

4.3.1.3 P/L decomposition

Ideally the P/L explanation should be in the same system as the risk information so that we can quickly flip from a risk view to a P/L view.

At a high-level node such as the ones of the last illustrations, the information would not be very useful as the net P/L would be a contribution of many individual positions. But at the individual book level or the position level, this information (together with the greeks) would give a lot of insight into where the P/L had come from on the day concerned.

Consider one of the desks deep within the global equity derivatives hierarchy:

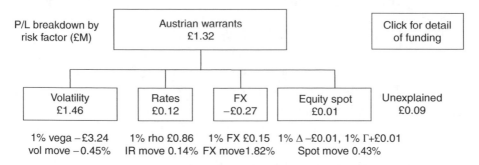

Obviously in a situation like this, the index move and net deltas and gammas might not give a good picture of the individual stock moves and position sensitivities, so we might need to drill down into detail to understand this component of the risk. Similarly, the FX numbers are opaque (what crosses? what exposures?), and the volatility exposure again would presumably come from individual warrant vegas. Still this kind of information is a good start if we want to understand risk and P/L in the same system.

4.3.1.4 Analysing the market data

Consider an instrument whose P/L is monotonically increasing in a risk factor S, such as any long position in a linear instrument or a call. Suppose we revalue the position for a series 1, 2, … of changes in that risk factor, and we order those changes by size $\Delta S_1 < \Delta S_2 < \cdots < \Delta S_n$. Since the instrument is linear, the resulting P/Ls will have the same ordering $\Delta PL_1 < \Delta PL_2 < \cdots < \Delta PL_n$. If $n = 1,000$ and we are looking for the 1% VAR, then by definition it is ΔPL_{10} and this comes from ΔS_{10}.

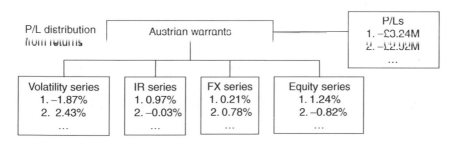

Matters are not as straightforward if more than one risk factor is involved; nevertheless it can be helpful to access both the whole P/L distribution VAR is estimated from and the individual risk factor returns which give rise to each P/L in that distribution.

4.3.1.5 Using aggregated risk information

It is only by allowing the user to really understand where the consolidated risk comes from, in terms of:

- Risk factors;
- Any peculiarities of local markets or particular books;

- Exposures;
- And market data points driving VAR;

that risk systems acquire true utility and robustness. No one number can capture every-thing (or arguably anything much), so the ability to remain sceptical, to dig down and to understand where the higher-level measures are coming from is key.* The risk manage-ment advantages in VAR may well lie as much in building and using the infrastructure, rather than in the aggregate risk information *per se*.

4.3.2 Market Data

The problem of finding accurate market data to drive a risk model is a difficult one. Even if you could get all the data you wanted, there is a choice between using a long history—and hence having more data and thus more accuracy at parameter estimation—and using a short history which gives less accuracy but is more responsive to market events.

There are no easy answers here: the only constraint is that for regulatory purposes there is a requirement in market risk models to use at least a year's data updated at least quarterly.

4.3.2.1 Weighting the data

One common solution to making a data series responsive to market events yet keeping a longer history is to weight earlier data less than more recent data. Thus, instead of using equally weighted data to estimate variance

$$\sigma^2 = \frac{\sum (x_i - E(x))^2}{n}$$

we can exponentially weight the data by a parameter λ

$$\sigma^2 = (1-\lambda) \sum \lambda^{i-1}(x_i - E(x))^2$$

This parameter controls how much later data are weighted over earlier: for the model to be well-formed, $0 < \lambda < 1$, and some practitioners use $\lambda = 0.94$, corresponding to roughly a quarter's effective data being used in the variance and covariance estimation. This approach is known as the exponentially weighted moving average or EWMA approach.

4.3.2.2 EWMA correlation estimation

We can also use an EWMA approach to estimate covariances. The usual approach first calculates the variances σ^2 for the tail end of the data, and then a recurrence relation is used to calculate today's variances for the kth return series $\sigma_i^2(x^k)$ and kl covariances $\text{Cov}_i(x^k, x^l)$ via

$$\text{Cov}_i(x^k, x^l) = \lambda \text{Cov}_{i-1}(x^k, x^l) + (1-\lambda) x^k_{i-1} x^l_{i-1}$$

---

* It is possible that the incentive structures within firms may sometimes be arranged so no one is encouraged to point out any deficiencies within the risk system, and this may especially be true for consultants or data vendors.

This gives a covariance matrix which reacts faster to changing market conditions than a linearly weighted one.

4.3.2.3 Data quality
Some of the market data supplied by data vendors may not be clean:

- It may contain clearly spurious data points.
- There may be a failure to adjust for corporate actions.
- The previous day's data may be repeated when there is a holiday in the market.
- Data are sometimes restated after a change in calculation method without documentation or any indication that you are not looking at the 'raw' numbers.

This means that many institutions have a group dedicated to *cleaning* market data before they are used for risk or pricing purposes.

Finally, it is worth noting the importance of keeping as much data as possible: you never know when market data, marks or other financial variables might be useful for modelling, so it is usually good practice to keep everything in case it turns out to have value. This is especially true for illiquid parameters: anything which cannot easily be sourced from a data vendor could potentially give the firm a modelling edge in the future.

4.3.2.4 Filling in missing data
Sometimes we have incomplete information on a market: a market data source may have failed for a few days, for instance. Also, there is a holiday in some market around the world on most trading days. Do we just omit the holidays from our return series? Or try to fill them in?*

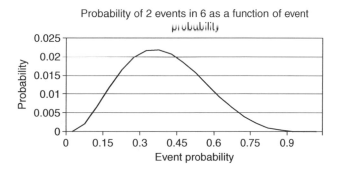

Probability of 2 events in 6 as a function of event probability

It might be helpful to be able to estimate the 'best guess' at missing data. There are a number of techniques for this which come under the general heading of *maximum likelihood estimators* (MLEs). Before we discuss the application of this technique, it is worth looking at it in a simpler situation.

---

* Note that if the previous close is repeated for the holiday then using this repeat data point will typically produce lower variances than throwing the repeated point out.

Suppose we play a simple slot machine six times, and win twice. We cannot see inside the slot machine to analyse the precise mechanism and hence deduce precisely how likely we are to win. But we can infer the likely probability of success from our experiment.

Let the probability of winning be p. The probability of having two wins in six events is $p^2(1-p)^4$. If we plot this function as a function of p, we can see that the answer we expect, $p = 1/3$, maximises the probability of 2 wins out of 6. $p = 1/3$ is the *most likely* probability: to determine p accurately, we would need to test the machine many times. $p = 1/3$ is the best estimate given what we know after six tests.

4.3.2.5 MLE for the normal variance

Suppose we have some data which we assume come from a normal distribution with mean zero but unknown variance. The probability of having an event x from a normal distribution with mean zero and variance σ^2 is

$$\frac{1}{\sqrt{2\pi\sigma^2}}\exp\left(\frac{-x^2}{2\sigma^2}\right)$$

So the probability of the independent events $x_1, \ldots, x_n$ is

$$\prod_{i=1}^{n}\left[\frac{1}{\sqrt{2\pi\sigma^2}}\exp\left(\frac{-x_i^2}{2\sigma^2}\right)\right]$$

If we maximise this function for σ, we get the most likely value for the variance. Relatively simple algebra gives the usual formula for this most probable value of σ.

$$\sigma^2 = \frac{n\sum x_i^2 - \left(\sum x_i\right)^2}{n(n-1)}$$

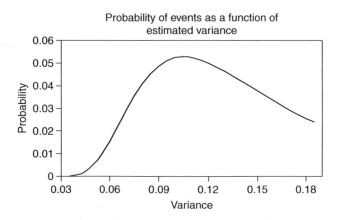

Probability of events as a function of estimated variance

This value for the variance is maximally likely. The data *could* come from a normal distribution with a different variance; it is just that any other choice of variance is less probable than this one.

If we plot the probability of seeing the actual data as a function of variance, then it is maximised for the value of σ above. This technique is known as an MLE.[*]

4.3.2.6 Maximum likelihood estimators

MLEs can be used in the missing data problem.

- We have m return series of length n, typical point x_i^j, with some points missing

$$x_1^1, x_2^1, ..., x_n^1$$

$$x_1^2, x_2^2, ..., x_n^2$$

$$\vdots$$

$$x_1^m, x_2^m, ..., x_n^m$$

- We assume that the data come from a multivariate normal distribution with covariance matrix Σ.

The probability of day i's return vector $x_i = \langle x_i^j, j = 1 \ldots n \rangle$ being from a series with average vector $E(\mu) = \langle E(x_i^j, j = 1 \ldots n), i = 1 \ldots m \rangle$ and covariance matrix Σ is just given by the multivariate normal distribution

$$\Pr(\underline{x}_i) = \frac{1}{(2\pi)^{m/2}} \left| \Sigma \right|^{-m/2} \exp\left(-\frac{1}{2} \left(\underline{x}_i - E(\mu) \right)^T \Sigma^{-1} (\underline{x}_i - E(\mu)) \right)$$

Therefore, the probability of each of these events occurring independently at times $i = 1, \ldots, n$ is

$$L = \prod_n^{i=1} \Pr(\underline{x}_i)$$

This is the likelihood function we must maximise.

Various software packages are available to find missing data; these aim to produce high values for this likelihood function. If you have more than one missing data item there are multiple choices of data. The search space increases fast with the number of data items to be synthesised, so it is sometimes unrealistic to aim for maximal L: sufficiently big is often good enough. This method is good for filling in a few missing items in otherwise complete series: the more data items missing, the less well it performs.

---

[*] See E.L. Lehman and Arthur Romano's *Testing Statistical Hypotheses* for more details.

Exercise. Decide on a method for selecting a data series to represent the risk of:

— A biotech stock spun out of an established pharmaceutical company;
— The rump company after the spin off;
— African countries' sovereign debt immediately after currency controls have been abolished;
— Those countries' FX rate versus the USD;
— Newly issued subordinated debt from a recently demutualised life insurance company;
— The 12-component stocks Telebras was split into. (*Context*: In May 1998, the Brazilian state telecom company Telebras was split up in anticipation of its privatisation to form 12 new holding companies. Virtually all the assets and liabilities of Telebras were allocated to the new companies. There are eight cellular companies, three fixed-line operators and one long-distance carrier. At the time of the split Telebras formed nearly 50% of the market capitalisation of the Brazilian market index, the Bovespa.)

CHAPTER 5

Credit Risk and Credit Risk Capital Models

INTRODUCTION

Lending is one of the oldest financial activities: you give me some money; I agree to pay you back with interest. You are exposed to my performance under this contract, and suffer a loss if I do not repay you. Credit risk, then, is the risk of loss from the failure of a counterparty to fulfil its contractual obligations. This failure is often due to default: the counterparty fails, and the lender may have to queue up with other creditors in order of seniority and present their claim to the liquidator.

In this chapter, we look at how credit risk arises, how it can be mitigated and managed and various models of a firm's net credit risk exposure.*

5.1 THE BANKING BOOK: INTRODUCING THE PRODUCTS AND THE RISKS

We begin with a discussion of some of the ways that credit risk is taken within commercial and retail banking activities. These positions are typically recorded in the banking book and historic cost accounted.

5.1.1 Retail Banking

The main activities of retail banking are well known: they include taking deposits, offering investment products, making loans either on a secured basis or in the form of a mortgage and providing credit card services.

---

* A good introduction to credit risk modelling is Christian Bluhm, Ludger Overbeck and Christoph Wagner's *An Introduction to Credit Risk Modeling*. See also Michael Ong's *Internal Credit Risk Models: Capital Allocation and Performance Measurement* for more details on credit risk modelling, or the references from the credit derivatives material [in Chapter 2] for further background on securitisation as a mechanism for transferring credit risk.

5.1.1.1 Deposits

These can take many forms depending on the facilities offered. In particular, deposit accounts offering full *chequing services* and paying either fixed or, less commonly, floating rates of interest are becoming increasingly common. Banks did not need to offer significant interest rates on standard retail deposit accounts until fairly recently, but rates on current accounts are now an area of competition between banks.

Structured deposits are also fairly common, where the bank offers an investment product packaged in a deposit account. Thus, instead of buying a note, the retail investor can purchase exposure to the same investment strategy via a (sometimes insured) deposit account.

5.1.1.2 The varieties of retail banking

Retail banking is a huge area with many segments depending on the type of client and the nature of the risks taken. Some areas, such as providing banking services to *high-net-worth* individuals, are often rather low risk. Here, there is typically a high barrier to entry, but once the infrastructure and brand are in place to conduct the business, these areas can provide a high return. Other retail areas such as middle market retail banking are highly competitive, and efficiency is key to maximising profits. Finally *subprime lending* is another niche area: the obligators here are of lower credit quality, so defaults are commonplace, but higher rates are charged to compensate for this.

One of the most common forms of retail exposure is the *overdraft* facility. This is usually attached to a deposit account with the bank providing access to additional funds up to some agreed limit in exchange for (often egregious) interest charges. In addition, banks often provide agreed-upon larger personal *term loans*: here the interest rate is significantly lower than the overdraft rate, but funds cannot be repaid early, or can only be repaid with the payment of a penalty.

5.1.1.3 Mortgage lending

One of the most common forms of secured retail exposure is the mortgage: here a property forms collateral for a loan. The degree of credit risk, just as in a repo, depends on the ratio between the value of the collateral and the value of the loan, known as the *LTV*, the creditworthiness of the borrower and the relationship between collateral value and borrower default.

Mortgages are usually long-term products, 25- to 30-year loans, and they might well have a complex structure such as a fixed rate for the initial 2 or 3 years, followed by a floating rate, possibly capped. They are often prepayable as clients' mortgage needs change as they move house. The ability of mortgage holders to prepay their mortgages means that this asset from a bank's perspective suffers *prepayment risk*: the principal balance can be repaid early and the bank loses the asset. [See section 10.2.1 for more details.]

5.1.1.4 Credit cards

Another form of retail credit exposure comes through credit card issuance. Here a financial services firm agrees to pay retailers for purchases made by credit card holders in exchange for an unsecured promise to repay the card balance. These exposures are typically *revolving*: any given debt has a short expected life, but new exposures roll in to replace old ones as card holders pay off some of their balances and then buy new goods.

5.1.2 Commercial Banking

Commercial banking services range from offerings that are very similar to retail for small businesses to much more sophisticated services provided to large corporates. Some of the main products are lines of credit (LOCs), loans, cash management and investment services, and asset finance.

5.1.2.1 Lines of credit

An LOC is a contractual arrangement where in exchange for the payment of an arrangement fee, a firm (usually a bank) undertakes to make funds available up to a fixed threshold for borrowing. The line may be initially *undrawn*, meaning that no funds are actually borrowed at the start of the contract, partially drawn down or fully drawn down.

Typically, LOCs are structured so that funds can be drawn down or repaid at any time or with only a short notice, so they act as *liquidity provision* for the user: if funds are needed quickly, the line can be used, and if the user has excess cash, it can be used to repay borrowing. Short-term LOCs used to be a very regulatory capital–efficient way of taking credit risk, so these were commonplace: post Basel II, the capital charges may be higher, but short-term LOCs are still an important banking product. Often these contracts are *rolling*, with the bank having the right to terminate the contract periodically.

5.1.2.2 Loans

Loans can take a considerable variety of forms:

- Interest on them can be fixed or floating.
- Often they are fixed term, but prepayable or even rolling loans are often possible.
- Principal is often repaid at the end of the loan, but alternatively it may *amortise* during the loan's life.
- The loan may be *syndicated* among a number of banks, so that they jointly lend the money, perhaps with a lead bank controlling most of the arrangements, and/or the loan may be *tradeable* from one bank to another.

5.1.2.3 Letters of credit and guarantees

A *bank guarantee* is a contract under which a bank undertakes to make good a client's obligations should the client fail to do so. Often the guarantee will be attached to a specific cashflow or transaction, so, for instance, the bank might agree to make a payment to a vendor should the client be unable to do so.

A *letter of credit*, in contrast, indicates the bank's willingness to make a payment or fulfil an obligation directly: rather than having first to demand performance from the client and then having recourse to the bank as in a guarantee, the letter-of-credit holder has a direct claim against the bank.

Corporates typically demand that their banks provide both forms of credit enhancement.

5.1.2.4 Cash management and investment services

These range from the simple provision of deposit accounts to comprehensive multi-currency outsourcing arrangements where a bank provides Treasury and back office services to a large corporate covering all its cash management needs and performing the accounting of these transactions too. Although not every bank is capable of offering this kind of rent-a-Treasury service, most offer a range of floating- and fixed rate deposit accounts in a range of currencies, securities trading and custody, FX spot and forward transactions, and payment services.

5.1.2.5 Asset finance

In *asset finance*, the bank lends against collateral such as property or equipment. A related situation is *commercial leasing* where rather than lending a corporate money to buy an asset and taking that asset as collateral, the bank buys the asset itself and leases it to the corporate.

For instance, suppose a firm wishes to fund a large, expensive asset with a long life such as a plane, a train or (less commonly) industrial equipment. Sometimes the expected life of the asset is longer than the term for which the corporate wishes to finance the asset. Train carriages can have a 40-year life, so it makes sense for them to be funded over a 40-year term, but a train operator might only have the franchise to operate a given line for 10 years. Therefore, it makes sense for the bank to buy and fund the asset and lease it back to the train operator. The operator does not have term debt on their balance sheet—just the ongoing lease cost—and the bank bears the joint risk that they will not be able to re-lease the asset at the end of the first lease term *and* that the secondary market value of the asset will be low at that point. This is often a relatively small risk, but it can be an issue if demand falls as in aircraft leasing for some time after the post-9/11 decline in air traffic, for instance.

The advantage of leasing is that the large amounts of cash needed to buy the asset are borrowed by the party with the lowest cost of funds: the bank. The disadvantage is that lease costs can be high to provide sufficient return against the *residual value risk* on the asset.

Exercise. Large transport projects are susceptible to political risk, in the sense that public disquiet over the performance of public transport can lead to political action which may change the effective value of a lease to a bank. How could a risk manager best highlight this risk to the senior management as part of the new product approval process for a proposed new locomotive and railway carriage–leasing operation?

5.1.2.6 Receivables finance

A *receivable* is a debt or unsettled transaction owed to a company by its debtors or customers. For a corporate, it typically occurs when it issues an invoice for goods or services: this is usually settled in due course, but there may be a significant period between the issuance of the invoice and the cashflow settling it. This gives rise to cashflow difficulties in some cases: the firm is profitable, but it needs the funds now to continue or expand its activities.

Financial services firms can serve their clients here in two ways:

- In *receivable financing*, a bank takes the receivable as collateral and lends some fraction of its value, perhaps 90%. In addition, it may help the client to service the receivable and collect the debt.

- Alternatively, in *receivable factoring*, a firm known as a *factor* buys the receivable for cash and collects on it itself.

In the first instance, the client retains the credit risk of the debtor, whereas it is passed on to the factor in the second.*

5.1.3 Forces for Change

Retail and commercial banking used to have the reputation of being a little sleepy. This might not even have been true at the time, and it is even less true now: several factors are causing banks to place more management attention on their retail and commercial banking activities, and to transform their risk profiles. These include the following:

- Portfolio credit risk modelling [discussed in sections 5.4.1 and 11.4.1] has made great strides forward in the past 10 years or so. Banks are now in a position to calculate the economic capital used by retail and commercial banking activities and to compare this on a uniform basis with the capital used by other areas. Performance measures based on return on capital are commonplace, and this in turn has led to an increased focus on risk-based pricing for retail and commercial banking products.

- Banks can now not just measure their portfolio credit risk; they can manage it using credit derivatives, securitisation and other tools. The ability to originate risk, keep some of it and sell on the rest has allowed some banks to move from a buy-and-hold model towards something that is closer to a flow paradigm for banking book activities. [See sections 5.4.4 and 10.3.2.]

- The new Basel II rules [discussed in Chapter 7] have changed the amount of regulatory capital required for the retail and commercial banks. Often retail has been a big winner here, potentially increasing the attractions of this area.

- Internet-based accounts have dramatically reduced the retail cost base for some players, allowing them to compete without requiring them to open branches or process large amounts of paper. In some countries, at least, this has made banking more competitive and has raised the rates paid on deposit accounts.

5.2 CREDIT RISK FOR SMALL NUMBERS OF OBLIGATORS

We begin the discussion of credit risk by examining its occurrence in various simple transactions with one or two obligators. Credit risk occurs whenever there is a cashflow in the future from a counterparty that might not be paid to us. Situations with *direct credit risk* therefore include:

---

* Receivables can also be used as collateral in a securitisation. [See section 10.3.4 for more details.]

- Loans and unsecured lending such as credit cards;
- Guarantees provided and written letters of credit;
- Committed LOCs which may be drawn down;
- Derivatives receivables;
- Other receivables including trade payments and perhaps your salary.

Settlement risk is another variety of credit risk: it occurs whenever there is a non-simultaneous exchange of cashflows between counterparties. The famous example of this was Bank Herstatt [discussed in section 1.3.1] where the two legs of spot FX transactions were not exchanged simultaneously.

Credit risk can be present before cashflows are due. [We have already discussed the presence of credit risk in IRSs in section 2.5.3.] The key issue here is the cost of replacing the cashflow or cashflows lost due to the failure of our counterparty. If that replacement cost can change through time—in the case of a swap perhaps due to interest rate movements—then we have potential future credit exposure (PFCE).

5.2.1 Single Transaction Exposure

One simple measure of credit risk is the cost of replacing a future cashflow or cashflows multiplied by the probability we will have to replace them.

5.2.1.1 Terminology

As with our previous discussion of credit derivatives, we will use PD to refer to the probability that a counterparty will not perform under an obligation, and *recovery* to refer to the percentage amount we receive if they do not perform. The *loss-given default* (LGD) is therefore the *exposure at default* (EAD) × (1 − recovery).

Our first attempt at a measure of credit risk is thus

$$PD \times LGD$$

5.2.1.2 Replacement value

The EAD for a simple loan is simply the PV of the future cashflows:* for other instruments, it is the PV *at replacement*. Here we assume that we are hedged, so if a counterparty does not perform, we have exposure on the hedge, and we will need to replace the original exposure to stay hedged. Imagine making a loan at 7% for 5 years and hedging it with an IRS: if the loan defaults, we still have the swap.

The PV of an exposure at replacement may be slightly different from the PV at default:

- Some contracts may contain *grace periods* where the counterparty has a period of time to correct their failure to perform, so we have to give them a short period to cure their failure before we can go ahead with replacement.

---

* A traditional banker might argue that the exposure is the notional rather than the PV of the remaining cashflows. The best answer may be that it depends on how the book is hedged: if the cashflows are PV hedged, then the credit risk is the PV; if the excess spread is not hedged or included in funding calculations, then the amount at risk is the notional.

- Some contracts may be so complex or illiquid that it takes us a significant period of time to replace them following a failure to perform.

- And in some cases, we may have to rely on the delayed information to determine whether a failure to perform has happened or not.

How much these effects matter depends on the volatility of the exposure: if a contract has a daily volatility of 1%, a 3-day grace period probably makes little difference; a large leveraged oil swap that might change in value by 20% over a week is much more dangerous.

5.2.1.3 Exposure in Basel I

The Basel I credit risk capital requirements [discussed in section 1.5.1] are based on the exposure at default (EAD) of a position. For a loan, the exposure for regulatory purposes is just the notional amount lent. For other positions, it is less obvious how to calculate exposure. Basel I introduced the idea of *pre-processing* positions into an exposure: this gives an estimate of EAD for regulatory purposes. The crude Basel I calculation of the exposure can be summarised as follows:

- For unfunded commitments such as LOCs over 1-year maturity, the exposure is defined as 50% of the amount committed.

- For unfunded commitments under 1-year maturity, the exposure is 0%.

- For guarantees, the exposure is 100% of the maximum claim under the guarantee.

- For transaction-related contingent items such as standby letters of credit, the exposure is 50% of the amount committed.

- For credit risk in derivative receivables, the exposure is defined as the current positive mark-to-market (i.e., the amount owing today) plus a fixed percentage of notional based on the underlying. This *PFCE add-on* is designed to capture the PFCE of the derivative, so these add-ons are higher for underlyings with potentially large volatilities such as commodities.

For many banks, this exposure calculation has now been replaced by the rules in Basel II [as discussed in Chapter 7].

5.2.1.4 Credit risk mitigation

The higher the PD, the riskier the exposure: this much is clear. But that risk can be mitigated in various ways:

- There may be *collateral*, which can be sold to reduce or eliminate the exposure in the event of non-performance. For large derivatives dealers, this can be a significant balance sheet item.

- There may be *covenants* in the contract, whereby the counterparty agrees not to engage in behaviour which is likely to lead to non-performance, such as increasing leverage, or which will decrease our expected recovery, such as selling assets.

- We may have support from a third party, either through a purchased credit protection on a credit derivative or through a *bank guarantee* or *letter of credit* where a bank undertakes to perform if the counterparty fails to do so, perhaps up to some maximum amount or under some conditions.

- Features of the contractual process might reduce credit risk. These include *gross real-time settlement*, where cash and assets in spot trading are exchanged simultaneously or nearly so; *netting*, whereby two counterparties agree to net various exposures to each other; or the use of off-shore jurisdictions to remove or reduce the exposure to unhelpful local legal risks.*

- The use of a *workout* group. This is a specialist part of an institution responsible for minimising the consequences of non-performance. Workout groups lead negotiations with failing creditors, manage the sale of collateral (which can be rather complex if it is non-financial—think of the potential issues in selling a power station) and may even *buy* further assets after default in order to gain extra voting rights in the liquidation process.

5.2.1.5 Exposure and seniority

The liquidation process in many jurisdictions proceeds by each creditor presenting their claim. These are ranked in order of seniority, with the most senior paid first, then the next and so on until there is no cash left. If there is not enough cash to pay all claims at a given level of seniority—as may well happen for senior debt, all of which is at the same level—all claims at that level get the same fractional amount, and lower claims get nothing. Therefore, though on average over many defaults we might find $R = 50\%$ for senior debt and $R = 20\%$ for subordinated debt, it is (almost) impossible in any particular default for the subordinated debt holders to get anything if the senior debt holders have not been paid in full.

This *waterfall* of money down the queue of seniority makes it particularly important to understand where an amount owed to you stands in the queue.

Insurance companies have sometimes taken a rather less careful approach to the management of credit risk than the best banks, so it is worth noting:

- General insurance claims and reinsurance receivables are typically unsecured senior claims. Thus, reinsurance typically involves credit risk to the reinsurer.

- On-shore life insurance companies are typically required to maintain a statutory fund to protect those they have written policies to. This fund is liquidated separately from the company in the event of default and is used for the benefit of policyholders. Retail life policies, therefore, have lower credit risk than non-life ones.

Finally, funds with a broker/dealer are often *client money*: this means they must be segregated, and are not available to help fund the firm or as a creditor asset in the event of

---

* A good example here would be if the local bankruptcy law did not support netting of exposures or did not permit *perfection* of collateral (that is, the right after non-performance of a counterparty to take and sell the collateral). A counterparty in such a jurisdiction wishing to engage in capital markets activities may find significant benefits in setting up and capitalising an entity not subject to these legal difficulties.

default, unlike deposits at a bank. Bank deposits therefore bear credit risk (which may be partly or wholly mitigated by deposit insurance), whereas broker/dealer client money only bears the minor joint risk of ineffective funds segregation and default.

5.2.1.6 Drawdown

For an LOC, the EAD is the drawn amount. As firms' credit quality declines, they tend to draw more on LOCs, so this amount might be significantly more than the drawn amount in ordinary conditions.

5.2.2 Potential Future Credit Exposure

Suppose we have entered into a contractual derivatives agreement with a counterparty. Ignoring any credit risk mitigation, our risk is that the contract has positive mark-to-market *and* the counterparty fails to perform. The EAD is therefore the positive part of the mark-to-market of the contract.

5.2.2.1 The evolution of mark-to-market

As market levels move, the mark-to-market of a contract evolves. If the $\sqrt{t}$ evolution of time rule holds for the variables concerned and the instrument is roughly linear, then we can expect the mark-to-market to migrate in a similar fashion, the uncertainty in value growing roughly as the square root of time.

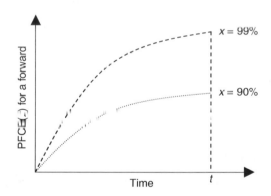

Define PFCE(x) to be the amount that the positive mark-to-market cannot exceed at x% level of probability for a fixed time horizon. Then for a forward to maturity t on a normally distributed underlying, we would find a picture as to the right. Increasing volatility would increase the size of the PFCE, as would increasing the time horizon.

For an IRS, we have a similar-shaped profile, but the exposure falls noticeably on resets towards the end of the swap, with the peak exposure for a typical swap coming roughly two-thirds of the way through. This maximum PFCE at a given confidence level is a measure of the exposure to the counterparty. (Obviously, the size and location of the maximum depends on the shape of the yield curve and the precise structure of the swap concerned.)

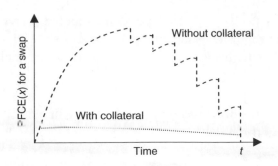

5.2.2.2 The effect of collateral

To reduce credit risk, many derivatives contracts include *collateral agreements* requiring one or both parties to post assets if the mark-to-market of the contract exceeds a threshold amount. The agreement also includes a remargining frequency which determines how often the position was measured to see if collateral is needed. Assuming that any collateral posted is effective, and that the remargining is frequent, we can see the significant reduction in PFCE as in the sketch above.

It is worth remembering that the collateral is only effective if we know what it is worth and what to do with it. Given that many counterparties demand the right to substitute one piece of collateral for another, and since our exposure can move quite quickly, it means that there is significant operational risk in collateral management. When a counterparty fails to perform, we have to notice that it has happened, present any notices required, wait for any applicable grace period, perfect the right collateral and sell it.

5.2.2.3 Portfolio PFCE

We can analyse a single transaction with a counterparty fairly easily. When there are multiple transactions on different underlyings, however, the situation becomes more complex. Here we would need to:

- Simulate a future path of all the underlying risk factors for the portfolio;
- Revalue the portfolio at each point in time for the new market variables;
- Calculate the net PV of the portfolio at each point in time;
- Repeat for many paths to obtain a distribution of portfolio PVs at each timestep;
- Calculate PFCE(x) as the xth percentile of the distribution at each point in time.

This calculation assumes that netting applies, that is, we are only concerned with the net amount owing under a portfolio of exposures, not the gross: this is usually the case for derivatives done under an ISDA master. [See section 10.1.1 for more details.]

5.2.2.4 Reporting credit risk

Credit risk reporting is usually based on the spot exposure—what we are owed now—and the PFCE at various time horizons. Typically, PFCE is calculated through time, and the maximum value it takes in any time bucket is used for that bucket. Limits may then be imposed by bucket.

Any proposed derivatives transaction with a counterparty C would then be analysed, and if the result of adding it to the firm's existing portfolio would blow the PFCE limit, then the transaction could not proceed without some mitigation to reduce the exposure to within limits.

Risk Factor	Credit Exposure to Corporate C
Sensitivity measure	Current exposure and 95% PFCE by time bucket
Limit	$10M spot, $20M PFCE to 1 year, $15M beyond 1 year
Application	Total net bank exposure to corporate C

5.2.2.5 Options PFCE

The choice of safety threshold for PFCE is sensitive if options have been purchased. Consider an 80 strike put on an underlying at 100 with a 10% volatility. At 90% PFCE, there is little chance we will be owed money at maturity since 90% corresponds to 1.3 S.D.s or spot at 100% − 1.3 × 10% or 87. A 99% threshold, however, is 2.33 S.D.s which corresponds to spot at 77: here the option is in the money, and so there is significant PFCE at this degree of confidence. In this instance, the PFCE(90%) is low, but the PFCE(99%) could be significant. Unless credit risk management is sensitive to these effects, trading might be incentivised to purchase lots of fairly out-of-the-money options from low-quality credits as they are not caught by the system: this is probably not the behaviour we want to encourage.

5.2.3 What Is a Credit Spread Compensation for?

Previously, we introduced the fair value of a credit spread s as compensation for the possibility of a credit event. If the probability of default is PD, the recovery after the event is R, and risk-free rates are r, a slight modification of the previous argument gives $(1 + r) = PD \times R + (1 - PD) \times (1 + r + s)$ (and of course in reality we would use the instantaneous hazard rate setting discussed in section 2.5.3 rather than this finite approximation).

Think about holding a risky bond position paying this spread s. There are several other factors apart from the possibility of a credit event which we might reasonably demand compensation for:

- The potential illiquidity of the bond;
- The fact that we might not be able to fund the bond as cheaply as a risk-free instrument, for instance, because it does not repo GC;
- The volatility of the mark-to-market of the bond (which has to be supported by equity in a mark-to-market environment, after all);
- Understanding the underlying credit risk and the structure of the bond.

In addition, we are assuming *risk neutrality*, that is, the investor is indifferent as to the amount of risk they take, provided they are properly compensated for it.

5.2.3.1 Historical evidence

If we examine bond spreads for a range of bonds over time and compare them with losses that arise from credit events, we find that in general a corporate bond investor gets more compensation than is fair for credit event risk. The excess amount depends on the period we look at and the quality of bonds we pick, but typically the spread is several times the fair compensation. This supports the view that the credit spread is the compensation for more than the PD for a risk-neutral observer.*

5.2.3.2 Implications

If this conceptualisation is correct, PDs derived from spreads as above are not pure real-world PDs. They may be suitable for pricing credit derivatives, but they probably do not capture the market's expectation of future default rates. To see the difference, consider the following applications:

- *One-off hedging.* We have a credit exposure and we want to know how much it would cost to hedge it. Clearly, the answer is the market price of credit today, and using a market-spread-derived PD is appropriate.

- *Portfolio hedging.* We have a credit exposure and we are going to put it into a large portfolio of similar exposures, and then sell the lot. Here we care about the exposure's incremental contribution to the cost of hedging the portfolio.

- *Expected loss.* We have a credit exposure and we want to know how much we should expect to lose if we hold it in a historic cost accounted book for an extended period, probably to maturity. Here a PD derived from a market spread will significantly over-estimate the risk if the spread is the compensation for more than default risk.

5.2.4 Partial Credit Mitigation

U.S. government bonds are good collateral against an exposure in USD. There is no credit risk, there is ample liquidity and the framework for perfecting this collateral in the U.S. is legally certain.

At the other end of the spectrum, taking Peruvian government bonds as collateral against a Chilean peso NDF with a hedge fund specialising in Latin America may be worse than useless: worse because it might appear superficially that the collateral is effective, whereas in many situations when the counterparty is in trouble, the collateral may well be worth much less too.

5.2.4.1 Collateral with correlation to the counterparty

One solution to this problem is never to take correlated or *wrong-way* collateral, and indeed some firms only accept collateral of a certain quality, typically cash, AAA government bonds or similar instruments.

---

* See, for instance, John Hull et al.'s 'Bond Prices, Default Probabilities and Risk Premiums' (*Journal of Credit Risk*, Vol. 1, No. 2).

However, for some firms, this approach is difficult from a business perspective. In *prime brokerage*, banks and broker/dealers seek to be the (only) provider of risk to hedge funds, trading securities and engaging in derivatives and repo transactions with them. Prime brokers often deal with hedge funds which specialise in a certain investment area, and the only collateral they have often comes from their area of expertise: a Russian specialist fund has little reason to buy U.S. Treasuries.

Similarly in corporate banking, clients may only have assets to pledge which are highly correlated with their business: copper for a copper miner, for instance, or even worse a copper smelter. In both of these instances, it is necessary to make an estimate of the likely value of the collateral in the event of default, and only to give credit for that fraction of the collateral's value.

One alternative way of dealing with large, profitable trades with correlated collateral is to price in a hedging programme. For instance, if you want to make a 10-year loan to a copper producer and the only collateral available is a copper smelter (which acts much like a call on copper struck at its cost of production), then ensure that the loan spread covers the cost of purchasing the puts on copper necessary to keep the deal within the firm's risk appetite.

5.2.4.2 Structural mitigants

Various contractual means of reducing credit exposure in addition to netting are becoming popular. These include:

- *Early termination agreements.* Here one or both counterparties have the right to terminate the transaction early, paying the other the mark-to-market value.

- *Downgrade triggers.* Here the right to terminate for a party comes into existence if its counterparty is downgraded below a certain level.

Both of these mitigants can significantly reduce exposure for long dated transactions.

5.2.5 Introducing Basket Credit Derivatives

Suppose we have three risky bonds, A, B and C, and we invest £10M in each bond. There is a range of outcomes of this investment:

	Number of Defaults		
0	**1**	**2**	**3**
	ABC	ABC	
ABC	ABC	ABC	ABC
	ABC	ABC	

Here we have indicated a default by striking through the name of the bond, so ABC is the state where none of the bonds default, and ABC the state where they all do.

5.2.5.1 nth-to-default products

Consider three securities structured from this portfolio:

- The *first-to-default* note will pay out just when there are no credit events on any of the bonds. It therefore only pays out in the state ABC, otherwise it returns whatever the recovery is on the first bond to default.

- The *second-to-default* note pays its scheduled coupon and returns full principal provided there is no more than one default. If a second bond suffers a credit event, it returns the recovery on that asset.

- Finally, the *third-to-default* note takes the risk that all three bonds will not suffer a credit event, so it pays out scheduled coupon and full principal in all states except A̶B̶C̶.

Clearly, the first-to-default note has a fair value coupon higher than any of the individual assets since it is at least as risky as the riskiest asset. The third-to-default note has a low coupon since it is only at risk if all three bonds default.

Collectively, these products are known as *n*th-to-default notes.

5.2.5.2 A very simple model

To give some insight into *n*th-to-default product, consider a highly oversimplified model:

- The credit event correlation between the assets is zero, that is, the assets are independent.

- The recovery on each of the assets is zero.

- Risk-free interest rates are zero.

- We only consider a 1-year time interval, and if a default happens, no coupon is received before default.

- All assets trade at their (risk neutral) fair value.

Suppose the credit event probabilities of A, B and C are 2, 3 and 4%, respectively.

The probability of each of the states using the same layout as above is as follows:

	1.8624%	0.0576%	
91.2576%	2.8224%	0.0776%	0.0024%
	3.8024%	0.1176%	

(This follows from the assumption of credit event independence: for instance, 2% × 3% × 4% = 0.0024% for state A̶B̶C̶.)

Each bond either pays out its coupon and returns principal if there is no credit event, or gives us nothing. Assuming that each bond's coupon is a fair compensation for this risk, the payout of the portfolio per state is as follows:

	£20,725,945	£10,416,667	
£30,930,027	£20,620,748	£10,309,278	£0
	£20,513,360	£10,204,082	

You can check the bonds are at fair value by noting that the probability-weighted sum of the payouts is £30M, the initial investment.

The fair value coupons for the nth-to-default notes in this setting are as follows:

- For the first-to-default note, 9.58%;
- For the second-to-default note, 0.256%;
- For the third-to-default note, 0.0024%.

Again, you can check this by calculating for each note the payoffs in each state, and checking that the probability-weighted payoff is £10M. For the second-to-default note, for instance, the payoff by state is

	£10,025,585	£0	
£10,025,585	£10,025,585	£0	£0
	£10,025,585	£0	

And the total payoff of all the tranches is

	£20,025,825	£10,000,240	
£30,930,817	£20,025,825	£10,000,240	£0
	£20,025,825	£10,000,240	

Although the assumption of credit event independence is unrealistic, this model does demonstrate several interesting features of nth-to-default notes

- The sum of the coupons of the three notes is larger than the total coupon on the bonds. This might seem bizarre: how can the fair value of protection on the three bonds be more than the carry they give? The answer is that if one of the bonds defaults, then we do not have to pay the (high) coupon on the first-to-default note, but we do receive the coupons on the two bonds that have not defaulted. In state ABC, for instance, we receive £20,725,945 from the underlying bonds but only pay out £20,025,825 on the tranches, so we make money. This profit in a fair value world is offset by a loss in another state: that state is ABC where we pay out more on the notes than we receive on the bonds.

- The timing of payments versus default affects the valuation considerably. For instance, if the notes paid a semi-annual coupon rather than an annual one, and the bond did not default until after the first note coupon had been paid, the fair value first-to-default coupon would fall to 9.14%.

> *Exercise.* Write your own version of the model discussed above.

5.2.5.3 Improving the model

A less naïve model for nth-to-default products would need to address at least the following:

- Interest rates and discounting (and possibly even stochastic interest rates or credit spreads, or both);

- The term structure of default, introducing fine time steps so that the timing of defaults, bond coupons and note coupons can be studied as in the hazard rate framework;

- Some notion of credit event correlation so that events on each of the bonds are no longer independent.

5.3 AN INTRODUCTION TO TRANCHING AND PORTFOLIO CREDIT DERIVATIVES

There are many sources of information on the historical performance of credit risk loss distributions: ratings agencies have tracked the corporate bonds for more than 40 years—although these data are rather U.S.-centric—and banks have their own internal loan databases. Typically, if we examine these data, we find the shape of the loss distribution discussed earlier [in Chapter 3] with a small EL and a much larger UL in the tails of the distribution.

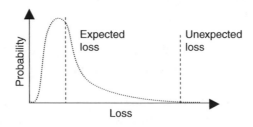

To make the discussion concrete, suppose a vehicle owns a $1B diversified portfolio of 100 different investment-grade corporate loans, all of 5 years' maturity. This portfolio will have a loss distribution like most other credit portfolios, as shown. Suppose that the loans within it pay a *weighted average coupon* (WAC) of Libor plus 150 bps, say, and that this portfolio is the only asset in the company. How might we fund the vehicle?

5.3.1 Funding and Loss Absorption

We need $1B to pay for the portfolio. Our first thought might be to fund this mainly with debt to enhance our leverage: say issue $990M of debt and have $10M of equity. But then two or three defaults on the portfolio would eliminate our equity buffer, so this debt would be fairly risky. The debt holders are bearing a significant amount of the credit risk from the portfolio.

Clearly, we do not know which loan will be the first to default—if we did, we would not have bought it—but it is likely that over 5 years, we will see several defaults. Therefore, we need enough equity to support a handful of defaults: $50M would support five (and more if recoveries are non-zero), so that is a reasonable buffer.

Now suppose the economic cycle turns over the 5-year life of our loan investments; then we might see a rise in defaults—as we did around 2001—and so even a $50M buffer might not be enough. If we have more than $50M of equity, our ROE will suffer, but if we stick with just a $50M loss buffer to *credit enhance* the debt we issue, it might not get a good rating due to the possibility of unexpected losses. Therefore, we split the required $950M of 5-year debt into two *tranches*:

- A *mezzanine*, or mezz, tranche of $70M, which is subordinated and so bears the risk of the next losses should the equity be exhausted;

- A *senior* tranche of $880M, which now has sufficient credit enhancement thanks to the junior and mezz to achieve a good rating since it would take at least $120M of losses on the portfolio before the repayment of it is at risk.

5.3.2 Securitisation and Tranching

This capital structure illustrates the basic ideas of the securitisation and tranching of credit risk:

- *Securitisation* is the process of moving a portfolio of assets into a vehicle and issuing securities from that vehicle which pass the risks and returns of those assets on to some other party. The party that initiates this process is known as the securitisation *sponsor*. Often—as in the example discussed above—the sponsor is also the *originator* of the risk. The assets are known as the securitisation *collateral* or *collateral pool*.

- The risk transfer can happen either by the sale of an asset, for instance by the sale of a loan to the securitisation vehicle, or by the vehicle writing CDS protection on each underlying asset. The former situation is known as a *cash securitisation*: the purchasing vehicle has to find cash to pay for the assets, so it, rather than the originator, bears the cost of funding them. The latter case is a *synthetic securitisation*: in this situation, the assets are funded by the originator, and it is just the risk of a credit event happening on them that is passed on to the securitisation vehicle. In both cases, the resulting securities are a kind of ABS.

- *Tranching* is the process of splitting up the loss distribution of a securitised asset pool into various tranches. Typically, there is the lowest tranche which bears the risk of the first losses on the portfolio, called the equity or *junior* tranche; higher tranches with some credit enhancement bear the next losses; and above them there may be one or more *senior* tranches which are rather unlikely to suffer losses. The junior tranche acts as *credit support* for the rest of the structure: if it is thick enough—so that it can absorb sufficient losses—all the tranches above it can be rated. (In *reinsurance*, a reinsurer takes some risk from a primary insurer *attached* between two points in exchange for a premium: in risk transfer terms, this is similar to the tranche of a securitisation. The same terminology is used so for instance in the example above we would speak of a $70M mezz tranche attached at $50M.)

5.3.2.1 Buying protection

We can think of the securities issued by a securitisation vehicle as providing protection on progressively more and more defaults, regardless of which credits in the portfolio suffer.

- The first few losses in the distribution typically depend on underwriting standards, general economic conditions and luck. For most portfolios, it is fairly likely that at least a few credit events will occur, so we would have to pay a significant amount to the equity tranche holders in exchange for the protection they provide against the first losses. Often the amount paid to the equity tranche risk taker is whatever is left from the excess spread (ES) of the portfolio after the other tranches have been paid, in analogy with the equity holders in a general purpose company.

- The next losses in the distribution are less likely: therefore, the cost of mezzanine protection is cheaper than junior protection but still significant. Thus, the coupon on a mezz security might be a few hundred basis points over Libor, and the holder bears the risk that sufficient credit events will occur to exhaust the equity completely.

- Finally, the senior tranche is rather cheap to protect: perhaps a single-digit or tens of basis points spread in exchange for the tail risk of many credit events.

The total loss we could be exposed to if our entire example portfolio defaulted is clearly $1B, but the reality is that losses will probably be tens of millions of dollars. This EL is born by the junior tranche,* so the coupon on it must be sufficient compensation for the likely loss of principal. The junior tranche pays a high spread for a high risk. This part of the distribution is therefore typically suitable for either the originator—who has a reason to believe that their own underwriting standards will minimise the number of credit events—or a high-risk/high-return investor such as a hedge fund. In contrast, the senior tranche is a low-risk, low-return asset. Depending on the *attachment points* it may well be as good as a typical AAA credit, or even better.

The analysis of securitisation tranches therefore depends on:

- Understanding the behaviour of the underlying collateral, including its ES and its loss distribution;

- Understanding how this behaviour is allocated between the tranches.

---

* Note that this situation is slightly different from our discussion in Chapter 3 where only losses above the EL were supported by equity.

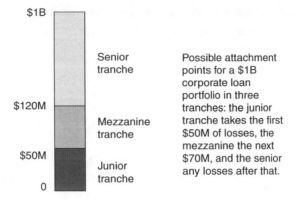

$1B

Senior
tranche

$120M

Mezzanine
tranche

$50M

Junior
tranche

0

Possible attachment
points for a $1B
corporate loan
portfolio in three
tranches: the junior
tranche takes the first
$50M of losses, the
mezzanine the next
$70M, and the senior
any losses after that.

5.3.2.2 Example levels

Consider our $1B loan securitisation again, as shown in the figure. How much can we afford to pay each investor to take the risks that they respectively bear? We assumed that the collateral portfolio has a weighted average coupon of 150 bps, so before any credit events, there should be $15M per year to allocate:

- First, we will have fees in the structure to the originating bank for administering the loans, to the parties running the securitisation vehicle for their labours and so on. If these amount to 25 bps, that is $2.5M.

- If we have to pay 15 bps to the senior investor, that is $1.32M spread over Libor.

- Suppose the mezz investor requires 150 bps to take the risk of credit events above the first $50M: their spread then amounts to $1.05M.

- This leaves $15M − $2.5M − $1.32M − $1.05M − $10.13M for the equity investor. Even if they have no return of principal at the end of the transaction, this corresponds very roughly to a 20% annual return, so the equity holder is well compensated for the risks that they run.[*]

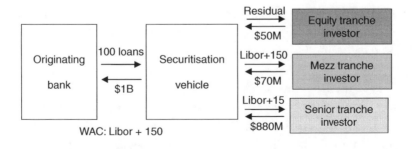

Residual

Equity tranche
investor

$50M

Originating

bank

100 loans

$1B

Securitisation

vehicle

Libor+150

Mezz tranche
investor

$70M

Libor+15

Senior tranche
investor

$880M

WAC: Libor + 150

---

[*] Note that even if ten defaults happen in the first year, the equity holder still receives the excess spread from the portfolio after all other parties have been paid, so they may still get a good return provided that there is enough excess spread. At least in simple structures like this, it may be the mezz holder that suffers most if there are many defaults.

This example demonstrates how the total spread from the portfolio can be allocated to the sellers of protection on the various tranches of the underlying collateral. The structure we have examined is too simple, though, so now we must make it more practical.

5.3.3 Collateralised Debt Obligations

The securities produced in a securitisation are known as *collateralised obligations*, since the vehicle's ability to pay on them only depends on the performance of the underlying collateral. If the collateral is a loan portfolio, we speak of CLOs for *collateralised loan obligations*; CBOs for bonds and so on. The generic term CDO refers to any debt collateral.*

The CDO market grew explosively through the 1990s: we review why and discuss some of the issues in CDO transaction structures.

5.3.3.1 The appeal of CDOs

One of the reasons for the growth of the CDO market in its early days was *regulatory capital optimisation*. A $1B loan portfolio on the bank's balance sheet might need $80M of capital to support it: in 1996, some banks could securitise the portfolio, retain a $100M equity tranche which contains nearly all the risks and rewards of the portfolio and reduce their capital requirement by a significant amount. Since those days, the capital required for securitised risks has become somewhat more closely aligned to the risks transferred, but in some sense, the genie is out of the bottle: securitisation and tranching technology is now a well-established way for risk originators to fund their activities, optimise capital requirements and hedge some or all of their portfolios.

Another key reason for the popularity of CDOs was that at least in the early days of the market, it was possible to buy a collection of bonds and then sell CBO tranches backed by those bonds for a lower PV: in effect, the sponsor could end up owning the equity tranche for nothing or very little. This is sometimes known as an *arbitrage CDO*.

5.3.3.2 Liquidity

The underlying collateral may pay coupons at various times of the year, so we will expect cash to accumulate in the securitisation vehicle. In contrast, the CBO tranche coupons will all be due at once, perhaps twice a year for a semi-annual pay structure. Therefore, we will need some cash management to invest collateral interest, and possibly the need to borrow for short periods to cover any gaps between collateral interest being paid to the vehicle and paying coupons on the issued notes. This is often done via a *liquidity facility* provided by the sponsor or a third-party bank.

In a synthetic securitisation, we might issue tranche securities as in a cash securitisation, but risk is taken via writing default swaps rather than by buying a loan or a bond. Therefore in this situation, the vehicle is long cash having been paid by the CDO tranche investors but not paying itself. As in a CLN, this cash would typically be invested in a high-quality security or portfolio of securities.

---

* Further information on CDOs is available in Gregory's or Schonbucher's books (*op. cit.*) or in Sanjiv Das' *Credit Derivatives: CDOs and Structured Credit Products*.

5.3.3.3 Prepayment

The collateral may pay early, either because contractually it is *prepayable*—as in some pre-payable loans or mortgages—or because we cannot find sufficient collateral with the same maturity, so we accept slightly shorter (or longer) maturity collateral too, such as loans between 4 and 5½ years' residual maturity in a 5-year CDO. We then have the choice of either issuing prepayable tranches, so, for instance, the senior tranche prepays first, then the mezz and so on, or trying to reinvest any amounts received early in similar collateral.

Securitisations where the underlying collateral is residential mortgages are known as RMBS (whereas CMBS refers to a *commercial* mortgage-backed security, i.e., one where the collateral is a corporate rather than a retail mortgage). In both cases, we speak of a collateralised mortgage obligation (CMO).

5.3.3.4 Fixed versus floating

In our simple example, we assumed that the collateral was entirely floating rate loans and that we could issue floating rate tranches against it. However, it may be that the collateral contains a mix of fixed and floating rate assets. Then we would need to control interest rate risk by swapping the fixed rate loans for floating. There is risk in this, however, as if we have a swapped fixed rate loan which defaults, the swap payments are still due. Therefore, we sometimes swap the *expected* fixed rate payments rather than the contractually due ones to reduce this contingent interest rate risk. In the presence of prepayable fixed rate assets in the collateral pool, this situation becomes considerably more problematic as now we can lose an asset through either default or prepayment, and prepayment rates themselves are interest rate sensitive, as we shall see later.

> *Exercise.*   What are the risks if we issue some floating rate CDO tranches and some fixed rate? Can they be fully hedged?

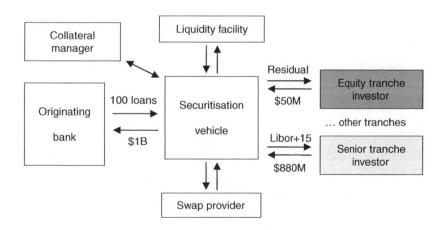

5.3.3.5 Collateral substitution and management

We may wish that the securitisation vehicle has the ability to *trade* the collateral pool rather than having it remain static. For instance, if any of the underlying loans is downgraded below investment grade, we might decide that the best strategy was to sell the loan, probably at a loss, and replace it with another investment-grade loan rather than have investors bear the increased default risk of the lower credit quality.

Therefore, we might need a *collateral manager*, either to exercise their discretion with trading the collateral or to trade it within certain preset rules, such as substitute-on-downgrade. If this collateral manager has particular expertise in the area, for instance, because they are a reputable fund manager, then their inclusion in the structure might add value to investors. If, however, there is a possibility of gaming—collateral being substituted because the manager does not want to own it themselves any more—then investors should be very wary.

Another situation where we may have varying collateral during the life of the securitisation is where the originator themselves has *revolving* exposures. For instance, the average life of a particular exposure within a bank's credit card book might be only a few months, but since the bank is continually originating new exposure as the old ones fall off, we may be willing to let the bank substitute new exposures for old ones, provided of course that they are underwritten to similar standards. This is not necessary for term assets, so we use the phrases *term* or *self-liquidating securitisation* to refer to a situation without revolving collateral: otherwise we have a *revolving securitisation*.

Exercise. If you have access to either internal credit card data or collateral performance data from a credit card securitisation, examine the default, weighted average spread and effective maturity behaviour of the portfolio. How does it vary through the economic cycle?

5.3.3.6 Finer tranching
We have simplified the graduation of losses into just three tranches. In real CLO or CBO structures, we would typically find more tranches corresponding to a finer graduation of loss transfer, perhaps 10 or even more.

5.3.4 Structuring and the Waterfall
The term *waterfall* refers to how the stream of cash coming off the collateral pool is allocated among the various tranches: we can imagine it falling down, first to the expenses, then to the senior tranche, to pay its coupon, and so on down until whatever is left after all other claims goes to the equity tranche holder. In general, the waterfall is just a set of rules determining how to divide the cashflows coming off the collateral pool into payments to the various securities (and possibly other accounts). The risk/return characteristics of the tranches can be modified by varying the waterfall. For instance, the following features are sometimes introduced:

5.3.4.1 Spread accounts
In the example earlier, we saw that there was considerably more spread coming off the collateral than was necessary to service the mezz and senior tranches. Instead of handing all of that over to the equity tranche, an account could be set up to capture some of that spread before handing the rest over to the equity holder. This is known as the excess spread or ES A/C.

The ES account would fill slowly over time as the collateral paid interest, and would act to effectively increase the attachment level of the mezz. If the ES account is not needed to absorb extra losses, it is paid out at the end of the transaction, often to the equity holder. The equity holder therefore only suffers a delay on receiving the ES if things go well: if they go badly, the account adds credit support to the mezz and senior.

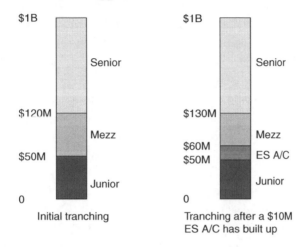

5.3.4.2 Coverage tests

Instead of diverting cash automatically, we could use a test to determine whether to start to divert the spread into an account. For instance, we might require that sufficient cash come off the collateral to pay 150% of the required coupon on the mezz and senior: if the spread falls beneath this level due to delinquencies or defaults on the underlying loans then spread would be diverted from the junior tranche to provide credit support to the higher tranches. This is an *interest coverage test*: similarly, we might have an *overcollateralisation test* which initiated interest diversion if the value of the collateral pool fell beneath 105% of the value of the mezz and senior tranches.

Another possibility is to use IC or OC tests as triggers for *early amortisation*. If the tests are failed, interest payments to the equity are suspended entirely and all the ES goes into amortising the principal balance of the rated tranches.

These features go some way to reducing the appeal of the equity (which is often an easy tranche to sell to investors, and is anyway often retained by the originator) and enhancing the security of the mezz (which is often rather more difficult to sell). The process of deciding which features to include in a securitisation is known as *structuring*: typically this is a balancing act between the conflicting needs of the buyers of the different tranches. Other parts of the puzzle are the ratings agencies that may rate the tranches, accounting policy staff who may well be keen to ensure that the securitisation vehicle does not consolidate on the originator's balance sheet, and tax specialists who wish to ensure a desirable tax treatment of coupons received on the collateral pool and coupons paid out on the tranches.

5.3.5 Index Credit Products

A number of *credit indices* have become commonplace over the past few years. These consist of a list of credits whose senior obligations are used to form the collateral pool for a standardised CDO. For instance, one of the well-known indices, the iTraxx Europe™, consists of 125 European investment-grade corporates diversified across a range of industry sectors and recalculated regularly.*

5.3.5.1 Traded tranches

Standard tranches on these indices are also defined: for the iTraxx, these are 0–3% equity, 3–6% mezz, 6–9% senior and so on. As with any index, the aim of standardisation is to promote liquidity and price visibility: an investor who wishes to take risk on a $1B portfolio of 125 names attached between $60M and $90M need only buy $30M of the 6–9% tranche to obtain the desired exposure.

Since tranches in the main indices are fairly liquid, an investor can be relatively sure of a secondary market if they wish to change their position during the life of the trade, something that is by no means assured for a customised collateral pool.

5.3.5.2 Single tranches

In the early days of CDOs, a dealer could buy a collateral pool, sell all of the tranches, book a significant profit and be flat risk. The available profits in this process are now somewhat smaller and investors are perhaps more cautious of the precise underlying collateral behind the CDO tranche they are buying, so the business of finding buyers for all the tranches is more difficult. Therefore, selling a single tranche can be attractive. If the underlying is an index, there is no problem as the dealer can buy back the tranche sold, and the ability to do this keeps bid/offer spreads tight on the index tranches.

But what if the underlying collateral pool is not an index with traded tranches? What exposure does a dealer have if they have sold the mezz on some collateral pool but not the junior or senior? Clearly, the risk of a tranche somehow depends on each credit in the underlying collateral pool, but how exactly—how would we calculate a *delta hedge* that replicated a tranche? This is the problem of how to model the pool loss distribution given an understanding of each piece of collateral in the pool, and we will examine various solutions to it in this chapter and later in the book. First we examine one of the major issues in solving this problem:

5.3.6 (The Problem with) Credit Event Correlation

One way of thinking about credit events is the tank analogy. We have a vehicle with some degree of armour plating and we are making our way through hostile territory. Depending on how much armour it has, the vehicle can take a certain amount of damage and proceed with impunity. However, if it is hit too many times, the armour fails. The equity tranche is like an unprotected vehicle—anything damages it—whereas the senior tranche is a massively well-protected tank which can take a huge amount of damage before failing.

5.3.6.1 Two routes to the same place

Now suppose we are faced with two possible routes for a journey that has to be made. On one, there are six enemy groups who cannot communicate with each other. Each

---

* See www.indexco.com for more details on the iTraxx indices and www.markit.com for the CDX™ indices of North American and emerging market credits.

independently makes a decision whether to try to ambush us. Given that an ambush is difficult and dangerous for the enemy, it is very unlikely that we will face more than two or three ambushes on this route: on the other hand, we probably will face at least one.

On the second route, there are again six enemy groups, but they are in communication with each other. Either there will be no ambushes, or all six groups will ambush us on the route, one after the other.

If we take the first route with the unprotected vehicle, one ambush is likely, so we will probably not make it back to safety. On the second route, however, the enemy may have decided to stay at home, and so no losses are possible and then the unprotected vehicle will be fine. In contrast, on the first route, the big tank can take the damage caused by two or three assaults without difficulty, so it will almost certainly make it home along that route. On the second route, though, if the enemies come out, they *all* come out, so even the big tank will be damaged along this route.

5.3.6.2 The effect of correlation on the tranches

The two routes are like two levels of credit event correlation: in the first, correlation is low, so if we have a few credit events in the portfolio, it does not mean we will have many more. The second route is analogous to a situation with high credit event correlation: if we know we are in this situation *and* we have a few credit events, more are highly likely.

Increasing credit event correlations makes senior protection more valuable as we are more likely to be in a situation where we need thick armour. Similarly, increasing correlation here makes first loss protection less valuable, as we might not need it at all. Also, of course, changing correlation can only move value between the tranches; it cannot change the value of the sum of the tranches, since that is the value of the original portfolio, so if protection on the senior tranche gets more valuable, that value has to come from somewhere.

To see the effect in a simple case, consider a basket of six credits. Clearly, situations are possible over a period: no defaults, one, two, etc. The following graphs show the changes in the cost of the first-, second- and third to default and senior [fourth, fifth, and sixth to default] basket credit derivatives as a function of credit event correlation.

The analogue of the equity tranche here is the first-to-default option: increasing correlation decreases its value. The values of two top tranches increase with correlation.

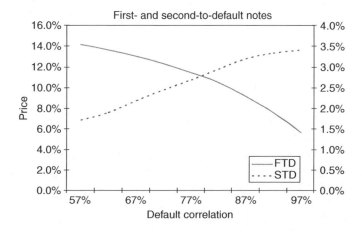

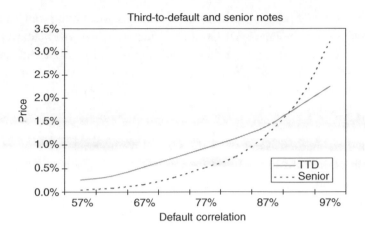

5.3.6.3 Implied correlation

One of the useful things about the traded tranches discussed earlier is that we can use their prices to infer average credit event correlations. These correlations backed out from tranche prices are known as *implied credit event correlations*, in analogy with implied volatilities: both are the value of a parameter which gives the right (market) price using a model. Also, as we will see later [in section 11.4.4], simple models of correlation usually give different implied correlations for different tranches—this is a powerful clue that there is something not being captured by the model, just as the implied volatility smile gives us a clue that the model of log-normal returns is not quite right when we deal with equity or FX options.

> *Exercise.*   Compare and contrast the market events needed for an investor who buys a single A-rated corporate bond to lose money with those needed if they purchase a single A-rated tranche of a corporate bond securitisation. If the two securities had the same PDs, how would they nevertheless differ?

5.3.7 Practical Credit-Adjusted Pricing

If we transact a single swap with a counterparty without any credit risk mitigation, then a good approximation to the fair market swap price including credit effects is gained by:

- Discounting payments from them to us in the future along the appropriate risk-free curve plus the counterparty's credit spread;
- Discounting payments from us to them along the appropriate risk-free curve.

[This will slightly overstate credit risk even here as we have the option to assign after default as discussed in section 10.1.1.]

The presence of a collateral agreement complicates matters. Here we actually have to model the behaviour of the swap and the collateral and only credit-adjust the residual owed to us after collateral.

Portfolios of transactions which net and where collateral is available add another level of complexity: as in the calculation of PFCE, we have to model the evolution of the entire portfolio of exposures and the value of the collateral over time. The net payments owing to us at each point are calculated, the effect of collateral is subtracted and any residual is discounted by the counterparty credit spread.

For a new transaction, the credit-adjusted price can then be obtained from the risk-free price by the difference between the credit-adjusted value of the counterparty portfolio with and without the new transaction.

Finally, CDO technology allows us to deal with this exposure in the context of a portfolio of exposures: we consider the whole portfolio of receivables from all counterparties as a CDO, and the credit adjustment is the incremental contribution of an exposure to the price of securitising the portfolio.

5.4 CREDIT PORTFOLIO RISK MANAGEMENT

Securitisation technology has focussed attention on the portfolio credit risk loss distribution and on techniques for measuring and managing it: the issues are fundamentally different from single-transaction credit risk since we have to consider credit event correlation as well as the risk of each obligator. We now look at several techniques for estimating the shape of the credit risk loss distribution and determining its behaviour under stress. This leads to a discussion of how portfolio credit risk is managed through active risk distribution.

A crucial tool in controlling credit risk for retail portfolios is the underwriting process, so we then discuss techniques for assessing retail and small corporate exposures.

5.4.1 The Portfolio Credit Risk Loss Distribution

There are many sources of information on the historical performance of credit risk loss distributions: ratings agencies have tracked the corporate bonds for more than 40 years—although these data are rather U.S.-centric—and many banks have their own internal loan databases which typically rate exposures using a scale internally developed by the bank.

5.4.1.1 Internal ratings

Many banks have an internal ratings system which aims to segregate obligations into a number of categories on the basis of their credit quality. For instance, there may be seven buckets, with bucket 1 representing an essentially risk-free obligation, and bucket 7 the riskiest.

Internal ratings systems should be:

- *Objective.* There should be a clear algorithm for deriving the rating of any obligation based on objective data.

- *Qualitative and consistent.* The rating will be based on observable financial properties consistently combined, and where these are over-ridden by the judgement of credit risk staff, this can easily be discovered.*

- *Accurate.* The ratings system should distinguish different levels of credit risk.

Various approaches to the design of internal ratings systems are discussed below.

5.4.1.2 Transition matrices

Typically, if we examine bank or ratings agency data, we find a distribution of losses similar to the classic credit loss distribution discussed earlier: a small expected credit risk loss and a much larger UL.

More detail on how a credit moves from being (presumably) good at the time of taking on an obligation to a credit event can be obtained by looking at the *migration* of corporate issuers around the ratings spectrum. For instance, we could take a portfolio and record the starting rating of every exposure within it at the start of every year. If we look again at the end of the year, we can record the probability of a credit reaching a given rating over the 1-year horizon. If the portfolio is large—and perhaps if we aggregate data over the whole economic cycle rather than just looking for a single year—then the resulting table of transitions gives a reasonable picture of how far a credit rating can move in a year. For instance, the table below gives one such possible *ratings transition matrix.*

The value .0555 in the 3 to 4 bucket indicates that there is a 5.55% probability, for the portfolio studied, of a credit that started the year rated as a 3 being a 4 a year later. Examining the table, we find the following:

- It appears as if the ratings process discriminates between different risks in that the probability of a credit event (in the rightmost column) uniformly rises with rating.

- Most credits, most of the time, do not move: by far, the largest probabilities are on the diagonal, indicating no movement in rating.

Start-of-the-Year Rating	End-of-the-Year Rating							Credit Event
	1	2	3	4	5	6	7	
1	0.9087	0.0913	0	0	0	0	0	0
2	0.02	0.8994	0.0806	0	0	0	0	0
3	0	0.0462	0.896	0.0555	0.0018	0	0	0.0005
4	0	0	0.0411	0.9157	0.0415	0	0	0.0017
5	0	0	0.0096	0.0578	0.6459	0.273	0.0039	0.0098
6	0	0	0.0124	0.02	0.0498	0.8367	0.0319	0.0492
7	0	0	0	0	0	0.0833	0.7238	0.1929

---

* Some banks permit credit risk specialists to over-ride the system if the result is 'clearly wrong'. Provided these over-rides are not under the influence of any business group and they are reviewed by senior credit risk management staff, this process can make the internal ratings system more accurate. One of the challenges in credit risk management is combining the considerable expertise of traditional loan officers with the techniques of modern quantitative credit risk management.

- For the two highest-quality ratings, there are not sufficient data to distinguish between them on the basis of credit events: neither of them had any. But 2-rated corporates move down the rating scale faster than 1-rated ones. Higher-rated corporates therefore get to default via a number of downgrades, whereas lower-rated ones are more likely to jump straight to a credit event, or to jump up to a higher rating.

The behaviour of the portfolio over multiple years can be modelled by multiplying transition matrices: the square of the matrix gives us 2-year transition probabilities, for instance the cube, 3-year, and so on.* A ratings system would then typically show the kind of behaviour shown below.

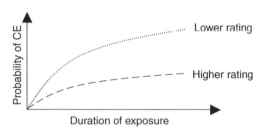

Once a firm is comfortable with the integrity of the ratings within its portfolio, these ratings can be used to model the portfolio credit risk loss distribution: this is the approach used in the first of the models we examine in the next section.

> *Exercise.* How many years would you have to wait to be sure of having a whole economic cycle?

5.4.2 Some Models of Portfolio Credit Risk

A number of models of the credit risk loss distribution have been developed: here we give a brief overview of two of them: CreditMetrics™ and CreditRisk+™. CreditMetrics is based on the transition matrices discussed above, whereas CreditRisk+ uses a parameterisation of the loss distribution:[†] both models estimate the shape of the portfolio loss distribution and hence provide a means of calculating the UL for credit risk portfolios.

5.4.2.1 The loss distribution from ratings transitions

Consider an exposure rated 4: using the table above, we have the following probabilities of migration to ratings at the end of the year:

3	4	5	Credit Event
4.11%	91.57%	4.15%	0.17%

---

* This only holds precisely if we make the assumption that transition matrices are *Markovian*, i.e., the history of a credit does not matter, but only its current rating does. This does not hold precisely—credits which changed their rating last year are more likely to move again this year—but it is not a bad approximation.

[†] See Christian Bluhm, Ludger Overbeck and Christoph Wagner's *An Introduction to Credit Risk Modeling* for a much more comprehensive discussion.

Suppose the average credit spread of each rating is:

3	4	5	Credit Event
125 bps	195 bps	280 bps	$R = 50\%$

Here the entry for a credit event is the recovery, and we assume that the credit event happens at the year-end.

If we have an exposure paying 100 in 1 year's time and risk-free rates are 5%, then the weighted average value is

$$4.11\% \times \frac{100}{1.0625} + 91.57\% \times \frac{100}{1.0695} + 4.15\% \times \frac{100}{1.078} + 0.17\% \times \frac{50}{1.05}$$

$$= 4.11\% \times 94.12 + 91.57\% \times 98.09 + 4.15\% \times 97.28 + 0.17\% \times 47.62$$

$$= 97.80$$

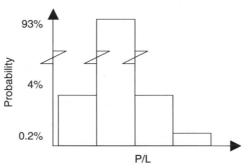

The model therefore gives us a distribution of future values for this position like the one illustrated. (Hopefully, a real credit exposure would pay us a coupon to compensate us for these possible losses, but we ignore this for the moment to keep the example simple.)

5.4.2.2 Transition matrix–based models

The CreditMetrics model of credit risk is based on transition matrices as discussed above. The whole portfolio of exposures is rated, and we use a ratings transition matrix to determine the distribution of future values at some chosen future time horizon, often 1 year.[*]

How should we measure the risk for a loss distribution? If the shape is asymmetric and fat tailed, there may be no simple relationship between the S.D. and the loss at 95 or 99% confidence as there is with the normal distribution for market risk. However, the S.D. σ is still an interesting measure of P/L volatility, and it can be estimated from transition probabilities $p_1, p_2, \ldots, p_n$ to states with losses $l_1, l_2, \ldots, l_n$

$$\sigma = \sqrt{p_1 \times (l_1 - E(l))^2 + \ldots p_n \times (l_n - E(l))^2}$$

where $E(l) = \Sigma_i\, p_i \times l_i$ is the average loss.

---

[*] Further details of CreditMetrics can be found in the *CreditMetrics Technical Document* available at www.riskmetrics.com.

5.4.2.3 What is the exposure?

Before we look at how to model more than one exposure in the CreditMetrics setting, it is worth commenting on the fundamental exposure data going into a portfolio model. Clearly, if we have a 5-year fixed rate loan which cannot be prepaid, we know precisely the cashflows we would lose if the loan defaults; in other situations, however, this is less clear:

- For a prepayable loan, we might estimate the effective average life of the loan to determine the maturity of the exposure.

- For a book of rolling unsecured retail loans such as overdrafts, statistical modelling may be necessary to estimate the effective maturity.

- For LOCs, the issue is EAD: as a corporate declines in credit quality on its way to a credit event, its credit spread for new funds rises, so it will tend to draw more on fixed-spread LOC funding. Thus, an LOC should be entered into the model with the predicted EAD, not the current exposure.

- Derivatives exposures are entered as their current exposure plus the PFCE at some chosen confidence level.

- Rolling exposures such as credit cards are probably best dealt with assuming a constant or slowly growing notional rather than the contractual maturity profile of the current portfolio: new exposures will usually come in to replace the old ones.

- Short-term exposures are particularly problematic since the ratings changes are assumed to occur over one time step, so if a time step is a year, a 3-month exposure is by definition risk-free in the pure version of the model. There are various solutions to this, including assuming all exposures have a minimum maturity of 1 year and using a fourth-root matrix (that is a transition matrix which, multiplied by itself four times, gives the original 1-year matrix) to model 3-month time steps.

5.4.2.4 Multiple obligations in CreditMetrics

In the setting above, the future distribution of credit exposures for various kinds of contract with one obligator can be estimated. However, banks have thousands, perhaps even millions of obligators rather than one: how do we deal with multiple exposures?

We need some way of modelling how the ratings transitions of one obligator comove with those of another. If there is no association between them, then we can simply model each loss distribution separately: but of course, this is unlikely as general factors, such as the overall state of the economy, interest rates and so on, tend to influence all debtors in a country at once.

There are several choices here. First we have to decide on a model of comovement—a normal correlation between the variables driving the transition is one of the simplest choices—then we have to calibrate it, which is not easy as there is little enough information on one 4-to-5 ratings transition without having to find enough 4-to-5s to decide how often they occur together. Given these data issues, some firms have chosen to use comovement information derived from equity or credit spreads [we discuss this approach a little more in section 11.4]. Another approach is simply to use one gross correlation: this means approximating millions of individual obligator versus obligator correlations by one (or a handful) of transition correlations. These can then be calibrated from the historical return distribution; for instance, in the CreditMetrics setting for a large number of obligators and assuming a normal correlation between the random variables driving credit events, the credit event correlation ρ is related to the average credit event rate μ and its S.D. σ by $\rho \approx \sigma^2/(\mu - \mu^2)$. This allows us to determine the correlation in this model from the actual observed credit event distribution either at a gross level or perhaps portfolio-by-portfolio (so we would have different credit card and large commercial loan book correlations).

5.4.2.5 Fixed-distribution-based models

Suppose we have a large number of credit-risky assets, and for each we estimate the probability of default, PD. For independent assets with a constant default rate, it is reasonable to assume a Poisson distribution of defaults, so if μ is the average default probability, the probability of n defaults is given by

$$\frac{e^{-\mu}\mu^n}{n!}$$

The form of the distribution is also fairly simple if we allow default rates to vary: the portfolio loss distribution is *negative binomial*, and σ is the default rate volatility; the probability of n defaults is given by

$$(1 - p)^\alpha \left(\frac{n + \alpha - 1}{n}\right) p^n$$

where $\alpha = \mu^2/\sigma^2$ and $p = \sigma^2/(\mu + \sigma^2)$.

The thinking in this style of modelling is much like the use of the normal distribution in market risk: the shape of the negative binomial distribution is broadly correct for any credit risk loss distributions, just as shape of the normal distribution on log returns is broadly correct for market risk; the distribution is relatively easy to work with, so we base our model on it and calibrate it to available data, in this case the average default rate and its volatility. The illustration shows an example from a 100-credit portfolio: this is the best-fit distribution for the portfolio data. The distribution, as we would expect, stretches far out to the right, with the probability of 30 defaults being 0.33 bp.

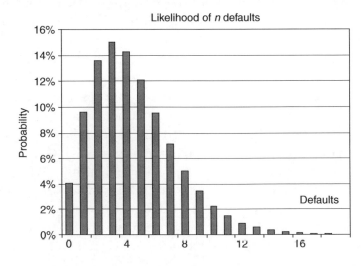

An example of this style of portfolio credit risk model is CreditRisk+. As with CreditMetrics, a structure of the default comovement can be introduced, for instance, by dividing the portfolio into independent sectors and using sector-wide credit event correlations.*

5.4.2.6 EL and UL estimates

One of the main reasons for implementing a model such as the ones discussed above is to calculate EL and UL: EL estimates are used for portfolio credit loss provisions and UL − EL for capital allocation.

5.4.3 Stress Testing Credit Portfolios

In market risk management, stress tests are used to measure exposure to extreme or unusual events. Similarly in credit risk management, historical, hypothetical and sensitivity stress tests are used to monitor portfolio sensitivity to changes in credit quality. Typically credit risk stress tests incorporate:

- Large moves in default probability and LGD;
- Breakdowns of credit event correlation assumptions.

Credit risk stress tests are often conducted together with the liquidity stress tests [discussed in Chapter 9], since periods of illiquidity often coincide with rising credit risks, so it can make sense to treat the two risks together. Various forms of credit risk stress test are popular:

- For firms with internal ratings models, a simple *ratings stress* test involves simply increasing the PD associated with each rating. Historical stresses can be used to estimate a reasonable move in the average PD for each ratings category.
- *Industry stress tests* apply larger moves to obligators from a particular industry group
- *Retail stress tests* analyse the behaviour of unsecured loan, credit card and related portfolios of a sustained economic downturn.

A good stress test should predict changes in:

- Expected losses;
- Provisions;[†]
- Delinquency;
- Economic capital.

---

* See *CreditRisk+ A Credit Risk Management Framework* (CSFB) available from www.csfb.com for more details of the CreditRisk+ framework in general and the use of sector correlations in particular.
† There is a regulatory requirement for a bank to be able to demonstrate 'across the cycle' capital adequacy. If a stress test suggests that it remains well capitalised even if a stressed level of defaults pertains for some extended period, then this may help to mitigate any regulatory concerns that the bank's regulatory capital calculations are too dependent on current market conditions.

5.4.3.1 Example

The example of Hong Kong may be helpful in thinking about the design of retail stress tests. Hong Kong suffered a significant fall in property prices from 1997 to 2004. At the low, prices fell approximately 70% from the peak, and a significant number of households were saddled with mortgage debt greater than their property's value. This period coincided with near-record levels of unemployment and political uncertainty: default levels rose very fast, and many kinds of retail exposure suffered significant losses.

It would be interesting to know how large the stress test losses on some bank's U.S. or Spanish* subprime mortgage portfolios would be if this Hong Kong event were used as the basis for designing the test.

Exercise.   How would you select a credit stress test for:

— A commercial leasing portfolio,
— An SME receivables portfolio and
— A commercial mortgage portfolio?

5.4.4 Active Credit Portfolio Management

A traditional retail and commercial bank's strategy might be crudely summarised as follows:

> Originate good assets, fund them by taking deposits and issuing unsecured debt, and hold them to maturity.

Here 'good' should mean good RAROC. The disadvantages of this approach are considerable:

- The take-and-hold proprietary portfolio can become a dumping ground for business groups, filled with low-yielding, illiquid and sub-par assets;

- It is a large user of capital, and there may be a lack of accountability for this and for the risks in the portfolio.

- The lack of ownership of the credit risk portfolio can result in over-exposure to cyclical industries, weaker countries or the counterparties the firm has strong relationships with.

5.4.4.1 Securitisation and the paradox of banking

Some of these problems can be mitigated with a sensible capital allocation strategy which charges business groups for the credit risks they originate. However, we cannot deal with the *paradox of banking*—that you naturally end up with more exposure to those counterparties you know best, giving rise to concentration risk—this way. With the advent of

---

* These are two markets which have recently been highlighted as having a retail property bubble which might burst if conditions change.

active portfolio management, however, these issues can be managed. Much of the risk of the credit risk portfolio can be passed on, and the banking paradigm becomes:

Originate good assets, securitise them, fund them by taking deposits and selling the mezz and senior tranches, and retain the equity tranche.

In this approach, the bank improves the velocity of its capital by securitisation, allowing it to concentrate on being an effective originator. The benchmark for originating risks also changes: now an asset is attractive if

Asset expected spread − Incremental contribution to the cost of securitisation > ROE threshold × Incremental contribution to the equity supporting retained risks

Thus, if we have an asset which adds 35 bps to our cost of securitising the portfolio it would reside in, our pre-tax ROE target is 40%, and on securitisation the retained risk requires 4% equity support; the break-even spread at which the asset is just earning enough to pay for its equity use is

$$40\% \times 4\% + 0.35\% = 195 \text{ bps}$$

5.4.4.2 Concentrations

Some assets dominate the credit risk portfolio to such an extent that they adversely affect our ability to securitise: a portfolio of 100 $10M loans is a good securitisation target; one with 99 $10M loans and one $500M loan is not. Here we need to assess the asset against its stand-alone hedge cost, for instance via CLN issuance or syndicating the loan, since it is not susceptible to securitisation. Soft limits forcing businesses to engage with the firm's risk management infrastructure before taking on concentrated exposures may be appropriate here.

5.4.4.3 Who owns the credit risk portfolio?

If we retain the equity tranche in a securitisation, depending on the attachment point, we may have most of the credit risk. How should this retained risk be managed?
There is no single agreed answer here:

- Some firms use a model whereby the originating business group sells all of its credit risk to a portfolio management group, which is then responsible for managing the credit risk portfolio. Provided this group has the authority to set a genuinely independent sale price, this has the advantage of separating origination functions and profits from credit risk management, and of separating the hedging decision from relationship management.

- However, it does mean that credits are not necessarily managed by the person in the firm with the most knowledge of them. Hence, some banks keep portfolio management functions within the originating business group.

Either approach can work well *if* the decision about hedging or selling an exposure is made on rational economic grounds.

5.4.4.4 Issues in active credit risk portfolio management

The active credit risk portfolio management approach can be good risk management, but it poses some challenges to the banking business model:

- By definition, active management means the ability to test a new credit risk and to refuse to accept it if it does not meet some hurdle criteria. This, in turn, has a potentially negative impact on clients. Banks will inevitably find themselves under pressure to take some uneconomic exposures: this is reasonable if the whole relationship with the client is generating a sufficient return, but where it is not, bank's shareholders should demand action.

- In times of spread narrowing, there may not be enough risk around to meet the hurdle rate in any area where the bank can easily originate risk. The right approach here is to reduce risk taking. However, this reduces revenue and may leave some capital unallocated. If the period of tight spreads lasts for years, a firm can find itself under pressure to do more business despite the lack of attractive risks.

> *Exercise.* Suppose you were concerned that a business group was originating loans which were not paying a sufficient spread, but due to its specialist nature, it is impossible to determine the correct securitisation price for the portfolio. How could you determine:
>
> — An appropriate level of provisions for the portfolio,
> — The economic capital for the portfolio and
> — The value of the portfolio to the bank?
> — Is your method for determining the level of provisions different from the one you would have chosen had you been asked to determine the securitisation price? If so, why?

5.4.4.5 Précis of the anatomy of credit pricing

The different styles of credit pricing can be summarised as follows:

- *Old style.* Profit on the credit risk portfolio comes from the carry after provisions, with these based on an EL. The portfolio is historic cost accounted, and large rises in provisions tend to effect earnings when the economic cycle turns.

- *Securitised.* Credit risks have to pay sufficient spread to cover their securitisation cost, a provision for EL on residual risk and the cost of equity supporting that risk. The portfolio is either mark-to-market or historic cost accounted, with the latter generating less earnings volatility.

- *Stand-alone.* Business groups are charged by a portfolio manager for credit risks an amount needed to pay for their hedging on a stand-alone basis, for instance, in the default swap market.

As we go down this list, the hurdle rate gets progressively higher, so banks which take the first approach will tend to hold assets rejected by firms using the latter approaches. The stand-alone approach has been used by some broker/dealers, but since relatively few credit risks, especially in the commercial banking space, actually pay sufficient spread to cover this cost, the effect of this is to reduce the amount of credit risk taken in the firm as a whole.

Exercise. If provisions are simply for EL, they will display an interest rate sensitivity, since defaults rise in high-interest-rate environments. Should a bank hedge this sensitivity? If so, how could this be done, and how would the accounting work?

5.4.5 Credit Scoring and Internal Rating

Consider the steps in the lending process:

- *Solicitation.* Either the firm solicits applications for loans, for instance via advertising, or a client comes to us with a new request.
- *Information gathering.* We find out something about the applicant via various means including interviews, visits, review of financial data or accounts, possible use of credit reference agencies, ratings agencies or other available data.
- *Recommendation.* On the basis of our information gathering and consideration of why the client wants the money an internal rating is assigned and a lending decision is recommended. This is then reviewed, perhaps by a loan committee in the case of a corporate loan or branch staff in the case of a retail exposure.
- *Closing administration.* Any collateral is perfected, documentation is finalised and signed and funds are made available.
- *Monitoring.* The performance of the obligator and their condition are reassessed periodically.

One aim of internal ratings is to allow many applications from different kinds of corporates in different countries to be assessed on an equitable basis. It also allows the bank to set break-even spreads for internal ratings classes which encourage lending to 'good' counterparties.

5.4.5.1 Corporate assessment

For a rated counterparty, we might be content to rely on the due diligence of the ratings agencies (although some commentators have suggested that there may be biases in the agencies' ratings, in favour of financials over industrials, for instance). For a corporate counterparty without a rating, though, we cannot do that. Instead, internal ratings need to focus on the traditional 6 Cs of lending:

Character, Capacity, Capital, Collateral, Conditions, Control

Character is a judgement call. For capacity and capital, we need to look at measures of financial performance. These fall into five rough categories:

- *Liquidity ratios.* These help us to see if the company is able to meet its short-term obligations, since without liquidity, default is possible regardless of solvency. For instance, one example here is

$$\text{Quick ratio} = \frac{\text{Cash} + \text{Marketable securities} + \text{Receivables}}{\text{Current liabilities}}$$

- *Efficiency ratios.* These tell us how effectively management is using the firm's assets to generate income. For instance, operating expenses/total sales is a measure of how much expense is involved in generating a unit of sales.

- *Leverage ratios.* As discussed earlier, leverage is a measure of the ratio of (non-deferrable) debt to (loss absorbing) equity. A crude example would be long-term debt/common stock; more sophisticated versions would adjust for off-B/S obligations.

- *Coverage ratios.* Here the idea is to measure the extent to which the firm's profits can pay interest expenses. EBITDA/gross interest expense is a commonly used coverage ratio.

- *Profitability ratios.* In addition to ROE, we might want to know the candidate's gross profit margin or return on assets.

Finally, despite the quality of ratios, big firms tend to be more stable than smaller ones, so we might have absolute thresholds, such as demanding that any firm rated 3 or higher has 5 consecutive quarters or 3 consecutive half years of EBITDA greater than £100M and a total equity greater than £500M.

5.4.5.2 Combining the data

How does a firm decide on what data to use for an internal rating, and how are they combined? Typically, it is helpful to resist the temptation to require too many inputs to an internal ratings system, partly because these may not be available for smaller or family-held firms or firms in countries which do not use IAS, and partly because it is important to know if an input actually adds any discriminating power to a ratings system, and with many inputs that is less clear. Therefore, leading firms often select between 5 and 25 pieces of data including financial ratios and absolute measures of the types discussed above, perhaps together with trend information.

These data are then combined in one of four main ways:

- *Multi-discriminant analysis* is a statistical technique which aims to explain the variance in one variable—the rating—by those in others—the inputs. This works well where the inputs are linearly related to the outputs, but this is not so in ratings, so we have to decide on the functional relationship between an input, such as EBITDA, and the output rating. (In that case, for instance, the relationship is sometimes assumed to be logarithmic.) Once we have set these relationships, the system is calibrated so that within-rating variances are minimised and between-rating variances are maximised, thus optimising the system's discriminating power.

- *Expert systems* are an attempt to systematise what a good human rater would do. The design process starts with interviews with experts. On the basis of these, the designer attempts to code a rule set which captures the expert's rating process. This is then run on a sample of corporates, and any errors discovered are used to improve the rule set. One of the advantages of this approach is that the rule set used in the system is usually comprehensible: we can see *how* the system is getting the final rating.

- *Scorecard approaches* are a simpler version of an expert system. Scores in a number of categories are allocated, and a corporate is rated on the basis of its total score. This method has the advantage of simplicity, but since many banks now use an automated ratings system, these methods are slowly being enhanced by more complicated rules which edge towards expert systems.

- Another systems-oriented approach is the *neural net*. Here we let the system itself discover an appropriate rule set through a training process. Although neural nets have been successful at certain data processing tasks, they tend to be less effective in high-dimensional problems where the inputs do not smoothly map to the output, so their use in internal ratings should probably be considered experimental.

5.4.5.3 Model benchmarking and use

Once we have a candidate internal ratings system, it must be tested. Here we need to consider several factors:

- All the available data on hand-rated obligations should not be used for developing the system, because then we will not have any data to test it.

- Testing against the ratings agencies is typically not sufficient as there is significant size bias in the available external ratings.

- Moreover, the agencies sometimes use a rather mechanical approach to subordinated or hybrid securities, such as an automatic one- or two-notch downgrade. We may not wish to incorporate that feature into our system especially if we wish to be sensitive to the precise structure of an obligation.

- The system needs to be robust across the economic cycle, so ideally it should be possible to recalibrate to old data and test the system's performance in different economic conditions.

- Finally, we need to look at the system's performance false positives versus its false negatives. The terminology here is from statistical inference:* first, a hypothesis is formed that the credit will not default. A *false positive* occurs when we incorrectly reject this hypothesis and hence decide that a good credit actually is not good. In contrast, a *false negative* occurs when we do not reject the hypothesis, but it is actually false, that is, we fail to spot a credit which is sliding towards default. False positives in credit models are usually better than false negatives: if we hedge an exposure we do

---

* See E.L. Lehman and Arthur Romano's *Testing Statistical Hypotheses* for more details.

not need to, it just costs us premium, whereas failing to spot a default about to happen is usually more expensive.

After a system has passed initial tests, it is rolled out under management controls. For instance, it might be used under limited circumstances, such as any corporate loan application for less than £1M is automatically granted for a corporate rated 4 or better, with larger loans being referred to loan officers for more detailed scrutiny. Once a history of the system's performance has been gathered, it should be reviewed for accuracy on a regular basis.

5.4.5.4 Credit scoring

The analogue of internal ratings for retail exposures is credit scoring. We gather data on the obligator and use them to assign a measure of credit quality. Obviously, the data gathered will be different—we will be asking about gross income rather than EBITDA—but the principles are similar: we are searching for a system which is broadly applicable and has discriminating power. This can then be used both reactively—to decide on whether to accede to a mortgage application, for instance—and proactively—to solicit application for a particular type of credit card.

> *Exercise.* Find your credit score. What could you do to improve it?

5.4.5.5 Standardised credit scoring processes

Some countries have established credit scoring agencies that sell credit ratings data on individuals to interested parties. There are advantages and disadvantages to the use of these data: on the positive side, certain criteria such as FICO scores are well understood by securitisation tranche buyers, so the use of these scores integrates smoothly into the securitisation process; on the contrary, the ability to underwrite good risk and reject bad is a core competence for a retail bank, so subcontracting that out to a credit scoring agency may be counterproductive.

5.5 POLITICAL AND COUNTRY RISK

Country or *political* risk is the risk of loss caused by uncertainty about political or policy changes. Typically, here the concern is the actions of governments, but other vectors of political risk include local legal systems, the military, or state-sponsored groups.

Examples of political risks include:

- The imposition or removal of taxes;
- The imposition or removal of exchange controls or exchange rate management systems;
- The repudiation or moratorium of government or central bank debt;
- The confiscation of assets including nationalisation;
- The imposition or removal of trade quotas or tariffs or both;

- The passage of legislation making previously acceptable business practices or ownership structures now illegal or subject to censure.

5.5.1 Examples of Country Risk

Country risks fall into three broad categories: sovereign defaults, convertibility events and broader country risks.

5.5.1.1 Defaults

One illustration of sovereign default is provided by China. The Xinhai Chinese Revolution occurred in October 1911. At the time, there were a number of debt instruments trading including a war loan of 1874 and a sterling loan participation issued by Baring Brothers. In 1912, an international loan was also granted. As Goetzmann and Ukhov* say, 'It is only reasonable to assume that an investor holding a promise by the Chinese Imperial Government would be concerned by the news that the government had been violently overthrown and replaced with a military strongman with an unclear popular mandate to rule.' These fears were well founded, and China had defaulted on all of these obligations by the mid-1920s. This is a good example of a default due to *repudiation*: the new ruling cadre simply did not see the debt as something they had an obligation to repay.

Another example of sovereign default is the Russian moratorium of 1998 [discussed in section 1.2.2].

5.5.1.2 FX events

Many Southeast Asian currencies suffered speculative attacks during the Southeast Asian crisis of 1997. The political reaction to this in Malaysia was stronger than most with Prime Minister Dr. Mahathir calling currency trading *unnecessary, unproductive and immoral*. The Malaysian response to the attack on the ringgit was to deliberalise FX markets. In September 1998, a number of measures were taken including:

- Prohibiting the transfer of funds into the country from externally held ringgit accounts except for investment in or the purchase of physical goods from Malaysia;
- Banning the provision of ringgit credit facilities to non-residents;
- Closing the off-shore market in Malaysian equity, and enacting various measures against short selling;
- Requiring prior approval for Malaysian residents to invest off-shore, and requiring repatriation of export earnings within a short period.

Together, these measures closed the off-shore equity and FX markets and effectively decoupled local interest rates from those implied by FX parity: the ringgit became for a period a (partly) managed currency, and those funds left in the country were locked in for a year or so, and then only repatriable at a managed rate.

---

* In *China and the World Financial Markets 1870–1930: Modern Lessons from Historical Globalization* (Wharton Financial Institutions Center Report 01–30).

5.5.1.3 Broader country risks

Country risk includes potential damage to a firm caused by terrorism, war or insurrection, damage to property or personnel inflicted by pro- or anti-government activity, and contract frustration or repudiation without possibility of local legal action. One obvious example is a revolution: this tends to lead to the seizure of property. Local firms, even if they wished to do so, might well be unable to perform on pre-revolution contracts.

Another situation occurs where there is sufficient instability in a region to cause risk: for instance, currently in Nigeria, a group calling itself the Movement for the Emancipation of the Niger Delta is fighting with government forces, sabotaging oil installations and kidnapping foreign oil workers. Since Nigeria is the world's eighth largest oil exporter and much of its oil infrastructure is in the Niger Delta, this is not just a tragedy for the firms and individuals concerned: it is also having an impact on the oil price.

Finally, a more unusual example: the Sarbanes–Oxley Act. This is a U.S. federal law of 2002 passed in response to a number of major corporate scandals including Enron and WorldCom: it contains a number of additional requirements for U.S. corporates and non-U.S. companies listed in the United States, of which the most onerous are requirements concerning the integrity and validation of a firm's internal controls. Meeting these requirements has proved most expensive, with one bank estimating its Sarbanes–Oxley compliance costs in excess of $25M. For some banks, this extra cost, incurred because the bank is listed on the NYSE, happened as a direct result of foreign legislation, and thus is an example of country risk.

5.5.1.4 Risk is in the eye of the beholder

The example of Sarbanes–Oxley makes it clear that country risk depends on where you stand: one man's strengthening of corporate governance is another man's costly and arbitrary imposition. This relativity of risk perception becomes even clearer in banks in developing countries: typically, it makes little sense for a bank to try to be safer than the country it is based in. Therefore, a AAA bank from another continent might find a single B country to be highly risky, whereas a BBB bank based in a neighbouring country with a much better understanding of local business practices and effective risk mitigation might see only a great business opportunity. They can both be right: the AAA bank might not have the resources or risk appetite to succeed, whereas the BBB bank does.

5.5.2 The Effect of Country Risk

Country risk introduces a risk premium into interest rates as investors demand compensation for bearing it. Thus, just as corporates pay an extra credit spread because they might have a credit event, sovereigns pay an extra spread because they might introduce currency non-convertibility or exchange controls.

5.5.2.1 Interest rate parity in the presence of country risk

Suppose a country risk event has probability p over a time horizon with local rates r_{local}. For an off-shore bank operating with home country rates r, it can either invest in the country and suffer country risk, or stay at home.

The forward rate for converting a unit of home currency into local is therefore related to the spot rate by the relationship shown in the diagram. Real interest rates in the risky country are higher than they would otherwise be to accommodate this *country risk premium*. In some cases, it may be possible to hedge some or all of this country risk, for instance via the use of NDFs. Just as firms control their exposure to credit event by the use of credit risk limits and credit risk mitigation, the country risk is controlled through country limits and country risk mitigation tools.

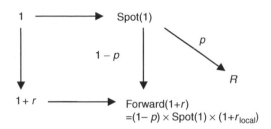

5.5.2.2 Is sovereign default rational?

Any obligator has the option to default. For (most) corporates, it is not rational to exercise this option as maximising returns for shareholders is not usually achieved by defaulting, although if entering bankruptcy protection under the Chapter 11 process allows a firm to shed pension obligations, then arguably it might maximise shareholder if not stakeholder returns. For a sovereign, though, the situation is different: default simply means failure to pay, perhaps even only failure to pay on foreign currency obligations together with a suspension of FX convertibility to reduce capital flight.

Clearly, this is rational if the cost of default, including the post-default increase in funding cost, is smaller than the cost of continuing to service the debt. If an emerging market leader wishes to act in his or her country's economic best interest, ignoring any political implications, this is the judgement that needs to be made. Some commentators have suggested that, in some cases at least, default may then be the rational decision.

> *Exercise.* How would you estimate the cost of defaulting versus the cost of continuing to service existing debt?

The markets sometimes give sovereigns incentives to default: if a left-wing leader is elected, debt spreads may go out, increasing the cost of future borrowing, and narrowing the gap between the pre- and post-default funding costs. Therefore, acting as if you think a country is likely to default increases the incentive for it to actually prove your worst fears correct.

5.5.3 Measuring Country Risk

Since country risk depends on a firm's location and predilections, measurement of country risk is inevitably arbitrary. There are several standard approaches, and they lead to rather different results.

5.5.3.1 Macroscopic approach

There are various indices of country risk. Typically, these are compiled using a scorecard approach: a range of factors is selected, each is allocated a maximum score and countries are scored on their performance.* For instance, the scorecard might be

Factor	Maximum Score (per 100)
1. Economic expectations versus reality	4
2. Inflation	4
⋮	
25. Expropriation of foreign capital	10

Economists and country experts would then complete this scorecard for each country.

5.5.3.2 Sovereign debt spreads

One obvious measure of the market's required compensation for sovereign risk is the spread of a country's sovereign debt to a risk-free instrument. If there is debt denominated in a risk-free currency such as USD, EUR or sterling, this spread can be observed. Note, however, that there is a significant risk premium effect in sovereign debt spreads: large spread widenings sometimes happen to many emerging market countries simultaneously. This occurs when the market withdraws risk capital from the risk class as a whole. An example of this phenomenon occurred in the Southeast Asian, Russian and Brazilian crises discussed previously.

5.5.3.3 Convertibility swap premiums

An off-shore investor can buy protection against a currency convertibility event via a *convertibility swap*. This is a product similar to a default swap where the protection buyer pays a periodic spread on a notional principal in exchange for the right in the event of a convertibility event to either:

- Deliver a fixed amount of foreign currency to an on-shore branch of the protection seller's bank and receive a risk-free currency from an off-shore branch: a *physically settled* convertibility swap or less commonly;

- *Receive* a sum in a risk-free currency: a *cash-settled* convertibility swap.

In both cases, the FX rate used for settlement is typically that which pertained to some short period, perhaps a few days, before the event. Thus, if the rate was 20 dodads per USD 1 day before dodad convertibility was suspended, the protection buyer on a $1M notional principal physically settled dodad convertibility swap would have the right to deliver 20M dodads on-shore and receive $1M off-shore.

---

* See, for instance, the *International Country Risk Guide*.

One important feature of the contract is the definition of a convertibility event. Triggers included in the contract can include the suspension of formal convertibility, the declaration of war, the confiscation of foreign investor assets, or other features. Clearly, the premium payable on a widely drawn convertibility swap contract is another measure of the market's required premium for country risk.

5.5.3.4 *Political risk insurance premiums*

The insurance analogue of the convertibility swap is *political risk insurance* (PRI). Although there are a variety of events used for convertibility swap definition, PRI is even more diverse. Insurable risks can include revolution; war or civil unrest; state or regional government confiscation of assets, frustration of contracts or repudiation of obligations; and suspension of currency convertibility. Many PRI policies, however, have a limited definition of insurable risk based on suspension of convertibility for an extended period.

A version of PRI is also provided by certain export credit agencies: these are typically government bodies in developed countries that in exchange for the payment of a premium provide protection to exporters to risky countries.

PRI premiums provide another measure of country risk. However, it is worth pointing out that narrowly written PRI policies have an image problem in some quarters based on a number of situations where a PRI provider has not paid on a policy despite the occurrence of events which appear, perhaps to a naïve observer, to have been insured. This highlights one of the difficulties of specialist financial insurance: the claims adjustment process sometimes appears to be used in place of initial underwriting due diligence.

5.5.4 Country Risk Management

Country risk can be managed by a variety of means. First, a firm needs to articulate its country risk appetite, perhaps via a limit structure. Then country risks in excess of that appetite are hedged.

5.5.4.1 *Limits*

Country risk limits constrain a firm's total exposure to non-performance by a country (and so by implication by any counterparty domiciled there). Thus, included in the limit would be positive credit exposures, the PV of securities exposures to country issues and any FX position:

Risk Factor	U.S. Country Risk
Sensitivity measure	Sum of positive spot and 95% PFCE credit exposures to U.S. counterparties, net PV of bond, FX and equity exposures by U.S. issuers
Limit	$2B
Application	Firmwide

If the firm has a presence locally in the country—a branch or company, fixed assets, retained earnings and so on—then these should also be included in the country limit.

We may also wish to constrain the size of total shorts to a country as a convertibility event could affect a firm's ability to monetise the value of the short position:

Risk Factor	U.S. Convertibility Risk
Sensitivity measure	Sum of PVs of short bond, FX and equity exposures by U.S. issuers
Limit	$1.5B
Application	Firmwide

Exercise. Should there only be country risk limits for non-investment-grade countries?

5.5.4.2 Hedge instruments

The obvious hedges have already been discussed: convertibility swaps, PRI and CDSs on sovereign or regional debt. These can be used to provide some mitigation to country risks if attractive opportunities arise in excess of a firm's country risk appetite.

5.5.4.3 Strategic hedges

Capital markets or insurance hedges tend to be expensive: at the time of writing, 5-year CDS protection on Ecuador has an annual premium of around 5%, for instance. For a firm with a significant on-shore presence, hedging their capital using this route is unlikely to be economic. Therefore, firms tend to use other risk mitigants. For instance, good practices include:

- The use of local partners to minimise capital investment. The slogan here could be rent-a-branch.

- Funding local activities as far as possible in local currency. Where possible, for instance, taking deposits in local to provide funding can reduce overall convertibility risk.

- Policies which enforce regular repatriation of earnings or transfer of assets off-shore where possible.

- Funding of on-shore investments so that country risk is passed on to third parties. For example, we could issue a synthetic sovereign CLN whose repayment is contingent on our ability to repatriate funds from an on-shore subsidiary. Since we own the sub, we take the risk that it will generate profit in excess of funding: this is extracted via a claim on the CLN issuer subordinated to the CLN itself.

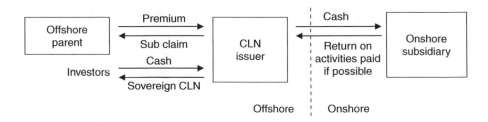

CHAPTER 6

Operational Risk and Further Topics in Capital Estimation

INTRODUCTION

Market risk and credit risk are highly visible risk classes: markets move every day and clients often default. Operational risk in contrast is more opaque and less easy to trace. It came to prominence as a result of its inclusion as a risk class attracting regulatory capital in Basel II [see Chapter 7 for more details]. Having been told by regulators that they needed to control, measure and allocate capital against operation risk, the industry has spent considerable effort in recent years trying to make sense of this risk class, and now it has a comparable importance for many institutions to its sister risk classes of market and credit risks. The first part of the chapter introduces some of these developments.

In previous chapters we have discussed models for market risk and credit risk: one reason for doing this is to estimate an economic capital requirement for these risks. The second part of this chapter discusses how to combine these capital measures together with operational risk estimates to produce an integrated capital requirement. Some issues relating to the accuracy of these capital measures are discussed and additional, even more difficult to quantify risks such as reputational and strategic risks are touched upon. Finally we say a little about the active management of capital requirements.

6.1 AN INTRODUCTION TO OPERATIONAL RISK

What is operational risk?* One definition that has attracted support is

> The risk of direct or indirect loss resulting from inadequate or failed internal processes, people, and systems or from external events ... Strategic and reputational risk are not included, but legal risk is.

---

* Further reading in operational risk includes Carol Alexander's *Operational Risk: Regulation, Analysis and Management* and Thomas Kaiser's *An Introduction to Operational Risk: A Practitioner's Guide*.

The idea, then, is that operational risk is in some sense the risk of doing business. Some examples of operational risk would include:

- Risk of rogue traders or other fraudulent activity, as in Barings;
- Risk caused by failure to comply with regulation or law, as when a firm is fined or has to make restitution for mis-selling or illegal trading;
- Risk of loss caused by failure to correctly record, manage or settle positions, for instance caused by mis-booking or systems failure.

Thus, operational risk runs the gamut from institution-shaking and headline-making events to very minor losses caused by a process failure.

6.1.1 Operational Risk Classes and Losses

In market risk measurement we often classify potential losses by risk factor: for operational risk, a similar taxonomy is possible to some degree. This allows us to gather and classify operational risk loss data and hence to begin to understand (some of) a firm's exposure.

6.1.1.1 Definitional issues

Before we can begin to gather data, it is important to have a clear statement of what is and is not an operational risk. For instance, is a loan which defaults but which was granted on the basis of a fraudulent application operational risk or credit risk? Given the lack of a consistent industry-wide standard here, a firm should at least have an internal definition which is consistently applied between business groups.

6.1.1.2 Risk classes

Operational risk can be classified by risk types as shown in the following table.*

Loss Category	Definition	Examples
Fraud	Losses due to staff or external activity that is fraudulent, illegal, contrary to policy or otherwise mendacious	Rogue trader, bank robbery, cheque forgery, computer crime, extortion
Employment practices	Losses arising from failure to implement best employment practices	Payments to victims of harassment or discrimination, workers' compensation
Client relations	Losses arising from a failure to meet a requirement to a business counterparty	Mis-selling, money laundering, provision of incorrect valuations, market manipulation
Physical assets	Losses arising from damage to physical assets	Fire, hurricane, earthquake, riot, vandalism
Business disruption and systems failures	Losses arising from systems failures	Hardware, software or telecoms outages
Execution and business processes	Losses arising from failure processes or transaction management	Miscommunication, wrong data entry, failure to settle or deliver, failure to manage collateral correctly, incorrect documentation, accounting errors, loss of customer assets

---

* This table summarises a more detailed classification used in Basel II: see Appendix 9 of the Basel Committee on Banking Supervision's *International Convergence of Capital Measurement and Capital Standards. A Revised Framework, Comprehensive Version.*

Notice the importance of *process* here: operational risk management encourages us to look across the firm's organisation at the processes flowing through a firm. Thus, an operational risk loss event could be caused by a failure in the equity settlement process, and this failure might impact on trading, operations and finance functions.

> *Extended exercise.* If you work in a financial institution, try to follow a trade from initiation to settlement, including all of the areas of the firm which the trade touches (including finance, operations, regulatory reporting and risk management).
>
> List all the processes the trade is involved in, and map the flow of information between the steps in each process. Even for a simple securities trade this might be a large project, but it will contribute significantly to your understanding of the processes in the bank.
>
> Now think of the possible errors in each step, and classify them according to the operational risk categories above (or using your own firm's categories).

6.1.1.3 Operational risk loss collection

Operational risk runs through everything an organisation does, and all its functions. In the past, each of those areas independently managed these risks: indeed there is a sense in which operational risk management is just what a firm's management does. However, this segmented operational risk management by itself is not satisfactory for two reasons:

- First, the firm needs good, integrated risk information. There has to be a structure which allows senior management to look down into the operational risks of each area and see the losses caused by operational risks.

- Second, small operational risk losses may be an early warning sign of something more potent. Although many operational risks, including compliance risks such as failure to obey regulatory or HR policies, often have relatively small direct impacts via fines, their potential impact on the firm's reputation can be immense. Therefore, the firm needs to ensure that early warning signals of operational risk are visible.

Therefore, the collection of operational risk loss data is one of the keystones of operational risk management. Typically, all business areas are required to report all operational risk losses above a certain threshold to the operational risk management group in a standard format. For example, a simple loss report might be something like the example below.

Reporting BU	**Latam debt trading**	Client involved?	Yes
Loss type	Execution and business processes	Loss size	$102K
Systems/staff responsible	Bond trading system, market data group, Latam bond trading	External reporting required?	Yes
Description	Incorrect identifier in system leads to failure to settle Peruvian Bond trade with client. Trade details not checked by trading. Bond bought in at higher price resulting in loss		
Mitigation	Data changed in system, operations have initiated programme to review all Latam bond market data		

6.1.1.4 Incentive structures around loss collection

Comprehensive loss data collection is difficult for several reasons:

- Many people have an understandable reluctance to admit to mistakes, especially if that admission could have an impact on performance assessment. Hence, it is vital that the firm sets up incentive structures to encourage correct operational risk loss data reporting (regardless of how the firm decides to treat these events for management P/L purposes).

- The processes with the highest operational risk are sometimes sufficiently badly controlled that operational risk losses are not measured or understood, so data may be missing just where they are most needed. Internal audit has an important role here, together with operational risk management staff: there should be a regular review process across the whole firm to identify areas which are not providing accurate reporting and suggest improvements.

> *Exercise.* What steps could a firm take to encourage good operational risk loss collection? How could the loss data collection process be audited?

6.1.1.5 The use of loss data

Good loss data allow management to compare the performance of a process or a function against:

- Similar processes or functions in the same firm. For instance, does debt trading misbook trades more often than equity trading?

- Similar processes or functions in the firm's peer group. For instance, do we have a similar ratio of fraudulently obtained credit cards to other players in the same market segment?

If performance is bad enough, either in relative or absolute terms, mitigation includes:

- *Process re-engineering.* Here systems changes may be needed, extra controls might be introduced or communication improved between groups. Making effective changes here can be one of the simplest, cheapest and most effective ways of mitigating operational risk.

- *Education/training.* Staff may not understand the desired process, or the culture of one or more groups may require management intervention.

- *Re-staffing/hiring.* The area may be under-staffed or mis-staffed.

In this sense, operational risk loss data are just another piece of information available to management in helping them to do their job.

6.1.1.6 Operational risk: why now?

There are two explanations for the growing emphasis on operational risk. The positive one cites:

- Increasing size of firms, especially the increasing dominance of certain large liquidity providers in some markets, which gives a systemic risk dimension to certain operational risks;
- Increasing use of large volume service providers whose failure could again introduce systemic risk;
- Increasing globalisation, the growth of e-commerce and the automation of business processes, resulting in systems of growing complexity and potential risk;
- Increasingly litigious business climate resulting in an increased impact of some operational risk events.

The negative one cites the decreased capital requirements for many institutions in Basel II as a result of the changes in the credit risk capital requirements [see Chapter 7 for more details]. Perhaps supervisors became uneasy about the extent of the reductions so it was decided that extra capital could be required to even things up. Operational risk was the easiest target.

These suspicions are perhaps unworthy: it is certainly the case that operational risk has caused significant losses, so it seems *a priori* reasonable that firms should have some capital against them. The difficulty comes in trying to work out how much.

Exercise. Try to find a service level agreement documenting the conditions under which a service is provided by an outsourcer. Analyse it for operational risk transfer: how much operational risk is borne by the outsourcer, and how much is retained by the client? What would the client's position be should the outsourcer fail?

6.1.1.7 Impact versus frequency

Suppose we try to classify losses by their severity on one hand versus their frequency on the other. For market and credit risks we could plot risks as in the figure below. There are two important things to notice here:

- First, for market and credit risks we can typically easily assign a position to a location on the diagram.
- And that allows us to devote attention to the high severity events, perhaps with a weighting of attention towards the higher frequency ones.

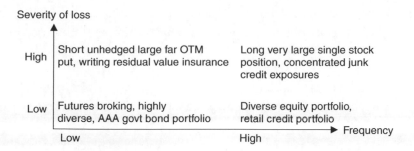

For operational risk, this typically does not hold. We can easily measure the low-severity high-frequency events, but that might not tell us much about the other operational risks facing the firm. In particular, high-severity low-frequency events, such as rogue trader losses, cannot be usefully quantified. Once we know such an event has happened, the issue(s) which allowed it to take place are addressed but almost by definition if we knew beforehand it could happen, we would attempt to stop it.

6.1.2 Scorecard Approaches to Operational Risk

There is more data potentially available to us in managing operational risk than simply losses. After all, we can run a risk without having a loss: we might just be lucky (for a while). Therefore, it makes sense to look at quantifiable factors which might indicate the presence of operational risk.

This gives rise to the idea of a *scorecard*: this is a compendium of indicators of possible operational risk in an area, some of which might give insight into the risks that have not yet manifested themselves as losses.

6.1.2.1 Scorecard indicators

The measures used in a scorecard are sometimes called *key performance indicators* (KPIs). Classes of KPIs include:

- *Indicators of process failure.* This is probably the most difficult category as the most useful measures will probably be highly process specific. For operations risk in capital markets, measures such as fails as a fraction of volume or percentage of trades requiring manual intervention are often used: other areas will similarly require the development of specialist KPIs.

- *Signs of policy failure.* This risk class covers situations where the process follows firm policy, but the policy is wrong, out of date or inapplicable to the current situation. One indicator might be the results of a regular review of documentation: even documentation age since last revision could be useful.

- *Legal risk indicators.* Here a firm could look at pending litigation, derivatives documentation disputed or unsigned and 'opinion risk' in highly structured transactions.*

---

* This occurs where the legality or effectiveness of a transaction depends on a legal opinion unsupported by applicable case law. It is relatively common situation in some areas of structured finance, and it can be quite insidious, in that the establishment of an unfavourable precedent or a change of policy by tax authorities can have a large impact.

- *Technology risk indicators.* These could include unscheduled systems downtime, indicators of code complexity or maintainability, usability metrics, security analysis and data quality indicators. Also of interest might be disaster recovery (DR) related measures, such as any issues related to backups, implementation of the DR plan or lack of systems redundancies.

- *Personnel-related measures.* Here personnel turnover, ratio of staff in each performance category, psychometric test results, training undertaken and complaints received could be used as metrics.

- *Modelling and valuation risk.* Reserves taken for model or valuation risk, percentage of models returned to the business group for changes by the model review team and model risk–related remarks (whether positive or negative) could be used as performance indicators here.

- *Project risk.* Measures in this area include on time and on cost completion statistics. The number of current projects or new products versus budgeted revenue might also give an indication of the extent of product or business innovation here.

- *Fraud and theft and damage to physical assets.* Insurance premiums may form a useful measure in this area.

6.1.2.2 Developing scorecards

There is a wide range of candidates for KPIs in most areas so selecting an effective set is difficult. This is especially so as any set of measures will likely introduce gaming: the measure rather than the risk will be managed. Therefore, some firms select a relatively wide range of KPIs to reduce the risk of gaming and to encourage good behaviour in a range of dimensions. KPI reporting should also be regularly audited to ensure that the figures are broadly correct. Finally, regular feedback between actual losses and KPIs is needed to check that significant risk indicators are not being ignored.

Exercise. Select an area of business you are familiar with. Develop a series of KPIs that could indicate the presence of operational risk. For each one discuss how easy it would be to gather an accurate measurement and whether the KPI could be readily manipulated.

The table below shows an example of a high-level KPI report: in a practical situation this summary would be supplemented by further detail of the indicators, trends, etc.

6.1.3 Some Issues in Operational Risk Management

In this section we discuss the use of KPIs and operational risk scenarios and place this in the context of sound operational risk practices.

6.1.3.1 Uses of KPIs

Trends in KPIs may be more useful as risk indicators than absolute levels: if we know that an average swap confirm is taking twice as long to sign now as a year ago, it is probably worth finding out why even if we are comfortable with the current time gap.

Operational Risk KPI Report: European Structured Finance				
Loss Category	**KPI**	**Level**	**Traffic Light**	**Trend**
Firmwide indicators				
Fraud	Experienced losses	0	Green	Flat
Employment practices	Total staff	14		
	Temps or staff with firm <1 year	2	Green	Flat
	Staff left in the last year	3	Orange	Up
	Number of disciplinary actions or complaints pending	0	Green	Flat
Client relations	Client complaints in last year	1 (no litigation)	Orange	Up
	Number of errors in info supplied	1	Orange	Down
Physical assets	Number of incidents with gross loss	0	Green	Flat
Business disruption and IT	% key systems uptime	99.2%	Green	Up
	Successful DR test?	Yes	Green	Flat
Execution and business processes	Internal audit action points	2	Orange	Flat
	All valuation models approved?	No	Orange	Flat
	Recent docs review	No	Orange	Flat
Business specific indicators				
Client relations and legal risk	Average age unsigned confirms	4.2 days	Green	Down
	Number of unsigned confirms	24	Orange	Down
	Opinion risk estimate from legal	High	Red	Flat
Execution and business processes	Accounting policy risk estimate	Medium	Orange	Flat
	Un'rec'd GL A/Cs	0	Green	Down
	Collateral management errors	0	Green	Flat
Business disruption and IT	% revenue from new products	31%	Orange	Up
	% not in system trades	22%	Red	Up

Similarly, even if a KPI does not seem to correlate well with losses, it does not mean it should be removed from an operational risk scorecard. Personnel turnover might well be irrelevant up to a certain level, for instance, but once a critical threshold is reached—roughly of there being enough people in the department who know what they are doing to actually get the work done—it can become important. Thus, though it is important to understand which KPIs seem to correlate with losses, keeping a few others which might give insight into high-severity low-frequency events is good practice too.

Typically, KPIs are not aggregated into a single overall risk measure: any such aggregation would be arbitrary. Instead one common approach is to use traffic lights: each indicator is assigned a level indicating red (dangerously high) and orange (problematic but not very high): anything else is green. A quick glance down the traffic lights can give a summary of status.

6.1.3.2 Scenario approaches
Scenario analysis for operational risk is similar to that for market or credit risks, although inevitably more judgemental. A scenario is selected, and experts estimate its impact on the

institution. Typical scenarios include:

- External shocks such as earthquakes, epidemics or terrorist events;
- The failure of major IT systems such as market data feeds, Treasury systems or the firm's links to external settlement agents.

One or both of these scenarios might well be linked with the firm's DR planning. Other scenarios include:

- High visibility events such as mis-selling retail products, distributing incorrect valuations on mutual funds or regulatory intervention accompanied by litigation;
- Fraud on a large trading desk or in areas responsible for cash management.

Once a range of scenarios has been selected, either a single loss estimate and probability are estimated for each or a full loss distribution is estimated. The impact of any mitigation such as insurance is factored in, and a loss estimate is obtained.

6.1.3.3 Non-quantitative issues in operational risk management

Some of the principles of good operational risk management were articulated in a BCBS publication.* A number of them relate to non-quantitative issues. For instance:

- The firm's board should be aware of operational risk as a distinct risk class, and they should approve and periodically review the firm's risk management framework in this area.

This authority from the board then flows down to managers:

- Management should have the responsibility for implementing the approved framework consistently throughout the firm via the development of appropriate standards and processes, and by ensuring staff are aware of their responsibilities in this area.

For most firms of any size this will mean that a specialist operational risk group is needed, probably within risk management. There are various approaches to the mission of these groups. Three extreme positions—probably none of which are held completely by any firm—which give an idea of the diversity of views are[†]

- *View 1.* The operational risk group is simply a sop to the regulator: the group is there because it has to be.
- *View 2.* The mission of the operational risk group is the calculation of economic (and ideally at the same time regulatory) capital for operational risk. The group's main functions are accurate data collection and modelling.

---

* *Sound Practices for the Management and Supervision of Operational Risk.*
† Views 2 and 3 are roughly the 'Calculative Idealism' and 'Calculative Pragmatism' of Michael Power's *The Invention of Operational Risk.*

- *View 3.* Capital allocation for operational risk is interesting but essentially arbitrary. The ORG may calculate capital as an ancillary function, but the main value to the firm comes from the detailed analysis of loss data and KPIs, understanding the causes of losses, and recommending cost-effective improvements to management.

Owing to this wide range of views, it is particularly important that firms articulate, albeit probably in more diplomatic language than the above, their attitude towards operational risk management. Once this has been elaborated, policies, procedures and culture which support the desired outcomes can be fostered:

- Firms should identify and review the operational risks inherent in all areas of the business. New products and businesses should be assessed for operational risk and appropriate controls put in place before they are undertaken.
- Operational risk losses should be measured and regularly reviewed by management, with mitigatory action taken as necessary.

This is just another example of a pattern we have seen before: the board has authority and responsibility, risk management implements processes to monitor and manage risks within that delegated authority, and the board monitors performance and takes action where needed.

6.1.3.4 Mitigation of operational risk
Market risk can be mitigated by hedging so it makes sense to set limits: if we are close to the limit, exposure can be reduced. But setting an operational risk loss limit of £50M would not have done Barings any good by itself. How then can operational risk be mitigated?

- For some forms of risk such as fire or earthquake, insurance can be bought.
- For others—as stressed by View 3 in the previous section—improving policies, processes or staff can be an effective mitigant.

6.1.3.5 Business features and operational risk
Creeping exoticism is not just an issue of model risk: it can also introduce other kinds of operational risk. For instance, American options can sometimes give rise to operational risk as they require either a trader to be notified or the system booking changed if the

counterparty has exercised; if the right is ours, we have to monitor whether early exercise is worthwhile. Similarly, considerations apply to the monitoring of barrier events, range accrual events, etc.

Another good slogan is *do not try to catch a falling knife*: if the operational risk involved in a trade or business seems too large you always have the option of waiting for others to sort out the wrinkles before you trade. [See section 11.5 for a further discussion of new products.]

6.2 THE TAILS AND OPERATIONAL RISK MODELLING

Fat tails are ubiquitous in financial distributions. For instance

- Historical market returns are fat tailed [as discussed in sections 2.3.2 and 6.2.1];
- The return distribution implied by option prices in many markets is also fat tailed;
- The operational risk loss distribution is usually assumed to include a small probability of very large losses.

Therefore, it makes sense to look at the mathematics available for modelling fat-tailed distributions. For market risk, useful results have been provided by *extreme value theory*,* so we begin by introducing that. The use of these techniques in modelling operational risk is then discussed.

---

* See Paul Embrechts, Claudia Klüppelberg and Thomas Mikosch's *Modelling Extremal Events for Insurance and Finance* or Alexander McNeil, Rüdiger Frey and Paul Embrechts' *Quantitative Risk Management: Concepts, Techniques, and Tools* for a more comprehensive introduction to EVT.

6.2.1 The Tails and Extreme Value Theory

The further we go into the tail of market risk return distributions, the worse the assumption of normality is. For instance, looking at the CAC-40 and expanding the negative tail, we find that there are many more days where the return is more negative than 2 S.D.s than predicted by the normal distribution: Figure 6 shows the issue. EVT often provides a better tool for studying this part of the return distribution.

6.2.1.1 The generalised Pareto distribution

Suppose x_i are identically distributed random variables with some distribution function $F(x) = \Pr\{x_i \leq X\}$. These could for instance be daily log returns from an equity index. Pick a parameter u which represents large losses: for instance, u might be 2 S.D.s. Then the *distribution of excess losses over u* is defined as $F_u(y) = \Pr\{x - u \leq y | x > u\}$. This represents the probability that a loss x exceeds the threshold u by at most y, contingent on it exceeding the threshold at all.

The *generalised Pareto distribution with $\xi > 0$* (GPD) is a two-parameter distribution given by

$$G_{\xi,\beta}(X) = 1 - (1 + \xi X/\beta)^{-1/\xi}$$

where ξ is the *shape parameter* of the GPD and β the *scale parameter*.

The reason the GPD is interesting is that for a large class of financial return series as u increases, $F_u(y)$ tends towards $G_{\xi,\beta}(y)$ for a suitable choice of ξ and β.[*] Thus, sufficiently far into the tails many distributions, including nearly all of the ones of interest in modelling financial returns, look like the GPD.

6.2.1.2 Fitting the GPD

The good news, then, is that we know that it is reasonably likely, at least, that the GPD will offer a good fit sufficiently far into the tails. The bad news is that we are not told how far we have to go before the GPD fits: is it 2, 3, 4 or 20 S.D.s?

There are various tools for determining the threshold u and fitting the parameters of the GPD to the returns beyond u, most beyond the scope of this book. One easy approach is simply to note that beyond u, if we plot natural logarithm of the empirical probability of seeing a return more negative than x versus $\ln[(x - \mu)/\sigma]$, the slope of the best-fit straight line is a crude estimator for ξ. Better techniques, sometimes based on MLEs, are discussed in the references in the previous footnote.

6.2.1.3 Tail fitting and VAR estimates using EVT

The GPD usually provides a much better empirical fit for the tails of market return distributions than the normal distribution: for instance, Figure 7 shows a GPD fit of the same CAC-40 data used previously. One caveat here is that EVT techniques are typically rather data hungry: we started with a little over 5,000 CAC observations, nearly 20 years worth of data. However, it was determined that u was approximately 1.75 S.D.s, and so

---

[*] This is a special case of the Fisher–Tippett theorem: see (*op. cit.*) for more details.

threw away all the returns that were more positive than that, leaving just 187 data points to fit the GPD. Obviously with only 2 or 3 years' worth of data we would have found it very difficult to fit the distribution with any degree of accuracy.

Exercise. Pick 10 positions at random from a bank's trading books. Try to get 10 years' worth of good, clean market data on *all* the variables needed to revalue the position (including implied volatilities where necessary). On the basis of the difficulty of this task, how sensible is it to set a confidence interval for risk management purposes beyond 99%?

Threshold (%)	Move under 1-Day VAR, Normal Distribution (%)	Move under 1-Day VAR, GPD Distribution (%)
97.5	2.8	2.7
99	3.3	3.9
99.9	4.4	9.1
99.99	5.3	19.3

The table shows the sensitivity of risk estimates to the choice of modelling distribution. Specifically, we show the size of the move needed to cover a given confidence interval using a normal distribution–based VAR and the parameterisation of the GPD which best fitted the CAC data. The VAR estimates using the GPD are much higher than those produced by a more conventional VAR model.

6.2.1.4 Multivariate EVT

Given enough data, determining if an extreme value distribution might offer a good fit to the returns of one market and, if it is, fitting it is often fairly routine.

Unfortunately, the situation is not nearly as straightforward for multiple markets. The concept of correlation—which is so useful in the theory of the multivariate normal distribution—does not carry over to multivariate EVT. [Instead the concept of a copula is needed, and a plausible one has to be selected: see Section 11.4.4 for a further discussion.] This means that, at least at the moment, EVT analysis is less readily applicable in the multivariate than the univariate case.*

6.2.2 The Case of Long Term Capital Management

A good example of the perils of making unjustified assumptions about the tails of market risk return distributions can be found in the failure of Long Term Capital Management (LTCM).

6.2.2.1 History and strategy

LTCM was a hedge fund (and an oxymoron) that operated from its founding in 1993 to its failure in 1998 and subsequent liquidation. Its traders were regarded as among the best

---

* See Paul Embrechts et al. (*op. cit.*) for more details of the multivariate case.

in the business, some of them having originated at Salomon Brothers' famously profitable and aggressive arbitrage group. The combination of well-known principals and what was for the time a large and sophisticated investor base led to LTCM being seen as a 'smart money' firm.

LTCM's strategy was a lot more mundane than its reputation would have suggested. There were three main components to it:

- *Mean reversion.* The fund took positions which involved the assumption that the spread between rates often mean reverts to an average value. Thus, when a credit or swap spread went above historical average levels, LTCM would short it on the assumption of mean reversion.
- *Leverage.* LTCM leveraged its positions by borrowing large amounts of money from various banks, not all of them its prime brokers. Moreover, limited disclosure meant that this leverage was not easily visible to the market.
- *Funding by writing options.* In addition to its borrowing, where it could LTCM took positions which resulted in significant cash inflows, allowing it to further leverage itself. Thus, a position on equity implied volatilities being over historics was taken by writing straddles, resulting in option premium income.

6.2.2.2 Position
In early 1998 a belief in mean reversion would have suggested the following positions:

- Long credit spreads;
- Long liquidity premiums (for instance, long off-the-run government bonds versus short on-the-run ones of the same maturity);
- Long swap spreads (that is receiving fixed versus short treasuries);
- Short long-dated equity index implied volatility;
- A yield curve position short the front end versus long the back end.

6.2.2.3 Portfolio risk
Portfolio optimisation is the process of producing the portfolio with the best-predicted return for a fixed-risk tolerance. Just as VAR estimates depend on distributional assumptions, so does the optimum portfolio. Phillipe Jorion has given a cogent analysis of this,[*] showing how LTCM's both the optimal portfolio and its risk estimate are sensitive to the distributional assumption. If we calibrate a model using a normal returns assumption and data for a few years before 1998, the optimal portfolio is close to LTCM's actual position (leveraged long corporate bonds versus short treasuries) leading commentators to suggest that LTCM used this assumption. As we saw in the last section with EVT versus normal risk estimates, using a different distribution can dramatically change our view of

---

[*] See Phillipe Jorion's 'Risk Management Lessons from Long Term Capital Management' (*European Financial Management*, volume 6).

potential risks. LTCM was presumably confident in its risk estimation or it would not have been so leveraged: this confidence was misplaced.

6.2.2.4 What happened?

By the end of 1997, LTCM had been hugely successful: investors had seen impressive rates of return, and the fund's assets had grown to over \$7B. This much capital is difficult to deploy effectively, so LTCM returned some to investors, at the same time using its growing fame to pressure banks into increasing lending to it. The net result was increased positions and increased leverage.

Unfortunately for LTCM and its investors, 1998 was an eventful year in the markets. The Russian default caused considerable volatility and a *flight to quality*.* Investors reassessed a range of market risks and risk capital was withdrawn from the markets. Thus, what started as a purely Russian event turned first into an emerging market crisis as investors sold emerging market bonds across a range of markets, and then into a broad credit market downturn.

This impacted LTCM's position in a number of areas:

- Long credit spread positions lost money as investors sold risky bonds;
- Long liquidity premium positions also lost money as investors came to value liquid investments more than hitherto;
- Long swap spread positions lost money as treasuries tightened.

At the same time equity index implied volatility went up as a large player started to buy back the short vega position in its equity derivatives book. LTCM was losing money on many of its positions. However, if the positions looked attractive on the basis of LTCM's model before the Russian event, they would have looked even better during it: some spreads were at historic highs, encouraging LTCM to double up rather than cut the position.

6.2.2.5 Failure and bail-out

Contrary to LTCM's assumption of mean reversion, market spreads continued to widen. LTCM's losses became so large that the banks which had provided leverage became increasingly concerned about the fund's credit worthiness, and they were on the brink of forcing the firm into bankruptcy when the Federal Reserve Bank of New York brought the lenders together and brokered a bail-out. The FED acted because of the size of LTCM's positions. For instance, one estimate is that LTCM was a counterparty to approximately 2.5% of all swaps traded globally in 1997. Hence the FED was concerned about the disruption its failure would have caused to the market and potential systemic risk.

The FED therefore 'invited' the principal creditor banks to contribute to a rescue operation. This was not the kind of invitation that was easy to refuse, and so the banks lent further

---

* This is an investor preference for less risky instruments, such as government bonds, compared to risky ones. It usually happens suddenly, in times of market crisis: investors sell risky instruments and buy safer ones. Thus, government bond traders sell emerging market bonds and buy U.S., U.K. and German government instruments, equity traders sell growth stock and buy value and so on.

funds to prop up LTCM, allowing it to be wound down in a controlled manner: they also contributed staff to a committee which managed the process. Orderly liquidation was complete by early 2000, with the original investors receiving a recovery of roughly 8%.

6.2.2.6 The lessons of LTCM

Different observers, inevitably, have drawn different conclusions from the failure of LTCM. Eight years after the event, it is worth highlighting:

- *Smart money sometimes is not so smart.* LTCM would not have gotten into the problems it did without the use of large amounts of leverage. That leverage was provided by banks lending it money without, it appears, understanding the fund's risks and positions. Certainly, had the market been generally aware of how simple LTCM's strategy was it is unlikely they would either have had the reputation they did or that the same degree of leverage would have been provided. Clearly, enhanced disclosure from hedge funds to their prime brokers has a role to play here.

- *Systemic risk does not just reside in banks.* The fact that the New York FED felt it necessary to intervene is evidence enough of this.

- *The ubiquity of model risk.* Sophisticated risk managers—whether on the risk taking or risk control side of the business—usually look at their risks using a range of different tools. They question the assumptions of their models and hence where the model might fail to give accurate predictions. It seems LTCM had so much confidence in its approach that it did not believe it could be wrong.

- '*The market can stay irrational longer than you can remain solvent*', as John Maynard Keynes is supposed to have said. LTCM's position would, after considerable P/L volatility, have made money had they had enough capital to survive that long: arguably the same is true of Nick Leeson's position at Barings. It was the combination of a position that was much more volatile than predicted and leverage that caused the failure of the fund.

- *The difficulty of measuring risk in the presence of asymmetric beta.* A high beta position is one that responds more than the market [as discussed in section 2.1.2]. The problem with some hedge funds is that they display asymmetric beta: when the market is flat or going up, the fund's returns are not highly correlated with the market return. But in a market crisis the fund's return are highly correlated with the market. Thus, what appear to be good returns with little market risk are in fact generated by taking risk in the tail of the return distribution. All of LTCM's positions could be seen as writing a put since they all earned good returns while the credit market was not in stress. The common undiversified risk factor was an assumption of normal market conditions. This is characteristic of asymmetric beta situations.

Exercise.   Read the President's Working Group on Financial Markets Report on Hedge Funds, Leverage, and the Lessons of LTCM. How many of the recommendations have been implemented?

6.2.3 The Stable Process Assumption

Before we look at the use of statistical techniques in modelling operational risk, it is worth stepping back to look at the overall framework in which statistical predictions are made. One important idea is that of a random process which can be observed. An early application of EVT was the Dutch dyke problem: how high do you need to build a sea wall such that it is not breached more than once in a hundred years? Given several hundred years worth of data and the assumption that there was an underlying random process generating storm surges (which produce big waves that could breach the dyke if it is too low) EVT was used to determine how high to build the walls.*

6.2.3.1 Random processes for market risk

In market risk quantification, we assume that there is a process generating random market moves in a risk factor such as an equity index. A simple approach might use the assumption that this process is log-normal in returns whereas in a more sophisticated analysis other choices might be made: but the first assumption is that there *is* a single underlying process which we can observe repeatedly. For an index like the FTSE 100, we might be a little nervous about this assumption: after all, the FTSE's composition potentially changes every 3 months, and the economic environment in which the FTSE 100 companies have to operate changes over time. However, broadly, at least for shorter periods of time, the assumption that there is a (probably fat tailed) process generating returns does not seem obviously wrong.

6.2.3.2 Random processes driving operational risk

Now consider operational risk. Does it make sense to talk about a single process generating a firm's operational risk losses? It is not clear that it does. The firm is changing all the time: processes are being re-engineered, new products are being invented and volume in old ones is changing, staff change jobs. Therefore, this author, at least, has some doubts about the wisdom of attempting to model the operational risk loss distribution.

6.2.3.3 Data requirements

The estimation of process behaviour at large confidence intervals requires a huge amount of data. Moreover, even if the operational risk process is stable, it must be firm specific, so there is no substitute for internal loss data. Therefore, even without the concerns above, a minimum requirement for operational risk modelling is a large amount of loss data gathered on a uniform basis.

6.2.4 Operational Risk Modelling

With that preamble, consider how we would calculate capital for operational risk on the basis of a model of the operational risk loss distribution should we consider such a thing possible.

---

* Recently there has been a change in the properties of the random process driving storm surges—sea levels are rising and weather patterns are changing due to global warming—so the predictions made using the old process calibration are no longer valid. Dutch dykes are not as safe as they were.

6.2.4.1 Risk factors approach

Just as a range of indicators from liquidity ratios to EBITDA are used to derive an internal rating in credit scoring, so KPIs can be combined to give an indicator of operational risk. The basic technique is the same: we seek to isolate a number of KPIs which drive the operational risk losses of a given business, then use the current levels of those indicators to estimate the current level of operational risk. This method has the advantage of being forward looking, but the disadvantage that it is difficult to gather enough data to be confident the KPIs really are capturing all of the material drivers of operational risk.

6.2.4.2 Loss distribution approach

Instead of looking forward using risk factors, a firm could look backward using past operational risk losses and model their distribution. This method is called the *loss distribution approach* (LDA).

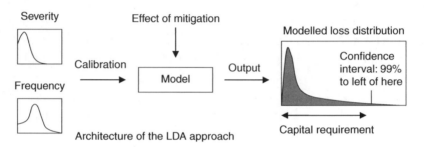

Architecture of the LDA approach

An outline of the architecture of a typical LDA is shown above. In the sketch, the firm:

- Gathers internal data on losses* in each of a range of categories.
- Perhaps supplements these data where they are sparse with external loss estimates.
- Analyses the data to find a possible distribution. Some firms use an EVT distribution such as the GPD discussed above.
- Estimates parameters of the chosen distribution using the data.
- Uses the distribution to predict the loss at the desired confidence interval x. Unlike market risk, however, where $x = 95$ or 99%, for operational risk x is typically a high threshold such as 99.9 or 99.97%.

6.3 ALLOCATING CAPITAL AND OTHER RISKS

This section is about two things: the assignment of capital to businesses and those risks for which capital allocation is difficult or impossible.

---

* Some practitioners suggest that it is best to gather data before the impact of risk mitigation such as insurance than to factor in the mitigation separately. The reason for that mitigation can introduce non-linear effects. For instance, if a bank has a fire insurance policy that covers it for up to £50M of losses per year, it is important to use the small fire losses (which do not give rise to a net loss) to gain insight into the likelihood of a large fire (which would).

Firms spend considerable effort in estimating capital requirements for market risk and credit risk, and many either have or are developing operational risk capital measures. Here we look at how these estimates are put together to produce an overall economic capital estimate for some activity, and what those capital estimates for quantifiable risks are used for.

Some risks that firms run are very difficult or impossible to quantify: a good example is reputational risk. A discussion of these 'other' risks, how they are managed and how they are supported by capital is presented. Finally, the problem of optimising the firm's risk/capital balance is considered.

6.3.1 Capital Allocation and Portfolio Contributions

The techniques of this chapter and the previous two give us some insight into the P/L distribution resulting from the market, credit and operational risks being run in a business.

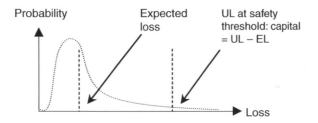

There is obviously some diversification between the different risk classes, so it is conservative to estimate the total capital requirement as the sum of the requirements for market, credit and operational risks. To make these capital requirements consistent, though, the same confidence interval should be used for all three estimates. Therefore, a firm might use, for instance, 1-day 95% VAR scaled to 99.9% for market risk [as discussed in Chapter 4], an internal measurement approach estimate for the 99.9 percentile of operational risk loss distribution and an estimate of the 99.9% of the credit risk distribution [modelled using one of the approaches of Chapter 5]. It might chose to model economic capital as the sum of these three contributions: this would be one choice of *risk-based economic capital model*.

6.3.1.1 Example

Consider a bank with three BUs as in the table below.

BU	Market Risk Capital Requirement (£M)	Operational Risk Capital Requirement (£M)	Credit Risk Capital Requirement (£M)	Total Allocation (£M)
Retail banking	9.2	17.2	156.3	182.7
Investment banking and capital markets	82.4	12.1	24.3	118.8
Corporate banking	16.2	7.4	247.1	270.7
Whole bank	100.2	36.7	383.7	520.6

Notice that the different BU's market and credit risk positions diversify each other, so the firmwide capital requirement is smaller than the sums of the individual positions.

The total capital allocated to each business is just the sum of the capital amounts required to support each of the risks in the business.* Following our earlier discussion [in Chapter 3], the following revenues would generate the risk-adjusted returns as shown below:

BU	Net Revenue (After Expenses and EL Provisions) (£M)	Capital Requirement (£M)	Risk-Free Return on Capital (£M)	Pre-Tax RAROC (%)
Retail banking	58.4	182.7	9.6	37.2
Investment banking and capital markets	73.1	118.8	6.2	66.7
Corporate banking	79.2	270.7	14.2	34.5
Whole bank	210.7	36.7	27.3	45.7

6.3.1.2 Performance measurement

In this example, the BU with the highest net pre-tax revenue, corporate banking, is actually the worst performing in risk-adjusted terms: it is earning good revenues, but not as much per unit of capital required as the other two groups (at least on the basis of this particular capital model). The best performing business group is investment banking and capital markets, so this unit should have the largest allocation of the bonus pool.

The RAROC approach allows this bank to compare performance across rather different business groups and hence to consistently reward units which add value for shareholders. However, it does depend on the capital model providing a meaningful risk assessment. Consider the new BU in the following table.

BU	Market Risk Capital Requirement	Operational Risk Capital Requirement (£M)	Credit Risk Capital Requirement	Total Allocation (£M)
Investment management	—	21.0	—	21.0

Obviously this BU's RAROC is heavily dependent on the operational risk capital allocation, and this is the most arbitrary of the three risk elements. If the firm is going to compensate staff in this BU on the basis of RAROC, it needs to have considerable confidence in the accuracy of its operational risk capital model.

---

* Some economic capital models include a diversification benefit between, for instance, market and credit risk: others consider such a benefit to be so dependent on unproven correlation assumptions that they do not consider it. There may also be other refinements in the model such as extra charges for concentrations or SR (where not captured by the market risk capital estimate). Finally, note that there is an implicit assumption that funding/balance sheet usage is being correctly charged, as discussed in section 8.2.1.

6.3.1.3 Risk capital or actual capital allocation

The firm's economic capital model for the three BUs above suggests the firm should have a capital of £520M. If it actually has £500M, then it is taking too much risk for the capital base and the board needs to take action to reduce risk.* But what if the firm actually has a capital of £600M? Should we allocate the full £600M to the BUs in the same ratio as the capital model suggests—£600M × 182.7/520.6 to retail banking for instance—or should the remaining capital be allocated to corporate centre to cover other risks not modelled by the capital model?

6.3.1.4 Diversification benefit

The sum of the capital requirement for the three BUs individually is £572.2M but the whole bank requirement is only £520.6M: who does this diversification benefit belong to? Again practices differ: some banks allocate the benefit in proportion to the BUs, so increasing their RAROCs: others keep it in corporate centre to balance the costs that inevitably accrue there.

6.3.1.5 An alternative approach: P/L volatility

Suppose a firm sets a threshold for capital calculation of 99.9%, as above. This sets an incentive structure: it encourages risk taking that is not charged for by the capital model, namely risk beyond the 99.9% threshold. Roughly speaking (portfolio effects make it more complex than this) it encourages the BUs to take credit risk with a PD < .1% or market risks more than 3.1 S.D.s from the mean (since $\Phi^{-1}(0.999) = 3.09$, where Φ^{-1} is the standard inverse cumulative normal distribution function). This may not be a helpful incentive structure since selling penny puts—far out of the money put options with small premium—is a notoriously risky business. Some firms prevent this via their limit structures, but some others take the view that this issue highlights a problem with risk-based capital models and so take an alternative approach.

Capital is required to support P/L volatility. Therefore, a capital model attempts to predict possible future P/L volatility on the basis of its drivers: risks. Clearly, instead of attempting to predict P/L volatility from risk measures, a firm could directly measure actual experienced P/L volatility. This suggests the idea of an *earnings volatility–based economic capital model*.

Here earnings volatility is used as the fundamental risk measure. This depends on earnings being an accurate measure for each BU, and so care is needed due to the potential for historic cost accounting—or a 'creative' use of provisions—to smooth earnings. Suppose that a firm implements the following controls:

- Rigorous and independent review of valuations of market-to-market instruments with remarking and revision of reserves and mark adjustments each month end;
- Model-based recalculation of EL provisions for credit risk each month on the basis of up-to-date data;
- Tight controls on the use of general provisions against operational risk losses.

Then the monthly P/L should reflect current earnings, and its volatility is a risk measure.

---

* In reality for most firms, regulatory capital is likely to constrain the firm before this situation occurs.

An earnings volatility–based capital model uses this data:

- First we calculate how much return shareholders demand for extra P/L volatility using a risk/return model such as CAPM.
- Then we relate each portfolio's P/L volatility to the return volatility of the firm's stock.

This allows us to estimate the extra return required from a business given its observed earnings volatility.

6.3.1.6 Pros and cons

The table below presents a summary of the advantages and disadvantages of risk- and earnings-based economic capital models.

Issue	Risk-Based Models	Earnings-Based Models
Nature	Bottom up estimate	Top down estimate
Ease of estimation	Requires considerable modelling	No modelling required
Quality of comparison	Good for comparing risk taking BUs	Can compare very disparate activities
Coverage of future volatility	Forward looking on the basis of modelled risks: *ex ante*	Backward looking on the basis of risks that caused P/L: *ex post*
Far tail coverage	Depends on model accuracy	Poor due to lack of data
How to reduce capital requirements	Hedges suggested as the risk drivers are modelled	Unclear as the risk drivers not modelled
How capital requirements are calculated	Required capital is quantified via EL–UL at a fixed soundness threshold	Capital indirectly quantified via model of P/L distribution or via CAPM
Incentive structure	Take tail risk	Smooth the P/L
Assumptions	Risk measurement and modelling is correct	P/L volatility and modelling is correct

6.3.1.7 Return on regulatory capital

Typically, firms focus on measuring return on economic capital because they believe this provides the most accurate assessment of performance. However, this performance is not unconstrained: firms must remain capitally adequate. Thus, having enough regulatory capital is a hard constraint that must be satisfied.

Economic-capital-based performance measurement therefore makes sense if you have more than enough regulatory capital—as most firms do. However if regulatory capital requirements are close to the available regulatory capital, the situation changes. Then it makes more sense to measure businesses on the basis of their return on regulatory rather than economic capital. Some firms even directly charge businesses for their regulatory capital use, encouraging position taking which optimises the firm's regulatory capital position.

6.3.2 Reputational and Other Risks

Two classes of risk are explicitly excluded from the definition of operational risk discussed earlier:

- *Reputational risk.* A firm's reputation is the collection of perceptions of it held by stakeholders including clients, the broader market it operates within, supervisors and others. Reputational risk is the risk that events damage its reputation, possibly leading to loss of clients, regulatory action or adverse publicity.

- *Strategic risk.* This is risk that a firm's strategy leads to losses or a deteriorating market position. Strategic risk includes the risk of acquisitions and disposals, business development risk and the risk of bad market positioning.

6.3.2.1 Responsibility for reputational risk

The responsibility for reputational risk, like all other risks, sits with the board. However, reputational risk management is more difficult to delegate because management decisions on particular issues here inevitably involve more judgement. Therefore, it is vital that significant reputational risk decisions are made by senior management: for the board to decide that the firm should commit to a trade which subsequently generates highly negative publicity is unfortunate; for a business group manager to do it without reference to higher authority suggests a failure of control.

6.3.2.2 Consequences of reputational risk management failures

A firm's reputation is an intangible asset: in accounting terms, it is part of goodwill. For some banks it is worth billions. It influences not just clients' desire to do business with the firm but also its cost of capital, the ease with which it can attract staff and how regulators perceive it.

A reputational risk event typically has an impact far beyond an immediate loss. For instance, if a firm mis-sells a product to retail investors and subsequently receives a fine from regulators, the fine is not usually a material fraction of earnings. Even if it has to make restitution to those investors, this sum is often immaterial too. But if the bank's credit spread goes out even 2 bps as a result, the impact on a hundred billion dollar funding base can be highly significant.

One of the problems with managing reputational risk is that these knock-on effects are difficult to measure: potential clients do not often tell you why they have decided not to trade with you and prospective staff at interview do not often openly express concern at your reputation. You will never know why that relationship did not prosper.

Exercise. Research some of the reputational risk losses suffered by financial services firms.

6.3.2.3 Reputational risk management

Reputational risk mitigants include:

- *The articulation of firmwide principles.* Firms should state their attitude towards business ethics, client service, responsibilities to stakeholders and the wider community, and conflicts of interest.* These principles should be realistic in the sense

---

* It is interesting that successful regulatory regimes often begin with principles and only then set detailed rules based on them. For instance, firms and individuals breaching FSA regulations are usually cited first for failure to adhere to one or more of FSA's 11 broad principles, and only latterly for a particular rule breach. See www.fsa.gov.uk for more details.

of genuinely encapsulating the best features of the behaviour of successful leaders within the firm.*

- *Consistent enforcement of principles.* It is counterproductive to articulate an over-arching principle and then not enforce it. Reputational risk is typically best managed when a firm can demonstrate that it does what it says it will do. Saying less but ensuring consistency between word and deed is often the best strategy. It is also vital that performance measures include compliance with the firm's principles: claiming to have the highest ethical standards but then paying managers based simply on how much they have made is likely to lead to significantly increased reputational risk.

- *Regulatory and legal risk management.* Reputational risk often arises from regulatory or legal failures, so it is vital that management monitors issues here with reference to their potential reputational impact. For instance, though the firm might judge that it is likely to win a particular legal case, the reputational damage caused by disclosure during the case and by the actions of lawyers representing the firm may not be offset by the benefit of winning.

- *Effective communication strategies.* Firms should ensure that stakeholders understand their principles and they should regularly highlight the consistency of their actions with those principles.

- *Crisis management strategy.* When there is a perceived failure to act ethically, or any other reputational risk issue, a firm should have processes in place to elevate the issue to senior management and to manage the crisis. This strategy will include potential disciplinary or other personnel-related issues, prompt action to address the problem and communication of the firm's understanding of the issue and attempts to resolve it.

Reputation has a peculiar dynamic: establishing a good one takes a long time, but losing it can take minutes and rebuilding a damaged reputation will certainly take years. Yet line managers are sometimes too focussed on immediate P/L to see the risk here: it is only processes and culture that ensure speedy elevation of reputational risks to senior management that will protect the firm.

6.3.2.4 Strategic risk: definition and examples

Strategic risk is the risk of loss arising from adverse business decisions including inadequate responses to market or industry changes or to institution-specific threats. It often arises when the bank's attempt to meet a business goal fails due to inappropriate or inadequate

---

* If a firm claims to be guided by principles which do not match the types of behaviour which are rewarded, staff dissatisfaction and reputational damage tend to result. Principles should not just be meaningless (or in social science jargon 'phatic') statements: they should be seen to encapsulate a standard to which the firm genuinely aspires, otherwise they can do more harm than good.

resources, failure to understand the changes needed or other project management failures. Examples of strategic risk include:

- *Inadequate responses to new technology.* A good example of this was the failure of some retail banks to offer efficient, easy to use on-line banking services in the late 1990s and early 2000s: these banks lost customers—and the ability to profit from the decreased costs of on-line transaction processing—to those competitors that were able to respond to the opportunity.

- *Bad business positioning.* There are long-term trends in business. For instance, in investment management at the moment hedge funds and PE are both attracting large investment inflows whereas traditional mutual funds are currently unfashionable. Firms need to respond to these trends to ensure that they are not stuck in an under-performing business segment.

- *Jumping on the wrong bandwagon.* This is the flip side of the previous issue: if you are going to enter a new business area then the process has to be properly managed. A traditional mutual fund would almost certainly need considerable resources to effect the technological, cultural and process management changes needed to offer hedge fund-based products, for instance.

- *Failure to manage cultural issues.* Successful exploitation of a new business opportunity or of an old one in a new geographical area often involves more than just good project management: business models are not necessarily portable from one location or business to another. History is littered with examples of expansions or acquisitions that have wasted massive amounts of shareholder's money due to these issues.

- *Failure to integrate an acquisition.* Even when staff on both sides of an acquisition are eager to work towards a successful integration, the project often fails to achieve the predicted synergies.

- *Resting on your laurels.* Managing change is difficult but there is no alternative: firms that do not innovate usually find their market share and profitability eroded by faster moving competitors.

6.3.2.5 The importance of strategic risk

Very few firms have been unlucky enough to lose $1B on a market risk event: large market risk losses tend to be measured at most in hundreds of millions even for firms with huge proprietary trading risks. Yet it is very difficult to think of a significant bank acquisition or divestment involving less than $1B, and even a medium-sized bank merger is a $10B or more transaction. Furthermore, a business line in a large bank or broker/dealer might well have a pre-tax budget in excess of $1B. Therefore, arguably strategic risk is much larger than market risk for most firms. It is much harder to measure so it tends to get less attention in risk reporting; but it should certainly not be ignored.

6.3.2.6 Avoiding group think

Should a firm have one strategy? It might seem a stupid question: how can a firm prioritise initiatives without a strategy? But the problem with having a strategy is that it can be wrong, and the more committed a firm is to pursuing it, the more wrong it can be. Therefore, some managers take the view that though they will encourage some initiatives by giving them more resources and discourage others by providing less, it is often worth having a range of projects simultaneously being pursued. Will the next big thing in the derivatives markets be insurance derivatives, emission derivatives or property derivatives? It is difficult to say. A given firm might feel it has a bigger edge in one area than another and so might chose to prioritise one business development over another: but doing *something* to progress the less favoured alternatives and regularly reviewing the state of the market is probably a good strategic risk hedge. A successful manager knows when to admit they have taken the wrong course and so to cut their strategic risk position. On the other hand, strategic risk is arguably biggest just when every member of the board wholeheartedly believes in a course of action: the lack of any dissent can make it very difficult to reverse course if things turn out badly.

> *Exercise.*   Read the equity research relating to a range of leading financial services firms. What are the analysts' views on each firm's strategic risks?

6.3.2.7 What risks should be included in a capital model?

Models can provide estimates of the capital required for market and credit risks, and there are benefits in at least addressing the question of what an operational risk capital allocation might be like.

Note that reputational and strategic risks are borne by equity holders when they buy the stock of a financial institution regardless of whether they are included in a capital model. But should they be included in such a model? There are two arguments:

- *Do not try to allocate capital for risks you cannot quantify.* Any capital allocation for reputation or strategic risk is bound to be arbitrary. Therefore, there is a strong argument for excluding risks that are essentially part of goodwill—such as strategic and reputational risk—from the capital model and letting shareholders decide whether they are sufficiently well compensated for bearing these risks.

- *Allocate everything.* Some boards keep their firm capitalised well above economic capital estimates for quantifiable risks, and all of this capital is allocated to BUs. The problem with this approach is that it can obscure the difference between necessary capital for goodwill risks from unnecessary capital that does nothing but dilute shareholder returns (and make the board feel warm and fuzzy inside).*

> *Exercise.*   Why is it rare for shareholders to demand the return of capital from a financial institution?

---

* Some of the extra capital may be to support acquisitions or to support volatility in the capital requirement.

6.3.2.8 Dialogue

The Finance Director Mr. P. Pincher is lecturing the asset/liability management committee on capital allocation, one of his favourite topics.

> We pay too much attention to "these numbers you know. Look what would have happened last year if we had listened to the capital model. Trading had a huge return on risk adjusted capital, and the model told us to put all our capital there. This year they've lost more money than you can shake a stick at and the retail bank has saved our bacon. All the capital allocation does is tell us where we should have put more resources last year. It tells us nothing about what's going to happen this year …"

6.3.3 Hedging versus Capital

Suppose after a careful analysis of all of our risks, we decide, somehow, that they require more capital to support them than we have. What alternatives are available to us?

- If we think all of our risks are adding economic value, then we should get more capital, for instance, via issuing a capital security.
- If not, we should reduce risk.

In this way we see that hedging is an alternative to capital: by buying a hedge, we reduce P/L volatility and hence the need for capital.

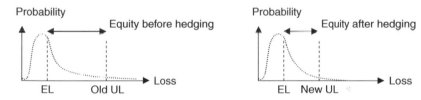

6.3.3.1 Available hedges

What hedges can we use to reduce P/L volatility? The routes that are available include:

- *Market risk hedges*. These are obvious: we reduce UL by reducing the tail of the market risk loss distribution, for instance, by buying puts.
- *Credit risk hedges*. Securitisation is the obvious answer here: we reduce the UL by paying someone else a premium for taking that risk, i.e., by selling the upper securitisation tranches.
- *Operational risk hedges*. The traditional method of reducing the risk of some operational risk losses is to buy insurance. Sophisticated insurance policies for large firms go well beyond natural disaster or liability insurance and allow firms a good deal of flexibility in deciding how much of each risk class they want to cover.

Notice that there is a trade-off between how broad a range of risks is covered and the premium paid for the hedge. If a firm really wants to cover all of the P/L volatility in a business, then the business must be sold: no other step completely removes future risk. If we do not want to go that far, the assessment of any hedge requires us to decide whether the premium paid for it is low enough given the effect it has on the business' capital.

6.3.3.2 What is worth hedging?

The result of this analysis can be slightly counter-intuitive. For instance, for most individuals the impact on our personal wealth of losing all of our goods in a fire is large. Therefore, it is worth paying a premium to reduce the UL, and so we buy fire insurance. For a large bank, though, it may make more sense to bear certain operational risks rather than to hedge them: after all, insurance companies have to make a profit, so they charge far more than the EL on a fire policy to provide cover. Thus, where a bank can afford to cover some fire-related losses, for instance those caused by retail branches burning down, it might decide not to hedge.*

6.3.3.3 Economic value-based decisions: theory and practice

The obvious way of evaluating a hedge is therefore to compare

- Shareholder value added before hedging, i.e.,

$$\text{Revenue} - \text{Expenses} - \text{Old EL} + \text{Old capital requirement}$$
$$\times (\text{Risk-free rate} - \text{Hurdle rate})$$

- EVA after hedging

$$\text{Revenue} - \text{Expenses} - \text{New EL} - \text{Cost hedge} + \text{New capital requirement}$$
$$\times (\text{Risk-free rate} - \text{Hurdle rate})$$

Of course, this assumes that *the only measure of risk is risk at the defined safety threshold.* In other words, under a strict application of this doctrine, risks would never be hedged which do not effect the UL at the specified confidence interval. Given that earthquake risk in many parts of the world is less than 0.1% annualised, on this basis a firm should not buy earthquake insurance in those areas. In practice, of course, firms do tend to cover some of these far tail risks, not least because of the difficulty of estimating their true probability and so of being sure that they are outside the loss threshold.

> *Exercise.* Try to find a document describing a firm's economic capital model in detail. Pick a business and deduce what hedges would increase EVA.

---

* The situation has been simplified here for ease of explanation: in reality, though the actual physical damage caused by branch fires might be a risk a bank might decide not to hedge, it would probably want cover against related losses caused by injury or damage to client assets. Therefore, it would include structured fire insurance with a fairly high deductible for property damage within a wider policy covering a range of risks.

CHAPTER 7

Bank Regulation and Capital Requirements

INTRODUCTION

There is fairly general agreement on the benefits of some regulation in the financial system [and a further discussion of the motivations for bank regulation can be found in section 1.5.1]. In this chapter, we turn to the structure of the international regulatory framework and discuss its main features, focussing in particular on regulatory capital. Before reviewing the precise rules, however, it might be helpful to discuss certain dilemmas in setting capital requirements.

HOW DETAILED AND RISK SENSITIVE SHOULD CAPITAL REQUIREMENTS BE?

Regulatory capital is intended to protect the stability of the banking system. *A priori*, it is not clear that the best way to do this is to align it with best practice in economic capital, for several reasons:

- First, if regulatory capital is highly risk sensitive, then capital requirements will rise in a given country as the economy turns into recession and default probabilities rise. This may cause banks to rein in lending, and hence make recovery from recession more difficult. Risk-sensitive capital requirements are by their very nature *procyclical*, and this may not be in the best interests of the economy as a whole.

- When regulators prescribe the details of risk-based capital requirements, they reduce the diversity of behaviour in the financial system. If every bank uses the same kind of model, they will all have an incentive to act the same way, potentially intensifying asset price bubbles and market crashes. [This phenomenon is discussed further in Chapter 9.]

- Detailed capital rules also stifle innovation in capital modelling since for them to be effective, regulators must continually revise capital rules to deal with the latest products and with market changes.

- However, if regulators do not give a detailed prescription of how capital is to be calculated, the same risk can receive a very different capital allocation in different banks, possibly resulting in competitive distortions.

- Finally, the higher the capital requirements, the safer the firms are, but the fewer firms there are since barriers to entry are high. Therefore, high capital requirements concentrate risk in a few, very large players. If one of these banks were ever to fail, the systemic impact would be enormous.

Thus arguably risk-sensitive capital requirements are not optimal from a systemic risk reduction perspective, and aligning economic capital with regulatory capital is inconsistent with having a level playing field between all regulated firms.

Moreover, higher capital requirements beyond a certain level do not necessarily give rise to a safer banking system:

- They encourage risk to leave the regulated banks for unregulated players not subject to capital requirements.

- They encourage bank consolidation leading to a lack of diversity in the banking system and increased systemic risk.

The right balance here is unclear.

7.1 REGULATORY CAPITAL AND THE BASEL ACCORDS

Regulatory capital forms what is nominally one of three key parts of the international regulatory framework for banks: the other two are *prudential supervision* and *disclosure*. Since regulatory capital requirements are at least to some degree public and meeting them is a major cost for banks, the section of the framework relating to capital requirements tends to attract the most attention, so we begin with that.

7.1.1 Before Basel II: Basel I and the Market Risk Amendment

The need for international capital requirements and the agreement of Basel I by the BCBS* has been discussed earlier [in section 1.5.1]. Here we consider the nature of these early capital rules and the subsequent Amendment to them for market risk.

7.1.1.1 The nature of Basel I

Basel I was the first set of capital requirements for all internationally active banks: it set crude capital charges for credit risk. Despite its simple nature, the magnitude of the achievement of Basel I should not be underestimated: an international agreement was made by all the countries which (at the time) had large banks active outside their home countries, and this agreement was sufficiently robust that no country subsequently withdrew. If this

---

* The definitive source for BCBS regulations is the BIS website, www.bis.org. Further discussion of regulatory capital can be found in a wide range of sources including Donald Deventer and Kenji Imai's *Credit Risk Models and the Basel Accords* and Chris Matten's *Managing Bank Capital: Capital Allocation and Performance Management*. Regulatory capital is a topic where both local implementation and detailed requirements can change rapidly, so a regulator's website such as www.fsa.gov.uk can also be a useful source of information.

situation is compared with the morass of international tax policy, perhaps supervisors can be forgiven for a certain pride in the international regulatory capital framework.

Basel I—and each subsequent Basel Accord—is an outline, not a law. National regulators have to make and enforce local regulations to implement the Accords, and that inevitably introduces differences between jurisdictions both in the details of the local rules themselves and in how those rules are applied within firms. In the EU, the process is even more complex as the Accord is first expressed as an EU directive then enacted in local law by regulators in each EU country. Thus, the Accords themselves should be thought of as a capital framework requiring local interpretation rather than a detailed set of prescriptions.

7.1.1.2 The Market Risk Amendment and internal models

One of the reasons that the 1988 Accord worked was that it covered most of the risks of most banks at the time: in the 1980s, credit risk was the dominant risk class in banks. By the early 1990s, however, banks were becoming more important players in the capital markets and those markets themselves were larger and more liquid. This meant that market risk was becoming a more significant risk in banks. Hence, there was a need to extend the Accord, and so in 1996 the Market Risk Amendment was agreed upon by the BCBS.

The 1996 Amendment gave firms the possibility of choosing one of several alternatives for calculating capital for market risk. All firms could choose to calculate capital using (one of a set of) *standard rules*. These were simple to apply and fairly crude but they tended to produce high capital requirements. Banks which wanted a cheaper and more risk-sensitive approach had to apply for permission to use an *internal model* such as VAR.* If a bank met the required standards in a number of areas, its supervisor could grant this permission, typically resulting in a significant capital reduction for the bank. The approach therefore incentivised banks to meet the standards required for the use of internal models.

The figure below shows the relationship between the risk sensitivity of the two approaches versus the requirements that have to be met.

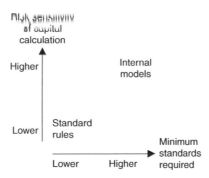

---

* The Market Risk Amendment does not *require* that for an internal model to be acceptable for market risk capital purposes it must be a VAR model, still less that it be a specific type of VAR model such as historical simulation. Nevertheless, the vast majority of internal models used for capital purposes are historical simulation VAR models.

There is some evidence to suggest that the pre-VAR market risk capital regimes of some regulators based on scenario matrices captured risk rather better than VAR models do. VAR is now so ubiquitous, though, that even if a firm believed that a different approach was more appropriate, the communications overhead of convincing regulators, ratings agencies, equity analysts and investors of that would be so large that it is unlikely they would judge the benefits worth the extra cost. VAR is now the *de facto* market risk standard.

Note that a more risk-sensitive capital charge does not *necessarily* imply a lower capital charge: typically, it will be lower for less risky assets but potentially higher than the standardised approach for very risky ones. Therefore, the ability to apply to use an internal model sets up the perverse incentive for more advanced banks—meeting the high minimum standards—to hold less risky positions, whereas less advanced banks using standard rules hold riskier positions. This also occurs in Basel II.

7.1.1.3 The internal models requirements of the Market Risk Amendment

The minimum standards in the Market Risk Amendment fall into three rough groups. Some of them relate to the control environment around the use of the model such as:

- The need for appropriately staffed and competent independent oversight of risk including senior management involvement in risk management;
- Appropriate control of valuations and of models [discussed in sections 1.3.4, 1.5.3 and 2.6] including the management of model risk and the use of mark adjustments where appropriate;
- External validation, perhaps by auditors, of the firm's internal risk model and its control environment;
- Policies and procedures ensuring that only positions taken with the intent of trading are booked in the trading book.

The model itself must also meet certain criteria:

- An estimate of loss at 99% confidence must be computed daily for the entire trading book portfolio.
- In calculating the loss, a 10-day holding period must be used. Banks may use loss estimates calculated using shorter periods and scaled up to 10 days by the square root of time [see section 2.1.5].
- A minimum of 1 year's data must be used to calibrate the model, updated at least quarterly. Where weighted data are used, such as in EWMA, the average time series length must be at least 6 months.
- To the extent that this is a material risk, firms must capture the non-linear characteristics of options positions and 'are expected to move towards' the application of a 10-day price shock.* Further, firms' risk management systems must have a set of risk factors which capture vega risk.†

---

* This would seem to de-bar the use of a simple delta VAR model for any bank with significant options risk. Whether a delta/gamma approach would be acceptable is unclear, but clearly full revaluation is a better technique.

    The requirement to move towards modelling a 10-day holding period rather than simply multiplying a 1-day VAR by the square root of 10 is another reason many banks prefer the historical simulation technique.

† Notice that this comes very close to saying that volatility must be a risk factor in the model without actually enforcing this requirement. It may be that some firms could persuade some supervisors that they are meeting this requirement without having a volatility risk factor. This is potentially a significant issue as the inclusion of a volatility risk factor can materially increase the VAR and hence the capital requirement.

Finally, additional requirements relate to adjuncts to internal models designed to highlight certain deficiencies:

- Regular comprehensive stress tests are conducted on the basis of both scenarios designed by the firm to capture adverse characteristics of the portfolio [as discussed in section 2.2.5] and scenarios supplied by supervisors.

- Firms must have information available on the largest losses experienced in the current period for review by supervisors.

- Backtesting must be carried out and backtest exceptions reported to the supervisor [as discussed in section 4.1.4].

The capital requirement for the market risks covered by the model is then the higher of:

- The previous day's VAR estimate; and

- The average of the daily loss estimates over the last 60 days multiplied by a factor which is typically 3.

This factor is known as the *VAR multiplier*. Since the square root of 9 is 3, the effect of this is to scale the VAR up to a 90-day 99% risk estimate.

7.1.1.4 Backtest exceptions

The VAR multiplier may be larger than 3 if a supervisor has significant concerns about the internal model or the control environment in which it is used, or if there are too many backtest exceptions.

There are a number of reasons that there might be a significant number of backtest exceptions:

- The positions used in the VAR model are not correct.
- The sensitivity of the positions to the risk factors used in the model is not correct
- The data series used either directly (in historical simulation) or to calculate volatilities and correlations (in var/covar approaches) are not correct or are not being used correctly.
- Clustering of extreme market moves, breakdown of correlation assumptions, or other forms of market movement not anticipated by the model.
- Loss making intra-day trading.
- Bad luck.

Clearly, supervisors will have more concern about causes towards the top of the list than those towards the bottom. Typically, here there will be a requirement to address the deficiency perhaps with the imposition of an additional capital requirement.

The Market Risk Amendment identifies three ranges based on the number of annual exceptions observed using the 99% 1-day VAR. We would expect in a 250-day year to see an exception roughly every 100 days, i.e., two or three per year.

- The *green zone* corresponds to between zero and four backtest exceptions per year. No further action is required here.

- A VAR model is said to be in the *yellow zone* if it has between five and nine annual backtest exceptions. Here the VAR multiplier is increased as in the table below.

- Finally, if there are 10 or more exceptions, the model is in the *red zone*. This requires a VAR multiplier of at least 4, and in practice, regulators are likely to be unwilling to let a firm use a model that is in this zone.

No. of Exceptions	5	6	7	8	9
Minimum multiplier	3.4	3.5	3.65	3.75	3.85

Part of the reason for this incremental treatment of more and more backtest exceptions is that a perfectly good 99% 1-day VAR model can produce significantly more than three backtest exceptions just due to statistical variation. Thus, there are four kinds of judgement we can make in reviewing the backtest exceptions of a VAR model. Suppose our hypothesis is that the model is correct. Then we have

	Model Is Correct	Model Is Not Correct
We decide the model is correct	Correct judgement	False negative
We decide the model is incorrect	False positive	Correct judgement

Here by 'correct' we mean that the model does correctly calculate the 1-day 99% VAR for the bank's portfolio. [The false-positive versus false-negative problem is similar to the one we discussed with the discrimination of internal credit ratings models in section 5.4.5.]

Unfortunately, even five backtest exceptions a year give a significant probability of a false positive: roughly 10% of correct models will generate five or more exceptions in a 1-year period. We simply are not getting enough information to accept or reject a VAR model on the basis of the number of exceptions to 99% 1-day VAR in 1 year:* our judgement will often be 'not proven'. One solution [as mentioned in section 4.1.4] is to look at the number of exceptions at different confidence intervals simultaneously. This is not, however, a requirement under the Market Risk Amendment.

7.1.1.5 The model review process
Supervisors typically make a decision on whether to grant a firm permission to use a model for capital purposes based on a range of information.

- First, a firm might have to submit a comprehensive application, containing details not only of the VAR model and its control framework but also of the firm's risk management organisation; processes and systems; the products traded; market risks taken and how these are measured; valuation and reserving policies; model risk control procedures; risk appetite; and limit structure.

---

* See Phillipe Jorion's *Value at Risk: The Benchmark for Controlling Market Risk* and Appendix 10 to the Basel Committee on Banking Supervision's *International Convergence of Capital Measurement and Capital Standards. A Revised Framework, Comprehensive Version* for more details.

- The supervisor will then typically conduct one or more review visits. It may also commission others such as auditors or third-party experts to review specific aspects of the model.

- Clarification on various aspects of the firm's risk control environment may be requested, and the firm may be encouraged to make various changes.

- Finally, the supervisor will make a decision on whether to grant permission to use the internal model for calculating capital on GMR alone or GMR and SR. If permission is granted, the supervisor must decide whether the VAR multiplier imposed should be larger than 3 and whether to place any limits on the products the model can be used for. If it is denied, typically a list of changes required before any re-application will be given to the firm.

- Permission to use the model will often include ongoing requirements, such as the disclosure of backtests. It may also require notification of new products, locations or markets; additional risk reporting; or other constraints.

> *Exercise.* If you work for a firm with permission to use an internal model for calculating market risk capital, see if you can obtain a copy of the application. What market risks can the model be used for, and what risks are excluded from the internal models approach?

7.1.1.6 General market risk and specific risk internal models
The market risk capital requirement for a position is composed of capital required for GMR and capital required for SR. Firms have the choice of:

- Using standard rules for both;
- Using an internal model for GMR and standard rules for SR;
- Using internal models for both GMR and SR.

To understand the trade-off, it is worth reviewing the SR rules.

7.1.1.7 Standard rules for specific risk
Where SR capital is not calculated using an internal model, the capital requirements below apply.

For debt securities and related positions,* the charges are as shown in the table below.

Comparing these with the Basel I charges for credit risk [discussed in section 1.5.1], they are rather similar: there are a few more risk categories, and a slightly wider range of

---

* These are the post-Basel II charges for specific risk, and they apply to bond positions and similar debt markets risks. Thus, being long a bond return via a total return swap will generate a specific risk charge just as if the firm held the underlying bond. See the Basel Committee on Banking Supervision's *International Convergence of Capital Measurement and Capital Standards. A Revised Framework, Comprehensive Version* for the details of the rules and their application.

charges, but it is clearly still a crude and relatively expensive framework for measuring issuer-SR.

Categories	External Credit Assessment	SR Capital Charge for Debt Securities
Government	AAA to AA−	0%
	A+ to BBB−	0.25% (residual term to maturity 6 months or less)
		1% (6–24 months)
		1.6% (more than 24 months' residual maturity)
	BB+ to BB−	8%
	Below BB−	12%
	Unrated	8%
Qualifying securities[a]		0.25% (residual term to maturity 6 months or less)
		1% (6–24 months)
		1.6% (more than 24 months' residual maturity)
Other securities[b]	AAA to AA−	1.6%
	A+ to BBB−	4%
	BB+ to BB−	8%
	Below BB−	12%
	Unrated	8%

[a] Qualifying securities include those issued by public sector entities and multilateral development banks, plus other securities plus (roughly) those rated investment grade or of comparable credit quality and listed on a recognised exchange.
[b] Strictly these are the charges for other corporate bonds: see BCBS *op. cit.* for further details.

> *Exercise.* Compare these capital charges with the credit spread volatility of an average bond in each rating category. If the SR capital charge for a bond is not for credit spread volatility, what is it for?

For equity securities the rules are even more crude as shown in the table below.

Category	SR Capital Charge for Equity Securities (%)
Liquid and diversified equity portfolios (typically index positions)	4
Others	8

7.1.1.8 Internal models for specific risk

The trade-off between the cost of modelling and capital requirements discussed earlier clearly pertains here: a firm that does not want to pay the heavy capital charges for SR under standard rules has the option of applying for an SR VAR model. But these models are complicated. Even after respecting regulatory requirements about the number of risk factors for interest rate risk, FX risk, equity risk and so on, a GMR model will typically only have tens or at worst low hundreds of risk factors depending on the bank's balance of business and the precise design chosen. An SR VAR model, in contrast, might have risk factors corresponding to each equity and credit spread in the portfolio, so the task of building such a model is at least an order of magnitude more difficult.*

In addition to the requirements for GMR internal models, there are additional regulatory hurdles a firm must jump before permission can be granted to use an internal model for SR:

- The model must explain the price variation in the portfolio—the usual backtesting requirement—and it must capture issuer concentrations.

- The model must be robust despite an adverse market environment. Therefore, some demonstration of adequate model performance during credit spread widening or an equity crash or both will be required.

- The model must fully capture name-related risk, so where proxies are used due to bad data quality or missing data, the firm must be able to show that the approximations introduced are immaterial or conservative;

- Finally, where a firm is subject to event risk that is not reflected in its VAR number, perhaps because it is a high-severity, low-frequency event [as discussed in section 6.2.1], the firm must ensure that it is factored into its internal capital assessment, for example, through the use of stress testing.

The decision regarding whether to apply for an SR model should be driven by business considerations: is the capital saving possible worth the extra cost of the model and the increased supervisory burden? If there is no or little reputational impact from not having an SR model, some firms may well decide that the game is not worth the candle. This is especially the case since there is now an additional requirement for the use of internal models: firms must demonstrate that they meet a soundness standard comparable to that for the internal ratings-based approach for credit risk in Basel II.

7.1.1.9 Example calculation of specific risk capital

Suppose a firm has permission to use a VAR model for calculating capital for GMR but not SR. We give a simple example of how a position contributes to the capital charge. Consider a 5-year S&P 500 call sold to a client:

- The position together with its hedges—perhaps long S&P 500 futures—will go into the firm's VAR model and will contribute to the VAR calculated using the equity risk factor. Assuming that the position is roughly delta hedged, most of the risk calculated this way will come from gamma.

- The position will also have interest rate risk since the call has rho risk. Unless it is fully hedged, this gives rise to an interest rate VAR.

---

* The Market Risk Amendment is notably unclear on the number of data series required for a specific risk VAR model to be acceptable. For instance, some national supervisors might view a model which has one series of bond returns per rating as acceptable; some might permit it for firms with simple trading activities but not for more complex firms; others may demand more risk discrimination from any applicant.

- There is also vega risk. If the firm uses a volatility risk factor, for instance using the VIX as a proxy for all volatility risk, the option will make a contribution here too.

- The model aggregates these contributions, and so the position contributes to the overall firmwide VAR.

- Finally, the position plus its hedges will contribute to the firm's SR position on the S&P 500. There is a 4% SR charge on the net position here.* The total SR charge is added to the GMR charge giving risk to the firm's total market risk capital requirement.

> *Exercise.* If you can, find a complete regulatory return for one period. Try to understand what each capital charge is for and how it was calculated. See how many different charges are imposed on a book you are familiar with and what risks each of them relates to.

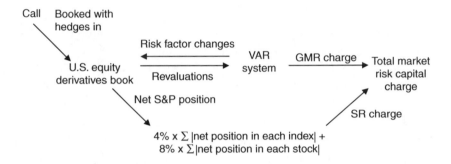

7.1.1.10 Dialogue

The head of risk management, Dr. R. Careful is having a quiet beer with the head of the legal and compliance department, Pru Lawyer.

'We have this new VAR model I've been meaning to tell you about. I want to schedule a trip down to see the regulator about it'.

'Oh yes? What's it do?'

'It's the work of that new hire of mine, got his doctorate in statistics. Extreme value theory. Very interesting stuff'.

'So this is a better model?'

'Definitely. It's based on the tails of the distribution. Gives us a much more accurate capital assessment'.

'High or lower?'

'Quite a lot higher actually'.

'Are the regulators making us use it?'

'No. They have no idea we have it at the moment'.

---

* Strictly, the ability to net the delta of the option with that of its hedges may require regulatory approval to use the firm's equity derivatives model for this purpose.

'Well for God's sake keep it quiet then. The last thing we need is that lot of hard arses down here thinking we've been underestimating capital for the last 5 years'.

'But it's a big leap forward. This is cutting edge technology'.

'We don't make leaps we don't have to, especially when they increase capital. The regulatory standard is what everyone else is doing. Just don't step out of line and no one gets hurt, OK?'

7.1.2 An Overview of Basel II

By 1999, the Basel I rules for credit risk had been arbitraged by some banks. Increasing liquidity in securitised credit risk allowed many institutions to reduce capital requirements by pooling assets such as loans, securitising them and selling the senior tranche (and so reducing capital) but keeping much of the risk by retaining the equity tranche. Something needed to be done. It was in this context that the BCBS met to develop a replacement for the 1988 Accord.

7.1.2.1 Aims and application of the new Accord

At the time Basel II was originally proposed, the supervisors' stated objectives were:

- To revise the 1988 Accord in a way that would strengthen the soundness and stability of the international banking system;

- To enhance competitive equality among internationally active banks;

- To promote the adoption of stronger risk management practices within internationally active banks;

- To maintain the overall level of capital within the international banking system (while accepting that there could be winners and losers within that system).

Some observers have suggested that, in addition, some supervisors had the objective of significantly increasing the capital charges for securitisation and other credit derivatives transactions in order to provide a robust barrier to the capital arbitrage provided under Basel I (by tranching, selling the senior, keeping the rest, and hence reducing capital without significantly reducing risk).

It is also worth nothing the prevalence of the term 'internationally active bank' in these objectives. Basel II was *not* initially designed for non-banks such as broker/dealers, nor was it focussed on the needs of smaller banks or banks primarily operating in a single country.

7.1.2.2 A range of approaches

Following the success of the system of alternatives provided in the 1996 Market Risk Amendment, Basel II also offers a similar series of choices: firms can choose to use simpler ways of calculating capital with lower entry requirements but likely larger capital requirements; or to apply for permission to use a more advanced approach which is more risk sensitive and will probably produce a lower capital requirement, but which requires proof of a certain standard of risk management.

7.1.2.3 The three pillars

Basel II defines a framework for bank supervision on the basis of three complementary safeguards:

- *Pillar 1*: A new regulatory capital framework.
- *Pillar 2*: The exercise of supervisory discretion. This requires regulators to visit the firms they supervise, gather information, assess the firm and, on the basis of that assessment, consider action including placing restraints on the firm's business, requiring control improvements or increasing capital requirements.
- *Pillar 3*: Disclosure. Firms will be required to publicly disclose more information than hitherto, presumably allowing the equity and debt markets to better understand and price the firm's risk profile.

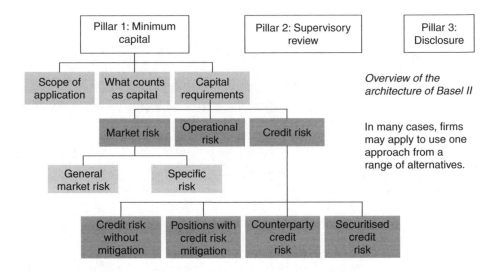

Overview of the architecture of Basel II

In many cases, firms may apply to use one approach from a range of alternatives.

7.1.2.4 Key features of the new Accord

Basel I had a very crude treatment of credit risk in the banking book, so it is no surprise to find that the centrepiece of Basel II is a more complex menu of approaches to credit risk. Other features include:

- Capital treatments of positions where credit risk is mitigated, including those involving lending against collateral, credit derivatives and guarantees;
- A revised capital treatment for securitisations;
- Capital requirements for operational risk.

7.1.2.5 The Accord development process

It is important to understand how Basel II developed in trying to understand the result. The process has been a long one and it has involved two groups, neither particularly united nor internally consistent: the supervisors and the banks.

The supervisors had somewhat disparate objectives partly because their constituencies are different. For instance, some European supervisors were aware that the Accord would be applied in the EU to a wider range of firms than simply internationally active banks; therefore, they were seeking a framework that did not introduce excessive distortions to that wider group of firms. Other regulators did not have that problem, and some were primarily concerned with retail and commercial banks reflecting the balance of activity in their jurisdiction. Finally, as ever in international negotiations, issues of competitiveness and the protection of national champions may not have been entirely absent.

During the process of developing Basel II, the BCBS conducted a number of quantitative impact studies (QIS). Here banks were invited to calculate capital using the new proposals. The rules were revised on the basis of the results of these calculations. Moreover, at various stages, comments from interested parties were sought, and these influenced the proposals too. The QIS and comment processes gave rise to significant changes in the Accord between the first proposal in 2001 and the final version; it is also possible that there will be further changes based on the effects of Basel II after implementation.

Firms almost certainly had different objectives in writing their comments on the proposed new Accord. Some may have sought the changes that resulted in the lowest capital requirement for themselves. Others took a broader view considering the health of the market or their client base as a whole. Where a number of large banks took the same position, they often had some success in altering the proposals. A good example here is interest rate risk in the banking book. As we discuss in section 8.2, banking books sometimes contain significant interest rate risk. The early draft proposed capital charges against this risk. Most large banks were opposed to this—not least because it would have significantly increased their capital requirements—and they successfully lobbied for these proposals to be removed.

> *Exercise.* Many firms' comments on Basel II are available on the BIS website. Read some of them and take a view of why the firm is saying what it is.

7.1.3 Basel II: Credit Risk without Mitigation

We begin an overview of the Basel II proposals with the simplest approach to calculating capital for credit risk in the new Accord: the *standardised approach*. A broad summary of the main features of this approach is given before we move on to a review of the methods for the calculation of capital requirements in the more advanced approaches.

7.1.3.1 The standardised approach to credit risk

Basel I had three risk buckets: free (0% risk weight); cheap (20% risk weight, i.e., a capital charge of 1.6% of notional); preferred (50% risk weight used for derivatives credit exposure and mortgage lending and producing a capital charge of 4% of notional); and standard (100% risk weight, i.e., 8% of notional). The standardised approach in Basel II has a few more categories.

For exposures to sovereigns, risk weights are assigned as shown in the table below. Here the credit rating is given in terms of the S&P system, but there is no implication that those are the only acceptable ratings: Basel II has the notion of an external credit assessment institution (ECAI), and a range of ratings agencies and other bodies are acceptable as ECAIs.

Claims on Sovereigns	
Credit Rating	**Risk Weight (%)**
AAA to AA−	0
A+ to A−	20
BBB+ to BBB−	50
BB+ to B−	100
Below B−	150
Unrated	100

Various multinational banks and other institutions such as the ECB, the European Community, good-quality multilateral development banks, including the well-known issuers IBRD, EIB and NIB, and of course the BIS are also explicitly 0% risk weighted.

There are two alternatives for the capital treatment of credit risk to banks, with the choice between the alternatives being made by the national supervisor and applied uniformly to all banks in their jurisdiction.

The first alternative simply rates banks one category lower than the sovereign of their incorporation, except for banks in countries rated BB+ to B− or not rated, which attract a 100% risk weight.

Claims on Banks: Second Alternative		
Credit Rating	**Risk Weight (Long Term) (%)**	**Risk Weight (Short Term) (%)**
AAA to AA−	20	20
A+ to A−	50	20
BBB+ to BBB−	50	20
BB+ to B−	100	50
Below B−	150	150
Unrated	50	20

The second alternative preserves some flavour of a Basel I distinction between short-term and long-term obligations from banks, as shown in the table, although 'short term' now means 3 months or less, and 'bank' includes broker/dealers provided they are supervised in a 'comparable' framework to Basel II. (Recall that undrawn LOCs under 1 year were 0% risk weighted in Basel I, making 364-day revolvers a preferred way of extending credit for many banks under the 1988 Accord.)

The standardised approach includes a different treatment for claims on corporates to those on other banks. The risk weights are as given in the table below.

Claims on Corporates	
Credit Rating	**Risk Weight (%)**
AAA to AA−	20
A+ to A−	50
BBB+ to BB−	100
Below BB−	150
Unrated	100

Some observers have commented that the distinction between banks and corporates is odd in that the ratings agencies themselves make no such distinction: why should there be a different capital charge for lending to a BBB bank from that for lending to a BBB corporate if we trust the rating's agency assessment of BBB?

The standardised approach recognises *retail claims*, which are uniformly risk weighted at 75%. To qualify as a retail exposure, the lending must be to an individual or a small business; it must be one of a limited number of products such as credit card-related lending and personal loans; the portfolio must be large enough so that no single exposure to any counterparty is more than 0.2% of the total; and the maximum exposure to any counterparty is no more than one million euros.

Retail mortgage lending is another distinguished category of credit risk in the standardised approach: this attracts a 35% risk weight, whereas commercial mortgages are risk weighted at 100%.

There are further detailed rules in the standardised approach for the treatment of past due loans and off-balance-sheet items such as revolving LOCs. Some of these, such as repo transactions, are discussed below when we touch on credit risk mitigation. In general, though, the standardised rules are not most banks' main concern; instead, most large banks will use one of the more advanced approaches.

7.1.3.2 Overview of internal ratings-based approaches to credit risk

Basel II gives firms the right to apply for permission to use an internal model as part of the calculation process for credit risk capital. In fact, there are two possible routes: the foundation internal ratings-based (FIRB) approach; and the advanced internal ratings-based (AIRB) approach. However, unlike market risk, neither IRB approach allows the bank to use its model to produce a capital figure; instead, the firm's internal ratings are used to estimate the inputs to a fixed formula which is then evaluated to give the capital required.

The process for calculating capital within these approaches is as follows:

- The firm first uses its model to estimate certain *risk components*. These are PD, LGD, EAD and maturity. Broadly in the foundation IRB banks, PD is calculated for each exposure and fixed supervisory values are used for the other components, whereas in the AIRB approach, banks are permitted to estimate all the risk components themselves.

- These components are put into a supervisor-supplied *risk weight function* which varies with the type of underlying exposure (sovereign, corporate, etc.).
- The risk weight function is evaluated to give a capital requirement.

7.1.3.3 Exposure types within the IRB approach

The exposure types recognised in Basel II are:

- Sovereign;
- Corporate (with a separate treatment for small corporates and for various forms of specialised lending);*
- Bank;
- Retail (with a definition similar to that used in the standardised approach and with separate treatments for residential mortgages; revolving exposures such as credit cards; and other retail exposure);
- Equity (meaning here equity in the banking book, that is, long-term non-traded positions: trading books positions are treated under the market risk rules).

7.1.3.4 Risk components

The PD and recovery of a credit exposure have been discussed [in section 5.2]. The EAD is an estimate of the amount owed at default: for a loan, this might be notional plus accrued interest, whereas for an LOC, it would be an estimate of the amount drawn. The LGD is $(1 - \text{recovery})$ times the EAD.

The basic idea of IRB approaches is that a good internal ratings system will associate an exposure with a bucket of PDs. Thus a rating of 4 for a given loan might mean that, on the basis of the bank's due diligence, the bank estimates the exposure PD of this borrower as between 1.2 and 1.7% annualised.

Similarly, a sophisticated credit risk system will distinguish between a fixed exposure (as in a conventional loan) and a variable one (as in a partly drawn LOC). Therefore, some banks will have EAD information available on each transaction too.

The last risk component M is the maturity of the exposure.

7.1.3.5 Risk weights: corporate, sovereign and bank exposures

For simple corporate, sovereign and bank exposures, the capital charge K is given by the maximum of zero and the result of evaluating the formula below.

$$K = \text{LGD} \times \left[\Phi \left(\frac{\Phi^{-1}(\text{PD}) + \sqrt{\rho}\,\Phi^{-1}(0.999)}{\sqrt{1 - \rho}} \right) - \text{PD} \right] \times \frac{1 + (M - 2.5)b}{1 - 1.5b}$$

---

* The distinguished classes of specialised lending are project finance, object finance (lending against physical assets such as planes or trains), commodities finance (structured short-term lending to finance ET commodities) and two forms of lending against commercial property.

where

$$\rho = 0.12 \left[\frac{1 - e^{-50PD}}{1 - e^{-50}} \right] + 0.24 \left[1 - \frac{1 - e^{-50PD}}{1 - e^{-50}} \right]$$

$$b = [0.11852 - 0.05478 \ln(PD)]^2$$

Here for corporate and bank exposures, PD is the greater of the 1-year PD associated with the exposure or 0.03%: this floor does not apply for sovereign exposures. The PD is defined as 100% for an exposure which has already defaulted. Φ is the cumulative normal distribution function for a standard (i.e., (0, 1) distributed) random variable, and Φ^{-1} is the standard inverse cumulative normal distribution function. Finally, ρ should be thought of as a kind of default correlation: this has been calibrated to lie between 12 and 24% for corporate, sovereign and bank exposures.

The figure below shows the variation of the risk weight (i.e., capital charge/8%) with the PD of the obligator: an LGD of 45% and a maturity of 2.5 years is assumed. The retail formula, discussed below, is also illustrated. Remember that for foundation IRB banks, the bank determines PD: LGD is set at 45% for senior claims and 75% for subordinated, and M at 2.5 years. AIRB banks are permitted to estimate PD, LGD, EAD and M.

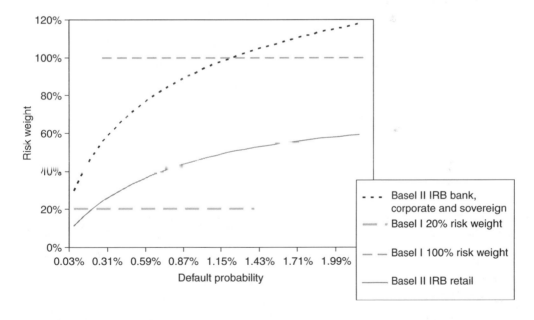

7.1.3.6 Motivations for the formula

The capital charge formula above is designed to measure the UL in a portfolio of credit risk: EL is supposed to be covered by loan loss or other provisions, and a different part of Basel II addresses the adequacy of these reserves. The formula was developed by the BCBS using a model of portfolio credit risk [related to the Merton model discussed in section 11.4.1] with one additional criterion: the capital charge for a portfolio should be the *sum* of the capital charges for the assets within it.

Therefore, the formula is designed to measure the *incremental* extra contribution to portfolio capital at a given confidence interval for a small additional exposure under the BCBS model. It is a large numbers' approximation, so there is an implicit assumption that the portfolio is large and diversified.

The formula can then be understood as[*]

$$K = \text{LGD} \times \underbrace{\Phi\left(\frac{\Phi^{-1}(\text{PD}) + \sqrt{\rho}\,\Phi^{-1}(0.999)}{\sqrt{1-\rho}}\right)}_{\substack{\text{Incremental contribution} \\ \text{to portfolio risk}}} - \underbrace{\text{LGD} \times \text{PD}}_{\text{Expected loss}} \times \underbrace{\frac{1 + (M - 2.5)b}{1 - 1.5b}}_{\text{Maturity adjustment}}$$

The BCBS decided on a confidence interval of 99.9% for credit risk, so the formula first calculates the contribution of the asset to the total UL at this confidence interval, and then subtracts the EL (which is LGD × PD). This result is then adjusted by a factor (typically close to 1) which adjusts for the maturity of the exposure: longer-term exposures are riskier than shorter-term ones.

As can be seen from the figure, the formula produces somewhat larger capital charges than Basel I for higher PDs. There is a concern that this might inhibit lending to smaller companies, particularly in jurisdictions where banking loans rather than the capital markets are a dominant form of corporate financing. Therefore, the committee introduced an adjustment to the formula for SMEs which reduce the capital charge somewhat.

7.1.3.7 Risk weights: retail exposures

There are three separate approaches for different types of retail exposure, all using variants of the corporate formula above.

- For mortgages, $\rho = 0.15$, there is no maturity adjustment, and so we have

$$K = \text{LGD} \times \left[\Phi\left(\frac{\Phi^{-1}(\text{PD}) + \sqrt{0.15}\,\Phi^{-1}(0.999)}{\sqrt{1-0.15}}\right) - \text{PD}\right]$$

- For qualifying revolving retail exposures, the same formula is used with $\rho = 0.04$, reflecting the fact that supervisors assume that the typical portfolio has more diversification.

- Finally for other retail exposure, again this formula is used with a correlation that varies between 0.03 and 0.16 (rather than between 0.12 and 0.24 for corporates) with ρ calculated using

$$\rho = 0.03\left[\frac{1 - e^{-35\text{PD}}}{1 - e^{-35}}\right] + 0.16\left[1 - \frac{1 - e^{-35\text{PD}}}{1 - e^{-35}}\right]$$

---

[*] See the BCBS' *An Explanatory Note on the Basel II IRB Risk Weight Functions* for more details.

For retail exposures, there is no distinction between the FIRB approach and the AIRB approach; in both cases, banks provide their own estimates of PD, LGD and EAD.

Notice how flat the retail curve is in the illustration compared to the corporate curve: Basel II requires comparatively little capital against retail exposures.

7.1.3.8 Risk weights: other exposures

There are further detailed rules for the calculation of capital for the specialist lending categories and for equity exposures. The equity rules are particularly complex, offering a range of alternatives each with their own floors, the possibility of national opt-outs, and extended transitional arrangements. This is an area where there seems to have been international tension: some countries have a long history of banks holding equity in their clients, and without the opt-outs some banks in these countries may have been faced with the unpalatable choice of selling their cross-holding or being capitally inadequate. On the contrary, in markets without this tradition of cross-holding, supervisors may have been unwilling to provide an incentive for banks to become more involved in providing venture capital.

7.1.3.9 Pre-processing within the IRB approaches

The exposure amount for various kinds of transaction is calculated within Basel using the idea of a credit conversion factor (CCF). This is just a percentage of notional which converts a transaction into an 'equivalent' loan. Thus, the CCF for most off-balance-sheet commitments is 75%. Uncommitted unconditionally cancellable lines, however, are 0% weighted: this is another echo of the old Basel I 0% weighting for short-term lines. One major change from Basel I, however, is the removal of the 50% CCF for OTC derivatives: these are now 100% weighted, although a models-based approach is permitted for the calculation of EAD instead of the simple PFCE add-ons of Basel I.*

7.1.3.10 Minimum standards required for the FIRB and AIRB approaches

Given the potential savings in capital available in the IRB approaches, particularly for retail exposures, it should be no surprise that supervisors have set fairly tough criteria for entry to the FIRB approach, and even higher ones for the AIRB approach. Broadly, these fall into the following headings:

- *No cherry picking.* Once a bank is permitted to use one of the IRB approaches for part of its holdings, it must extend it across the entire bank (although a phased roll-out is permitted).

- *Data requirements.* For the FIRB approach and for retail positions, the bank must use at least 5 years' data to estimate the risk components and to calibrate its assignment of internal ratings. For AIRB approach, at least 7 years' data are desirable.

---

* See Annex 4 of *International Convergence of Capital Measurement and Capital Standards. A Revised Framework, Comprehensive Version* (*op. cit.*) for more details.

- *Use.* Firms are not permitted to implement an IRB approach purely for regulatory purposes: the model must play a vital role in the lending decision, economic capital allocation and risk management.

- *Independent credit risk function.* Banks must have a group independent of the business with responsibility for the design, implementation and operation of the ratings system, and for the measurement of credit risk.

- *Stress testing.* A programme of stress testing credit exposures must be carried out which identifies plausible circumstances which could have a significant adverse effect on the bank's portfolio of credit risks.

- *Model validation and documentation.* A robust process must be in place to validate the assignment of PDs (and if applicable, other risk components) to individual transactions. The internal ratings model itself and the processes around its use must be fully documented and externally audited.

There are also certain quantitative criteria. These are designed to ensure that the internal ratings model provides a reliable and accurate measurement of credit risk: as discussed above, if a bank assigns a particular internal rating to an exposure, the supervisor needs to have confidence that it is a meaningful risk assessment. The model therefore must satisfy criteria relating to

- *Risk differentiation.* The model must have at least seven borrower grades for performing borrowers and at least one for defaulted borrowers. There must be no excessive concentrations of risk within these grades.

- *Separation of borrower risk and transaction risk.* PD measures the risk of borrower default. However, depending on transaction features such as seniority and covenants (and credit risk mitigation discussed below), the LGD may differ significantly between transactions with the same borrower. The system must reflect this.

- *Accuracy.* The process for assigning an exposure to a bucket (which in turn corresponds to a range of PDs) must have good predictive power, and similarly for the estimation of other risk components. A firm must have a process for assessing the statistical accuracy of the model.

There are no specific criteria concerning the methodology for assigning internal ratings, and indeed banks can use different approaches for different classes of exposures, for instance using scorecards for classifying retail credit card exposures, and a financial ratio-based model for large corporate loans [as discussed in section 5.4.5].

7.1.3.11 IRB approaches versus economic capital for credit risk

The diagram below illustrates the architecture many banks use to calculate regulatory and economic capital for credit risk.

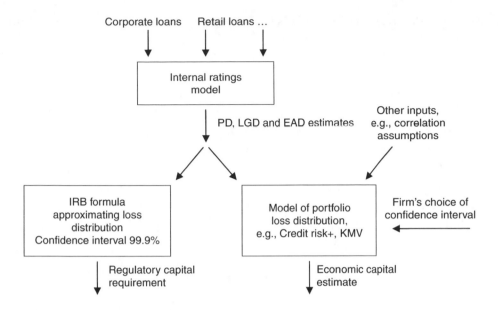

7.1.3.12 How accurate is the IRB formula?

Some banks will see a significant fall in the credit risk capital required between Basel I and the IRB approaches in Basel II. This has inevitably attracted the suggestion that the AIRB capital requirement is not sufficient to meet its stated aim: the estimation of capital required at a 99.9% confidence interval with a 1-year holding period.*

Some of the main issues that have been identified are as follows:

- The IRB formula was calibrated by the BCBS using correlations inferred from data supplied by banks. Thus, for instance, the endpoints of 0.12 and 0.24 for corporate ρ were set by the supervisors. It has been suggested that these are too low and hence capital requirements are understated.

- In particular, there is evidence that credit event correlations increase with PD: the IRB formula does the opposite.

- Moreover, history also suggests that there are systematic risk components in both LGD and EAD: LGD and EAD increase as the economy turns down, whereas the IRB formula simply calculates the incremental risk due to an exposure's PD. Of course, including this effect would have made the IRB approach even more procyclical than it already is.

7.1.4 Basel II: Credit Risk Mitigation in the IRB Approaches

Basel II uses the term 'credit risk mitigation' to refer to any situation where the presence of another instrument, such as a guarantee, a credit derivative or a collateral, reduces or may

---

* See, for instance, Paul Kupiec's *Basel II: A Case for Recalibration* (presentation at the 2006 FDIC-JFSR Basel II Symposium).

reduce credit risk. In general, Basel II is fairly tough on these situations, even within the IRB approaches.

7.1.4.1 The double default framework

If we have a loan to a corporate and buy a credit derivative of the same maturity from a good-quality counterparty, there is a natural tendency to assume that we are hedged. Basel II does not see things that way. Instead, there is the notion that this position—and any others where a bank has protection under a credit derivative or a guarantee written by a third party against a credit risk—is dealt with in the *double default framework*. This gives some benefit for buying protection, but not a complete elimination of the capital requirement.

First, a capital requirement K_0 for the underlying exposure is calculated using the IRB formula but using the LGD of the guarantor. The capital requirement for the position including the credit risk mitigation is then

$$K_0 \times (0.15 + 160\mathrm{PD_G})$$

Here $\mathrm{PD_G}$ is the probability of default of the guarantor. There is thus a maximum benefit for hedging of 85% even when the guarantee is written by a risk-free counterparty.

7.1.4.2 Collateral overview

In the IRB approach, taking various forms of collateral against credit exposure, gives some reduction in capital requirements. This *IRB eligible collateral* includes:

- Cash and gold;
- Certain debt securities including those rated at least BBB− (or slightly lower for sovereigns);
- Equities listed on a main index;
- Certain investments in UCITS;
- Certain investments in both residential and commercial property.

(The approach is similar in the standardised approach but without property being eligible.) There are two routes for calculating capital: in the first, capital is calculated by first adjusting the exposure amount for the effect of collateral in a crude way, whereas in the second, a models-based approach is permitted.

7.1.4.3 The simple approach to collateral

Here the BCBS has defined the effective exposure as

$$\max[0, E \times (1 + H_e) - C \times (1 - H_c - H_{fx})]$$

Here E is the exposure before collateral, H_e the haircut appropriate to the exposure, C the value of the collateral, H_c a haircut applied to the collateral and H_{fx} an additional haircut if the collateral is denominated in a different currency to the exposure. The real shock is the size of the haircuts: some of them are shown in the table below.

Collateral	Residual Maturity of Exposure	H_c (%) Sovereign Issuers	Others
Cash	Any	0	
AAA to AA− bonds	≤1 year	0.5	1
	1–5 years	2	4
	Over 5 years	4	8
A+ to BBB −	≤1 year	1	2
	1–5 years	3	6
	Over 5 years	6	12
Equities	Any	0	

The credit risk mitigation framework applies not only to lending versus collateral in the banking book, but also to repo transactions, prime brokerage and taking collateral against OTC derivatives exposures.

Therefore, banks are strongly incentivised to move towards one of the models-based approaches.

7.1.4.4 Models-based approaches to collateral

Here, subject to meeting certain standards, banks are permitted to estimate their net exposure to a collateralised exposure to a counterparty as follows:

- The bank must simulate the simultaneous migration of the value of the exposure and the collateral using a VAR model broadly meeting the standards required in the 1996 Market Risk Amendment.

- An exposure at a 99% confidence interval is estimated with different holding periods depending on the type of transaction as shown in the table below, and with particular operational requirements relating to the monitoring and management of exposure.

- Banks must account for the liquidity of lower-quality collateral within their model, and monitor and prudently manage 'wrong way risk'.*

Type of Exposure	Holding Period	Operational Requirement
Repo and reverse repo	Five business days	Daily remargining
Other capital markets (e.g., OTC derivatives exposure and margin lending)	Ten business days	Daily remargining
Secured lending	Twenty business days	Daily revaluation

This approach typically produces lower capital requirements than the simple approach, but it still threatened to cause considerable damage to the secured lending market since activities such as repo used to attract very low capital requirements under Basel I. Therefore, after a

---

* 'Wrong way' risk refers to the situation where collateral value is highly correlated with counterparty creditworthiness. [See section 5.2.4 for a further discussion.]

considerable volume of comments, the BCBS acknowledged the importance of the repo markets via the *repo carve out*. Where a repo is with *a core market participant*—sovereigns, banks and securities firms, mutual funds subject to capital requirements and pension funds—and where the collateral qualifies for a 0% risk weight in the standardised approach, the transaction may qualify for a 0% risk weight. Additional requirements for the carve out include:

- Daily mark-to-market and daily remargining;
- A short interval between failure to post margin and liquidation of collateral;
- Standard documentation of the transaction and a legally enforceable right to perfect the collateral.

At least within this safe harbour, the repo market is safe.

7.1.5 Basel II: Capital Rules for Positions That Have Been Securitised

After the range of alternatives discussed in the previous sections, it should be no surprise that there are several possibilities for securitised positions including a standard rules-like method known as the *ratings-based approach* and an IRB-style approach.

The risk weights for the ratings-based approach to securitised positions where applicable external ratings are available are broadly presented in the table below. Again, notice the implied distrust of ratings discussed earlier: an ordinary loan to a corporate rated BB− attracts a 100% risk weight, yet if a bank securitises its loan portfolio and retains a tranche rated BB−, it attracts a 650% risk weight. Not-rated tranches fare even worse: these constitute a *deduction from capital*, so if a bank securitises $1B of corporate risk and retains a $50M first-loss tranche, it must assign a full $50M of capital against it.

For most large corporate loan securitisations where the first loss tranche is a few percent of the total notional, this tends to discourage the common capital arbitrage trade of selling the rated tranches and retaining the first loss tranche: for other assets though (such as credit cards) where the rated tranches have much lower attachment points, capital arbitrage is still possible.

External Rating	Risk Weight for Most Senior Position (%)	Base Risk Weight (%)	Risk Weight for Non-Diversified Pools (%)
AAA	7	12	20
AA	8	15	25
A+	10	18	35
A	12	20	35
A−	20	35	35
BBB+	35	50	50
BBB	60	75	75
BBB−		100	
BB+		250	
BB		425	
BB−		650	
Below BB− and NR		1250	

The IRB approach has similar properties: the key idea used here is K_{IRB}, the amount of capital that the pool would have used before it was securitised. Suppose this is \$40M on a \$1B pool: then retained tranches below \$40M are 1250% risk weighted, and those above are subjected to a supervisory formula. This formula assigns a minimum risk weight of 7% even for supersenior exposures as in the ratings-based approach.

> *Exercise.* Does it increase or reduce systemic risk if large amounts of credit risk leave the banking system? What types of credit risk do the Basel II rules on balance encourage to leave the system?

7.1.6 Implications of the Basel Credit Risk Framework

The Basel II credit risk framework will dramatically change the capital required for some institutions' loan portfolio. This is bound to affect banks' behaviour, both positively and negatively. The following business segments seem likely to be affected in one way or another.

7.1.6.1 Lending to SMEs

One of the advantages of a non-risk-sensitive capital framework such as Basel I is that it does not discriminate against lower-quality borrowers. Basel II will make it more expensive for banks to lend to corporates with higher estimated PDs, and hence these firms are likely to find their borrowing costs increase. One area of particular concern is SMEs. Although the ability to use a revised risk weight function for SMEs goes some of the way to addressing these concerns, it is likely that smaller companies will be one of the losers in the new Accord.

7.1.6.2 Emerging market lending

Similar arguments apply to emerging markets, especially given that sovereign ceilings are built into certain parts of the Accord. Indeed arguably, at least from the perspective of having the best functioning banking system for their local economies, some countries with emerging markets might be best served by *not* implementing Basel II.* Politically, however, that would be problematic, so almost all countries have signed up the implementation of Basel II albeit in some cases with a delayed implementation schedule.

7.1.6.3 Banks' place in the venture capital market

Banks do not always compete with other banks; indeed, it is only in deposit taking that they have a unique place, and that is typically not particularly profitable. A good example of this is the provision of venture capital funding. During the technology boom of 1997–2000, some banks became more active in this area partly in the hope of winning lucrative investment banking business as the firms they funded went public. However, Basel II gives a considerable capital burden to holding equity in the banking book. This may well outweigh the funding advantage banks have versus non-bank venture capital players. Certainly, some banks appear to be retreating from venture capital as witnessed, for instance, by the spin-off of JPMorgan Partners in 2005.

---

* See Giovanni Majnoni and Andrew Powell's *Reforming Bank Capital Requirements: Implications of Basel II for Latin American Countries.*

7.1.6.4 *Smaller and less sophisticated banks*

Capital adequacy can be a competitive advantage. Consider a large bank with permission to use the AIRB approach for credit risk and a more advanced approach for operational risk. A smaller bank competing with it might only be on the FIRB approach and using a simpler approach for operational risk. Thus, the larger bank not only probably has a lower cost of funds but also has a lower capital requirement for many positions. Therefore, it can afford to undercut the smaller bank on good credit risks (i.e., those where the AIRB approach gives a lower capital charge).

This also extends to acquisitions: if the larger bank buys the smaller one, it can incorporate the target's book into its own, and so reduce capital requirements.

> *Exercise.*   In the previous chapter, we indicated that regulatory capital was a constraint rather than a scarce resource, so is banks' behaviour really changed by anything except the grossest changes in regulatory capital requirements? Does Basel II actually matter that much if most banks were capitally adequate before it and remain adequate afterwards?

7.1.6.5 *Pricing of liquidity lines*

Basel II has an extensive set of rules covering liquidity support. These are situations where, for instance, a bank provides an LOC to a securitisation SPV on a short-term basis to cover cash-flow mismatches between income on the collateral pool and coupons due on the tranches. Here the nature of the risk depends on the detailed form of the arrangement: a line that has first claim for repayment on any dollar coming off the collateral is a pure liquidity facility, whereas one that is not so senior may involve implicit credit support. Basel II has penal charges for lines which are not pure liquidity, focussing banks' attention on the risks here. [See section 9.3 for a discussion of some of the issues in the provision of liquidity to SPV.]

7.1.7 Operational Risk in Basel II

When the BCBS designed the capital charges for credit risk, they had a reasonably extensive literature of academic work, an extensive history of experienced losses on corporate and retail exposures (if not specialised lending), and a wealth of industry risk management practice to draw on. For operational risk, they had a lot less, so it is perhaps not surprising that the operational risk charges are neither time honoured nor risk sensitive. Three approaches are available:

- *The basic indicator approach* is a very simple way of calculating capital for operational risk.

- Firms may apply to use *the standardised approach*, a slightly more sophisticated method available to firms meeting certain standards of operational risk management.

- Finally, firms meeting higher standards may apply for permission to use an advanced measurement approach (AMA).

7.1.7.1 Basic indicator approach

Banks using the basic indicator approach must hold capital for operational risk equal to the average over the previous 3 years of a fixed percentage of positive annual gross income. This percentage, denoted by *alpha*, has been fixed at 15% by the BCBS. Thus, gross income is the operational risk 'indicator'.

7.1.7.2 Standardised approach

In the standardised approach, banks' activities are divided into eight business lines with a haircut, denoted by *beta*, applied to the gross income in each business line according to the table below. This gives a somewhat higher operational risk charge for activities the supervisor considers higher risk, and a lower one for those, such as the consistent beneficiary of Basel II, retail banking, that are considered lower risk.

Business Line	Beta (%)
Corporate finance	18
Trading and sales	18
Retail banking	12
Commercial banking	15
Payment and settlement	18
Agency services	15
Asset management	12
Retail brokerage	12

The qualifying criteria for entry to the standardised approach broadly include the following:

- Management involvement in operational risk management and approval of policies and procedures specifying the firm's framework for identifying, measuring and mitigating operational risk.

- The documentation and implementation of a sound operational risk management system, including the systematic gathering and reporting of operational risk loss data by business line and the allocation of adequate resources to operational risk management and other control and audit groups.

- The firm's operational risk system must have a key role in risk reporting and analysis.

- The firm must create an incentive structure to reduce operational risk, for instance by allocating capital to businesses on the basis of operational risk measures. There should be procedures for addressing non-compliance with operational risk policy or inadequate loss reporting.

- The operational risk management framework must be validated and regularly audited.

- A robust mapping system for allocating actual firm gross income into the business areas defined in the table must be implemented.

7.1.7.3 Why gross income?

It is easy to object to the basic indicator approach. For instance, it is not clear that there is much correlation between operational risk and gross income: indeed, at some level, one might even argue for a negative correlation, in that firms that tend to be highly successful in a given area presumably understand it rather better than the ones that are not. However, it is at least anti-cyclical, so viewed from a systemic risk perspective, there may be some justification for it.

In some ways, the standardised approach is less justifiable: it forces firms to artificially map their internal structure into the specified business lines with little obvious benefit. The entry criteria imply an incentive for better operational risk management, yet some of the betas are larger than alpha, so the benefit is in the opposite direction for some firms such as the investment banks.

7.1.7.4 Advanced measurement approach

AMAs give huge freedom to banks to innovate: if a firm is permitted to use this approach, the regulatory capital requirement is simply given by the bank's internal operational risk measurement measure, and there are rather few constraints on how banks derive this measurement. The (slightly ominous in historical context) motto here is, '*Let a thousand flowers bloom*',* that is, firms are encouraged to invent their own model suitable for their business.

What must an AMA model measure?

- Supervisors have specified that an acceptable AMA should capture the tail of the operational risk loss distribution at a standard 'comparable to' the IRB standard, i.e., a 99.9% confidence interval over 1 year.

- However, unlike the IRB approach, the sum of EL and UL must be calculated unless the firm can demonstrate that it has adequately accounted for EL. Presumably an EL reserve would help here, although that might be problematic from an accounting standpoint.

- This measurement must be based on at least 5 years' worth of internal data used either to calibrate the model or to validate it.

Various classes of operational risk model might be acceptable on this basis including:

- Scorecard approaches, provided the KPIs are used to estimate a 99.9% 1-year operational risk loss estimate;

- Loss distribution approaches.

As usual, there are further standards to be met, including all those required for entry to the standardised approach and the use of operational risk scenario analysis to evaluate exposure to high-severity events. An incentive, though, in addition to the reduction in capital potentially available through internal modelling, is the ability to recognise the risk mitigating impact of insurance subject to a 20% cap.

---

* See Roderick MacFarquhar (Ed.), *The Politics of China: The Eras of Mao and Deng*.

7.1.7.5 Dialogue

Pru Lawyer has finally caught up with the new EVT specialist in risk management, Fisher Tippett:

'I wonder if you can help me with something Mr. Tippett. You are an expert on modelling the tails of distribution, are you not?'

'I will do my best Ma'am'.

'The new Accord states, "There may be cases where estimates of the 99.9th percentile confidence interval based primarily on internal and external loss event data would be unreliable for business lines with a heavy-tailed loss distribution and a small number of observed losses." In your opinion does that hold for any of our businesses?'

'Off the top of my head I'd say it held for all of them'.

7.1.8 Floors and Transitional Arrangements

Like the text of this book, Basel II is coming to fruition in 2007. Some banks are already using the new Accord to calculate capital, although the most advanced approaches are in parallel running until the start of 2008. However, because of fears about the possible falls in capital required especially for banks using the more advanced approaches, the Accord contains a falling scale of floors to the capital benefit which can be obtained. The following table summarises these floors as a percentage of the Basel I capital charge.

From Year-End Approach	2006	2007	2008
FIRB	95%	90%	80%
AIRB or AMA	Parallel calculation	90%	80%

The QIS, particularly the last one before implementation, QIS 5, point towards capital reductions well below the floor levels for many banks, especially those with large retail activities. It will be interesting as end-of-the-floor period approaches to see if this remains true and, if so, what if anything supervisors do about it.

7.2 BASEL II: BEYOND THE CAPITAL RULES

Some regulators have always viewed capital as at best a third line of defence behind management and internal controls: others, especially in jurisdictions with little tradition of supervisors visiting banks, have tended to rely more on capital alone. Basel II requires regulators to exercise discretion with regard to their banks and hence suggests minimum standards of supervisory review. We begin by looking at these and then move on to another area of international convergence in Basel II: disclosures required from banks. Finally, we discuss some of the winners and losers from the Basel II process.

7.2.1 Pillar 2 in Basel II

Regulatory capital cannot substitute for inadequate people or processes: apart from anything else, if a firm cannot measure, understand and act on the risks it is taking, how can a supervisor have any confidence in its capital estimates? Pillar 2 in the new Accord therefore contains provisions aimed at enhancing bank's risk management practice by requiring regular dialogue with supervisors and the supervisory assessment of a range of issues not directly captured by capital requirements. In particular, it contains a range of principles which can be summarised under the following headings:

7.2.1.1 Banks' overall capital assessment

Banks must review their capital versus their estimate of capital requirements at a senior management level, and have a strategy for maintaining sufficient capital. Capital levels must be regularly reported and supervisors notified of any event which threatens capital adequacy. In turn, supervisors are required to intervene if this appears likely, and to enforce action to restore the capital position.

7.2.1.2 Supervisory review

Supervisors should regularly review the quality and veracity of banks' capital adequacy assessments and take action where these assessments do not meet minimum standards. The intent is that this is a continuing dialogue. Typically, the review process will include visiting banks; examination of regulatory returns, external audits and other assessments; and peer group comparisons. One aim of this process is to assess whether the bank's control environment is sufficiently robust that confidence can be placed in the bank's estimate pillar 1 regulatory capital: another is to determine if that capital requirement, assuming it is correctly calculated, covers all the bank's material risks. The Accord identifies three areas where supervisors are particularly encouraged to consider whether minimum capital requirements are adequate:

- Risks which are not fully covered by the pillar 1 rules such as credit concentration risk;
- Risks ignored completely in pillar 1, such as interest rate risk in the banking book;
- Broad market factors which may affect capital adequacy such as the economic cycle.

Exercise. If you are not a bank supervisor, imagine you are one. Consider the career consequences for you if a large bank you were involved in the supervision of were to fail. How could you prove that you had acted properly were this to happen? Does this help to explain some regulators' measure of interest in documentation?

7.2.1.3 Action including extra capital requirements

If a bank's current capital and controls are not judged adequate, one step supervisors can take is requiring banks to hold capital above the pillar 1 minimum. Others include demanding improvements in internal controls, extra reporting, restricting payment of dividends or recapitalisation.

Exercise. Which of these is most troublesome for the bank? Which is easiest for the supervisor to monitor?

7.2.2 Pillar 3 and Banks' Disclosures

The final buttress against systemic risk in Basel II is *disclosure*, also called *market discipline*. The basic idea is that by requiring firms to disclose details of their risk processes and exposures together with capital held, the market can better assess the risk profile of banks and so more efficiently price instruments issued by them.* The detailed disclosure requirements fall into two categories.

7.2.2.1 Qualitative disclosures

Basel II requires banks to disclose details of their risk management strategy, policies, organisation, processes and risk measures in major risk type. In particular, descriptions are required of key controls including:

- The internal ratings process and external ratings used where applicable;
- The valuation process for mark-to-market positions;
- Stress testing, backtesting and other model validation processes.

7.2.2.2 Quantitative and other disclosures

The principal requirements here concern the details of capital structure, available capital, capital requirements and method of regulatory capital calculation:

- Details of the group structure, accounting policy and consolidation;
- Information on the capital instruments issued, and the total amount of capital in each tier;
- Capital requirements for credit risk, split by portfolio, including information on which method is used to calculate credit risk capital;
- Breakdown of credit exposures by geography, type of obligator and maturity;
- Amount of impaired loans, actual losses due to and provisions for credit risk;
- Details of credit risk mitigation and securitisation;
- Information on off-balance-sheet exposures including counterparty risk on derivatives transactions;
- Capital requirements for market risk, split into risk type and with information on the scope of internal models permission granted for capital calculation;
- Details of VAR exposures and backtest results where applicable;
- Capital requirements for operational risk (but not details of operational risk losses);
- Measures of interest rate risk in the banking book [see section 8.2].

Although most of these disclosures are already commonplace among internationally active banks, there are some requirements which extend bank practice. For instance,

---

* Some commentators have noted that if the market really believed that Basel II required capital that captured all risks at a 99.9% confidence interval, then *all* senior debt issued by internationally active banks would have a maximum spread corresponding to a 0.1% PD. This is far from being the case.

banks typically do not at the moment disclose details of the scope of their VAR models permission.

> *Exercise.* Obtain the annual reports for a range of large banks and compare their current disclosures with the Basel II requirements.
>
>   How do you think equity analysts and investors will react if major banks are not on the AIRB approach for credit risk? What about the AMA for operational risk?

7.2.3 The Impact of Basel II

Basel II is a complicated edifice which will produce dramatic changes in the capital requirements for some banks. Its impact is only slowly becoming clear, and it seems likely that it will result in further market changes during implementation.

One major source of information on the impact of the new Accord is the QIS, so the results of the latest QIS are reviewed.* Then we move on to deduce further likely impacts on the basis of the nature of the new capital rules.

7.2.3.1 QIS 5 results

QIS 5 is the most recent large study on the effects of the new Accord. On average, it shows that the capital requirements for the banks surveyed will fall under Basel II compared with Basel I.

However, this trend masks considerable variations, and there are definitely institutions where capital requirements rise: these are likely to be more specialised banks.

Some of the largest falls in the fifth QIS are for smaller banks under the AIRB approach, with retail mortgage banks particular beneficiaries. Commercial banking portfolios do not, on average, see large changes in capital requirements, although there is a reallocation of capital towards riskier exposures. This is not a surprise: what might be is that on average, it seems that charges for securitised positions *fall* slightly post Basel II. There is a large reallocation of capital from senior positions, which become cheaper to hold, to junior ones, which should become more expensive.

(However, note that many of the banks which are active in securitising assets they originate are already under a capital regime which has punitive charges for holding junior tranches. This is the case, for instance, for many large U.S. players. Therefore, large capital requirements for retaining the riskiest tranche of securitisations may already be built into the 'before' figures.)

> *Exercise.* What do you think the results of a QIS which examined the impact of Basel II excluding large banks would reveal?

---

* See the QIS results on the BIS website www.bis.org for more details.

7.2.3.2 Winners and losers

The table below summarises some of the winners in the new Accord, with brief reasons.

Winners	Reasons
Ratings agencies	Need for ratings in the standardised approach
Public rated term bonds	Rating clear, no debate on capital charge
Unregulated financial services companies	Able to profit from areas that become more expensive for Basel banks
Advanced banks, especially in AIRB approach	Lower capital requirements versus their competitors
Banks with high quality low default correlation assets available for securitisation	Securitisations where the rated tranches form a very large percentage of the whole are cheaper
Retail banking generally and mortgage banks in particular	Broadly cheaper in the new Accord

There is one certain impact of Basel II: increased costs for banks and supervisors and increased revenue for consultants and other advisors.*

Indeed, the combination of changes to accounting standards, Sarbanes–Oxley and Basel II has been estimated by some large banks to increase costs by hundreds of millions of dollars a year. A detailed cost–benefit analysis would make interesting reading here.

Some losers should be obvious from the areas of Basel II where capital charges are higher:

- Some revolving structures and unrated transactions;
- Some forms of lending against collateral not meeting the standards for models-based capital calculation;
- Banks with high exposures to emerging market credits;
- Smaller banks unable to meet IRB standards;
- Banks with significant exposure to equity in the banking book or significant holdings of lower or non-rated tranches of securitisations;
- Banks with little or no market and credit risk, which suffer from the operational risk charges.

---

* It is worth noting that there is an unhelpful incentive structure in regulatory matters: supervisors want banks to raise standards, so they tend to articulate progressively higher 'minimum' requirements. However, it is difficult for the industry to respond even when the target is seen as unrealistically high: banks may find challenging their supervisor problematic. Moreover, supposedly independent advisors can sell more advice and more advanced systems, the higher the standards are, so they are inclined to over-estimate the level required too, at least in public statements.

Part Three

Treasury and Liquidity Risks

CHAPTER 8

The Treasury and Asset/ Liability Management

INTRODUCTION

The performance of corporations is judged in part by the results of their activities as revealed by their *financial statements*. Risk management is therefore partly financial statement management. For instance, much of trading book risk management focuses on activities that affect the *income statement*, as that is where the earnings from trading activities are disclosed. Many firms also have risk outside the trading book though: in particular, banks have both credit risk and interest rate risk in the banking book. Much of that credit risk comes from lending, and some of the interest rate risk from how that lending is *funded*. Thus, for instance, if funds to make a 5-year fixed rate loan are obtained by issuing a floating rate liability, the bank runs the risk that the floating rate will rise above the fixed rate and so the loan asset will become unprofitable. This will not produce a mark-to-market loss since the loan book is historic cost accounted: rather it affects the bank's *net interest income* (NII). Therefore, we will need to understand how this NII arises and how to measure its sensitivity to changes in interest rates.

This leads us to consider the assets and liabilities on a firm's *balance sheet* and how they behave as market rates move. The process of controlling the joint risks of assets and liabilities is known as *asset/liability management* (ALM).

The Treasury takes a key role in ALM, so the functions of this group are discussed together with some comments on the analysis of funding strategies and their risks. Finally, we consider a key organisational structure for many banks—the ALCO—and how it oversees the ALM process.

8.1 AN INTRODUCTION TO ASSET/LIABILITY MANAGEMENT

Financial institutions have assets, such as securities, loans, goodwill and office buildings, and they have liabilities, such as deposits taken and debt issued. Assets and liabilities are raised by a variety of BUs in the course of their activities. It is unwise to allow this process to happen in an unconstrained fashion: it may well be undesirable, for instance, to have the commercial bank and the retail bank both issuing bonds in the wholesale market on the same day. In this chapter we study the process of managing the asset/liability mix in the bank. We begin by looking at the structure of the bank's assets and liabilities, where the different elements come from, and how Treasury fits in.

8.1.1 The Trading Book, the Banking Book and the Treasury

We have already met the two main divisions of risk taking books:

- The trading book, which typically contains mark-to-market positions in marketable securities, derivatives and related instruments;
- The banking book, which typically contains historic cost-accounted positions such as loans, together with deposits and other retail and commercial banking instruments.

To be able to make a loan, the banking book needs to be able to advance cash. Similarly, the trading book needs cash to be able to buy a security or pay an option premium. Typically, that money comes from the bank's *Treasury*.

8.1.1.1 Functions of the Treasury I: liquidity management

One of the major functions of the Treasury is to ensure that the bank has enough cash or *liquidity* to function effectively. This involves first deducing the current cash position of the bank, and then estimating the likely future position at some time horizon by adding the cash expected from *positive liquidity* situations and subtracting that required for *negative liquidity* situations. For instance, the following *generate cash* and so are positive liquidity:

- New deposits;
- Issuing new securities including CP, MTNs and capital securities;
- Redeeming deposits made with other institutions or drawing on available lines;
- Liquidating securities positions.

The first two would be liabilities of the bank, whereas the last involves selling an asset. On the other hand, the following asset purchases *require cash*:

- Making loans including lending money against collateral;
- Buying new securities;
- Drawdowns on written LOCs.

Finally, declining liabilities require cash, for instance, when depositors call their deposits or borrowers repay their loans.

Treasury has to balance the sources and sinks of the bank and act to ensure that there is always enough cash. If cash is needed, it will determine the best source: if the bank has surplus cash, this will be invested. This process is called *liquidity management* or *funding the bank*.

8.1.1.2 Functions of the Treasury II: liability and capital structure management

Suppose that the treasurer looks at the bank's liquidity position and sees that tomorrow, if no action is taken, the bank will not have enough cash to meet the likely needs of its capital markets business. How should the decision be made about what to do to raise funds? There are a number of factors to consider:

- *Cost.* Some ways of generating liquidity are cheaper than others: for instance, the bank may fund in the CP market at 1-month Libor flat, but if it borrows by issuing long-term debt, it might have to pay a double-digit credit spread.

- *Term.* Typically yield curves point up. Therefore, borrowing for longer is typically more expensive than borrowing for a shorter period even if there is no term structure to credit spreads (and there usually is). However, if we are borrowing to fund the purchase of an asset, we need to balance the lower costs of having short-term funding versus the need to *roll* that funding: to replace it when the first borrowing matures. This is part of *liquidity risk management* [the topic of Chapter 9].

- *Volatility.* Some forms of funding are contractually short-term but behave as if they were longer-term: a good example is retail deposits. Here the depositor can come and ask for their money back at any time, but in fact they tend not to. Therefore, these sources of funding have a longer behavioural term than their contractual term. Alternatively, they are said to be *low-volatility funding* because they do not migrate very quickly.

- *Impact on rating/credit spread.* The more debt a firm issues for a given amount of capital, the more highly leveraged it becomes, and hence the more risky. Thus, raising funding has implications beyond the cash received: it is important to consider the impact on the bank's capital structure. At some point as more and more funding is required, extra capital will be necessary, or the bank's debt will be downgraded.

- *Regulatory capital position.* If funding is raised to buy risky assets, these assets will increase the bank's regulatory capital requirement, and so this too will eventually push a bank that is growing its asset base into issuing capital instruments.

Obviously, then, there is a need to centralise liability issuance so that these issues can be considered in the context of the firm's whole capital structure. This happens in Treasury.

8.1.1.3 Functions of the Treasury III: surplus cash

Some banks, particularly those offering attractive deposit rates, can end up with surplus cash. Obviously, if this is not invested then value is destroyed; so another Treasury role is to invest any excess cash at the end of each business day. Again here there are a lot of choices ranging from interbank depositors to buying securities into the *liquidity portfolio*—the portfolio of securities held by Treasury as part of their funding management activities.

8.1.1.4 Functions of the Treasury IV: capital

As the group responsible for the bank's capital structure, it makes sense for Treasury to have a role in capital management, perhaps jointly with risk management. This may extend in some firms to reporting regulatory capital as well as allocating economic capital.

Summary of Treasury Functions

Standard Functions	Treasury May Have a Role in
Provision of funding to business groups needing cash	Capital allocation
Acting on the bank's behalf in the interbank market	Regulatory reporting
Use of bank's LOCs	Managing relationship with ratings agencies
Issuance of securities and their management	Disclosures to investors
Investment of surplus liquidity	
Management reporting on funding and capital structure	

> *Exercise.* Which functions does Treasury carry out in an institution you are familiar with?

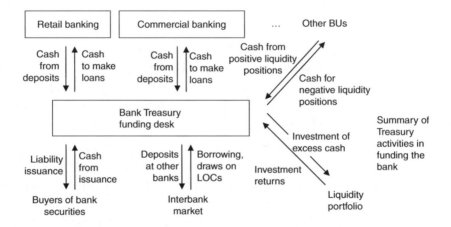

8.1.2 Accounting for an Old-Fashioned Bank

There is an old joke that the key to success in banking is the 4-6-4 rule: borrow at 4%, lend at 6% and be on the golf course at 4:00 p.m. That model, if it ever existed, is now very much out of date. However, it is worth looking at a simplified bank on the basis of this premise in some detail, partly to understand how it fits together, and partly to see how the business reality is reflected in financial reporting.

We return to the simple bank discussed earlier [see section 1.4.3]. This institution has £100M of equity capital. The capital supports three BUs. Suppose the balance of their activities is:

- *Retail banking.* This group takes deposits. Currently the total amount deposited is £500M.

- *Commercial banking.* This group makes loans to large corporates. The total notional lent is £900M.

- *Capital markets trading.* This group trades securities and derivatives. It is long £300M of securities and has a total notional exposure of £1B of swaps.

Retail banking is positive liquidity: it contributes £500M to Treasury.

Commercial banking and capital markets are both negative liquidity: they require a total of £900M + £300M = £1.2B of funding. Treasury therefore issues £400M of short term debt and borrows £200M in the interbank market.* The remaining £100m comes from the bank's equity.

8.1.2.1 The balance sheet

A *balance sheet* is a view of the financial state of a corporation at a fixed point in time. Although the precise details of what appears on a financial statement depend on the accounting standards required in the corporate's country of incorporation, the broad features are the same. Most balance sheets have three parts: assets, liabilities and shareholders' equity. For our revised version of the simple bank we have:

Assets		Liabilities	
Loans	£900M	Retail deposits	£500M
Securities	£300M	Issued debt	£400M
		Interbank borrowing	£200M
		Shareholders' Funds	
		Equity	£100M
	£1200M		£1200M

Note that the balance sheet balances: assets = liabilities + shareholders' funds.

This means that the liabilities are smaller than the assets, the extra funding coming from shareholders' funds.

8.1.2.2 Aside: accounting assets and liabilities

It is worth being clear about two uses of the terms *assets* and *liabilities*. In common parlance an asset is something with a positive expected PV, so, for instance, an in-the-money swap is an asset in the sense that it is worth something to the holder. But in accounting, derivatives do not appear on the balance sheet, so by themselves they are not assets or liabilities for accounting purposes.

8.1.2.3 The income statement

The *income statement* is a financial report giving a corporation's net income and indicating how that money was made or lost during a period. Equity analysts scrutinise the income statement to understand how repeatable earnings were and where they came from: a highly regarded firm will typically have positive, growing, low-volatility earnings from its core business.

The 4-6-4 indicates one of the key sources of income in banking: net interest income. This is simply the difference between the yield on assets and the cost of servicing the liabilities: if we borrow at 4% and lend at 6%, the NII is 2%.

---

* Obviously, this is a simplified picture intended to illustrate aspects of ALM rather than the details of a real bank's ALM: see Donald Deventer et al.'s *Advanced Financial Risk Management: Tools and Techniques for Integrated Credit Risk and Interest Rate Risk Management* for a more detailed discussion.

Suppose for our simple bank the NII on contractual items is the interest income from the loan portfolio minus interest paid on deposits and on interbank borrowing. To get the total NII, we need to add in income from the securities portfolio and subtract interest payments on issued securities.

Suppose this comes to a total of £32M in the previous year. Other elements of income are:

- Trading P/L of £30M;
- Net fees (for instance, from loan commitments and transaction fees) are £15M;
- Operating expenses are £40M.

The pre-tax part of the income statement is therefore as shown in the table below.

Income Statement	
NII	£32M
P/L on financial transactions	£30M
Net fees	£15M
Operating expenses	(£40M)
Net operating income	£37M

8.1.2.4 Leverage and ROE

The bank's pre-tax ROE for this year is £37M/£100M or 37%. On a balance sheet basis, it is leveraged about 12:1, as it has £1200M of assets for £100M of equity: the derivatives activities may add extra leverage which cannot be seen from an accounting perspective.

8.1.2.5 Memoranda and derivatives disclosures

In addition to the assets and liabilities appearing on the balance sheet, banks typically disclose some information concerning contingent liabilities and derivatives, such as the notional amounts of financial guarantees written and the replacement value of derivatives transactions, in other sections of the accounts known as the *notes to the accounts* or *memorandum items*. Although these disclosures are interesting, neither notional nor replacement value are particularly insightful measures for guarantees and derivatives.

> *Exercise.*   Review the balance statement, income statement and derivatives disclosures for a number of international banks. What do they tell you about the risks taken and the balance of activities? What questions are left unanswered?

8.1.3 Assets and Liabilities through Time

Suppose we add another dimension to the balance sheet: the *repricing* period. This is the period of time an interest rate is fixed for. Thus, a deposit which pays 3-month Libor reprices after 3 months: to keep the cash in the deposit we have to pay the then current 3-month Libor for the next 3 months.

8.1.3.1 Repricing periods for the simple bank

Consider a crude set of time buckets of 0–90 days, 91–180 days, 181–360 days, 1–2 years and beyond 2 years under a 30/360 convention.

For our simple bank we assume that its loans are 15-month fixed rate instruments which cannot be prepaid, and that the security assets are 4-month fixed rate notes. On the asset side of the balance sheet we have the entries below.

Assets	0–90 Days	91–180 Days	181–360 Days	1–2 Years	>2 Years
Loans				£900M	
Securities		£300M			

Now consider the liabilities. Suppose the bank's deposits pay 1-month Libor; they then go into the first bucket. The bank's issued debt is just fixed rate notes. If they have a residual maturity of 9 months, that puts them in the fourth bucket.

Finally, suppose the bank's interbank borrowing is for 1 month at 1-month Libor + 10 bps. Then we have

Liabilities	0–90 Days	91–180 Days	181–360 Days	1–2 Years	>2 Years
Retail deposits	(£500M)				
Issued debt			(£400M)		
Interbank	(£200M)				
Equity					(£100M)

The sum of each bucket indicates the amount of funding that the bank has to replace during that time interval at then current rates. For instance, adding down the 0–3-month bucket there are no assets expiring but £700M of liabilities, indicating that the bank has £700M of funding repricing between now and 3 months: a finer bucketing scheme would reveal exactly when.

8.1.3.2 The funding gap

The total amount needed is referred to as the *funding gap*: if it is positive, extra funding from some source will be needed, whereas if it is negative, the bank has surplus cash which can be invested.

It is also useful to keep track of the cumulative gap through time. This gives us

Gap	(£700M)	£300M	(£400M)	£900M	(£100M)
Cumulative gap		(£400M)	(£800M)	£100M	—

8.1.3.3 Net interest income

Now suppose that the security assets pay 4.2% fixed; the loans, 5%; our issued debt, 4.2%; and 1-month Libor is 4%. What is the NII? For the first month, it is roughly

$$\text{Asset net interest received} - \text{Liability net interest paid} = £300M \times 4.2\%/12$$
$$+ £900M \times 5\%/12 - £500M \times 4\%/12 - £200M \times 4.1\%/12$$
$$- £400M \times 4.2\%/12 = £1.05M$$

But now suppose 1-month Libor rises to 5%. The asset interest income is the same as they are fixed, but the cost of funding them goes up and so if we have to pay the new rate, the NII decreases

$$= £300M \times 4.2\%/12 + £900M \times 5\%/12 - £500M \times 5\%/12$$
$$- £200M \times 5.1\%/12 - £400M \times 5.2\%/12 = £133K$$

We can understand this from the gap report. The gap of £700M in the first bucket indicates that there is interest rate risk on that notional: we have £700M of assets funded by liabilities which reference short-term rates, so if these rise, NII falls.

NII is one of the most important sources of income for traditional banks. Therefore, the management of the interest rate sensitivity of NII (aka interest rate risk in the banking book) is a key concern.

> *Exercise.* What is the NII figure for the banks you studied in the last question? Can you estimate how much of this is due to interest rate risk in the funding mix and how much to taking credit spread risk?

8.1.3.4 Risks taken for the return

What risks is the simple banking taking to earn its NII?

- There is credit risk on the loans.

- As we have seen, there is interest rate risk.

- The firm has a significant amount of funding from deposits and from the interbank market. If these deposits were to be withdrawn, for instance, due to a loss of confidence in the bank, the bank would find itself having to replace that funding with a source that might be considerably more expensive. Thus, it is taking *liquidity risk*.

8.1.4 What Is ALM?

ALM is the process of managing the mismatch between assets and liabilities. Financial firms are often leveraged as we have seen in our simple bank. Therefore, a relatively small decline in the value of assets or a small increase in the value of liabilities can make a big difference to shareholders' equity = assets − liabilities. This can happen because asset and liability durations are not *matched*: often the bank has a funding gap as in the example above. There a small risk in interest rates increases the price the bank has to pay on its funding and hence causes a reduction in NII.

Banks may also have other risk exposures in their funding, for instance, via the issuance of liabilities in one currency to support assets in another, or through the use of liabilities which have uncertain durations such as deposits. Effective ALM involves managing these risks.

Historically, some banks lent long and funded short, exposing themselves to significant interest rate risk. The reason for this was twofold. First, for a traditional upward-pointing yield curve, the bank made money on the spread between the short-term funding rate and the long-term rate (even before credit spreads were taken into account). Even if short-term rates rose, compressing NII, the bank could respond by originating more long-term assets provided the curve did not invert. Second, the use of accrual accounting for the banking book meant that investors could not see the extent of the interest rate risk taken: NII includes both income made from taking credit spread risk and income from taking interest rate risk on funding.

8.2 BANKING BOOK INCOME AND FUNDING THE BANK

This section explores the process of funding the banking book in more detail. We look at how Treasury funds BUs and hence how it manages the interest rate sensitivity of NII. Non-interest income is also discussed briefly.

8.2.1 Transfer Pricing

Suppose the commercial banking unit makes a $100M 10-year loan at 6% fixed. Treasury advances the funds for this to the BU and adds $100M to the bank's funding requirement. How should we judge the commercial bank's performance on this asset? Clearly, they are taking credit spread risk on the loan obligator: it is their business to assess this risk. But they should not be involved in the decision regarding how to fund this asset, so regardless of how the $100M is actually raised, the BU should be funded on a *matched basis* reflecting the risks it is mandated to take. Since the loan is a 10-year fixed rate asset, Treasury should charge them with the bank's 10-year rate. The net credit spread income is 6% minus the bank's cost of funds for the same maturity. Hence, the BU will recognise a profit over the life of the loan equal to the accrued credit spread at 10 years minus any losses on the loan, and any profit or loss reflecting how the loan is actually funded will appear in Treasury's P/L.

8.2.1.1 The transfer pricing book

The process of deciding on the correct funding rate for assets, or interest income for a BU which raises liabilities, is known as transfer pricing (TP). In general, the transfer price of an asset or liability will be determined by the repricing period: thus, for instance, commercial banking might be paid the bank's cost of overnight funds for taking a corporate deposit, reflecting the fact that it is demand repayable. This incentivises the BU to attract deposits which are good for the bank, i.e., cost less than the bank's cost of funds on the same terms elsewhere.

TP is often achieved by booking an explicit funding instrument between the BU and Treasury. Suppose the bank's cost of funds through issuing 10-year bonds is 5.1%.

The commercial bank would book its 10-year loan at 6% fixed with the client, and it would fund it by booking a 10-year borrowing at 5.1% from Treasury. Treasury would book the other side of this loan in the *TP book*.

Exercise. To reflect the bank's real cost of funds, the TP book should update the rates it borrows from or lends to the BUs frequently. Is a daily update enough or could this give rise to funding arbitrages?

8.2.1.2 Transfer pricing for liabilities
Similarly, if the retail bank offers its clients a demand repayable deposit paying Libor −10 bps, this would be booked versus a deposit with Treasury at the rate the bank can fund itself overnight, Libor −5 bps say, recognising the profit earned by the BU in attracting that deposit.*

8.2.1.3 Loan profitability
The advantage of adopting the TP approach is that it separates the two components of NII—P/L from taking credit spread risk and P/L from taking interest rate risk on funding—and sends them to the parts of the firm that should be managing them. The lending BU's job is to assess the credit spread it can originate a loan at versus the risk in that loan: the Treasury's job is managing funding.

Another advantage is that the TP book accumulates all the interest rate risk on the bank's funding base; hence, any natural offsets within the bank can be exploited and the net position can be managed by a single group of traders. The illustration below shows how the TP book is used.

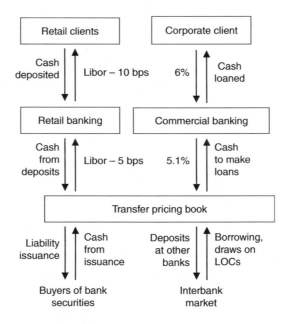

8.2.1.4 Transfer pricing for prepayable assets
Suppose the retail bank originates a 30-year mortgage. Following the logic above, this should be funded at the bank's cost of 30-year funds. But this mortgage might well be

---

* TP should be on a bid/offer basis, so assets raised are funded more expensively than the credit given from liabilities. See Donald Deventer et al. (*op. cit.*) for more details.

prepayable: the borrower can repay the funds early, and indeed the average length of the bank's 30-year mortgages might be as little as 7 years. Therefore, the funding rate needs to be adjusted for the embedded prepayment option: the correct rate is the bank's 30-year funding rate plus the value of the option to prepay. [The value of this option is discussed in section 10.2.1.]

8.2.1.5 Behavioural maturity
The effective maturity of retail deposits is typically much longer than their contractual maturity: people can withdraw their money at will, but mostly they leave a positive balance in their accounts, and this in turn is available to fund the bank. Therefore, the bank might give some benefit on the basis of the estimated *behavioural* maturity profile of a deposit book.

8.2.1.6 Structured liabilities
Suppose the credit derivatives trading desk issues a 5-year CLN. What credit should the TP process give for the funding raised? The answer depends on the structure. If the CLN proceeds are invested in third party assets, then the liquidity raised does the bank no good.

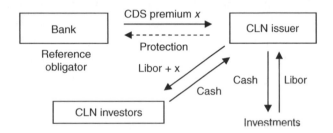

However, if the cash is invested in the bank's own paper, then the structure is helpful in funding the bank. Therefore, Treasury might well be prepared to issue debt customised to the needs of the structure, for instance, using a shelf program to create a special issue of a 5-year senior FRN for the CLN issuance SPV to buy. Finally, if no collateral is involved and the CLN is a direct liability of the bank, as in the version of the structure below, then the TP credit given should reflect the option-adjusted spread (OAS): the funding is not guaranteed term funding since if there is a credit event, the CLN desk will have to pay out the funds received on the hedge it has bought against the CLN from the default swap market. For a high-quality reference obligation, this option adjustment will typically be small.

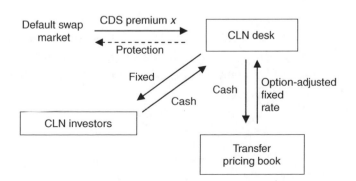

8.2.1.7 Capital consumption and net interest income

Suppose a BU makes a $1M FRN for 5 years at 6-month Libor + 90 bps and the bank's cost of 6-month money is 6-month Libor + 10 bps. The TP book funds this loan at this level, so we might assume that the NII attributable to the business for taking credit spread risk is 80 bps. (This is not quite true due to the provisions and capital allocated to the loan.)

Let us assume this capital allocation is 4% of notional, reflecting the incremental contribution of the loan's risk to the portfolio UL [as discussed in Chapter 5]. This capital is funding too: equity is sold for cash, after all. Therefore, the funding need is not $1M but $960K.*

If 6-month Libor is 4.1%, the components of NII in this case are:

- Loan interest at the cost of funds: $1M × 4.2% = $42K;
- Loan interest on the credit spread: $1M × 0.8% = $8K;
- Funding: −$960K × 4.2% = −$40.32K.

So the NII comprises $8K on the credit spread and $1.68K (=$40K × 4.2%) on the equity allocated. Since this earning on the equity support is just due to 6-month Libor, it has interest rate sensitivity: if Libor drops from 4.1 to 4%, the earnings on the credit spread are the same, but the earnings on the equity support drop to $1.64K (=$40K × 4.1%).

Finally, notice that if the loan was at fixed rate, the TP book would fund it fixed. Suppose the 5-year cost of funds is 5% and the credit spread is the same. The components of NII are then:

- Loan interest at the cost of funds: $1M × 5% = $50K;
- Loan interest reflecting the credit spread: $1M × 0.8% = $8K;
- Funding: −$960K × 5% = −$48K.

Again there is $8K credit spread income but the earnings on the equity support are $2K, and this does not have interest rate risk since the BU has had its funding locked in to term by Treasury. Notice that this analysis shows the BU's position as funded by the TP book in Treasury. The whole bank's position may be different due to the interest rate risk on the bank's actual funding. Therefore, we turn next to look at the whole bank picture.

8.2.1.8 Interest rate sensitivity of earnings on equity support

The idea that we only fund the fraction (1 − percentage of capital allocation) of an asset also makes sense in terms of the gap analysis of the simple bank. The £100M equity is also funding as the full gap analysis below shows: the cumulative gap only closes in the last bucket when the bank's equity is included.

---

* The loan will also make a contribution to the portfolio EL, so some of the gross NII will be used to increase the loan loss provision for the portfolio.

Assets and Liabilities	0–90 Days	91–180 Days	181–360 Days	1–2 Years	>2 Years
Loans				£900M	
Securities		£300M			
Retail deposits	(£500M)				
Issued debt			(£400M)		
Interbank borrow	(£200M)				
Equity					(£100M)
Gap	(£700M)	£300M	(£400M)	£900M	(£100M)
Cumulative gap		(£400M)	(£800M)	£100M	—

This reflects the fact that the balance sheet balances: liabilities + equity = assets.

8.2.2 Interest Rate Risk in the Banking Book

The gap of £700M above indicates that £700M of funding falls off in the first period: it gives an indication of the sensitivity of the funding base to changes in interest rates.

8.2.2.1 *Using the cumulative gap to estimate the effect of an interest rate rise*

Suppose that the gap occurs in the middle of the first bucket (i.e. 45 days in) and interest rates rise tomorrow by 1%. Then we have an increase in funding costs due to the first gap of roughly

$$£700M \times 1\% \times 45/360 = £875K$$

For the next bucket again we assume that the change in the funding position happens in the middle of the bucket. Then we pay interest for 45 days on the £700M before the securities mature, and then on £400M for the remainder of the bucket. The cost of a 1% rise in rates for this bucket is therefore

$$£700M \times 1\% \times 45/360 + £400M \times 1\% \times 45/360 = £1.375M$$

The illustration shows the position we are assuming we have via using bucket midpoints. In the 181–364-day bucket the issued debt expires and that funding will have to be replaced at the new rate; therefore, the contribution of this bucket to the cost of a 1% rise in rates is larger: it is £400M × 1% × 90/360 + £800M × 1% × 90/360 = £3M. Proceeding this way the effect of a change in interest rates on the bank's NII can be estimated.

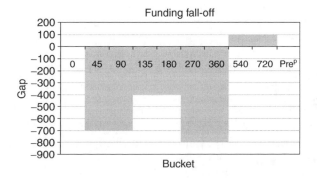

8.2.2.2 Improving the estimate

Suppose the loans in our simple bank pay interest quarterly, and the securities pay a semi-annual coupon at maturity. Then a more accurate view of the asset cashflows is

Assets	0–90 Days	91–180 Days	181–360 Days	1–2 Years	>2 Years
Loans	11.25	11.25	22.5	911.25	
Securities		306.3			

At Libor = 4% for the liability side we have the following cashflows (ignoring dividend payments on equity):

Liabilities	0–90 Days	91–180 Days	181–360 Days	1–2 Years	>2 Years
Retail deposits	(501.67)				
Issued debt		(8.4)	(408.4)		
Interbank	(200.68)				
Equity					(100)

This gives the following improved estimates for the gaps and cumulative gaps:

Gap	(691.1)	309.15	(385.9)	911.25	(100)
Cumulative gap		(381.95)	(767.85)	143.4	43.4

The positive ES from the loans at this level of Libor serves to reduce the funding need: the loan portfolio throws off cash.

The NII can now be estimated just by taking the sum of the asset and liability interest cashflows. A detailed analysis would not just use many more time buckets; it would also include a more detailed modelling of the cashflows from both assets and liabilities.

> *Exercise.* To estimate the NII, you need not just the current levels but also the expected replacement costs. For instance, £691M of funding is expected to fall off in the first bucket. How would you determine the rate at which this can be replaced?

8.2.2.3 Deposits

Consider a fixed rate demand callable retail deposit account. Since the funding from this account could be called by the depositor at any time, we could argue that this reprices overnight. But in reality only a small percentage of the total notional of deposits will be withdrawn even if rates rise and a better account is available elsewhere. Some banks have modelled this effect in detail and they can predict the approximate behavioural maturity of accounts like this, as shown in the figure.

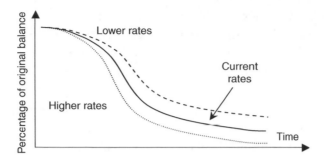

This type of *deposit migration analysis* would give a less conservative view of the cumulative gap as we would give credit for the 'sticky' nature of retail deposits.* Similarly, a bank could model the effect of rates on draws on LOCs, on interbank and commercial deposits and so on.

Exercise. What data would you need to model the behavioural maturity of deposits? What differences would you expect to find between the behaviour of basic chequing accounts and accounts designed for higher income customers here?

8.2.3 Non-Interest Income and Operating Expenses

Traditional banks have two other large items in the income statement besides NII: non-interest income—often mostly from fees—and operating costs.

8.2.3.1 Non-interest income
For completeness we briefly discuss some of the elements of non-interest income:

- Service and penalty charges on accounts;
- Fees for investment products;
- Fees for selling products from partners or insurance subsidiaries;
- Custody fees;
- Net gains on securitisation and servicing of securitised assets [see section 5.3.1];
- Revenue from long-term equity holdings such as venture capital.

Note that though many of these appear superficially to be free of the interest rate risk associated with NII and non-matched funding, they may not be. If interest rates go up and the equity market declines, retail investors tend to move money from higher-margin equity-linked investments into lower-margin deposits, reducing fee income.

8.2.3.2 Operating costs
Banks tend to go through cycles here: when times are good, the bank expands, increasing its cost base. Then when times are leaner, there is a significant focus on cost reduction,

---

* Here we have not considered new depositors that are attracted to the account; only the behaviour of current depositors is considered.

sometimes counterproductively. At the moment, for instance, some banks are feeling a customer backlash against the outsourcing of services to low-cost jurisdictions. This might reduce costs, but there is a perception it has reduced service too: as always, reputational risk management is needed in significant management decisions.

Exercise. Research cost-cutting exercises by financial services firms. What factors predisposed them to success, and when did they generate reputational damage?

Comment on the asymmetric phenomenon that when a service appears to be free, it is often not valued, but when it is removed, resentment usually follows.

8.3 ALM IN PRACTICE

The practice of ALM has evolved using a number of tools: the gap reports discussed in the previous section are one; this section discusses several more. First, we look at ALM in the TP book and how the risks identified earlier are managed.

Gap analysis is useful for looking at the bank's funding base over a fairly short-term horizon. Another perspective comes from looking at the total value of a bank's assets minus its liabilities, and how this measure changes with interest rates and FX rates. This leads to a discussion of risk reporting for ALM, FX risk and the role of the ALCO.

8.3.1 Risk in the Transfer Pricing Book

If the process of funding the BUs discussed in the previous section is followed rigorously, the TP book will end up with a significant amount of risk:

- All the loans needed to fund assets in the banking book to term;
- All the interest payments made to the BUs reflecting deposits gathered;
- All the interest rate risk from the bank's issued liabilities;
- And the prepayment options from prepayable assets and liabilities.

It may also acquire a significant amount of basis risk, since, for instance, a loan paying prime + 100 bps will be funded by the TP book at the bank's spread to the prime rate, prime + 10 bps say. The net risk position in the TP book therefore also incorporates the bank's total basis risk exposure, allowing Treasury to manage the position, for instance by issuing liabilities linked to prime rather than Libor.

8.3.1.1 Managing the transfer pricing book

The TP book will need all of the infrastructure of a sophisticated interest-trading desk to handle its risk: an experienced trading team; a system allowing it to monitor the interest rate exposure in real time and make accurate prices to both the BUs and the market; appropriate risk limits and so on. Treasury will manage the book within the board's appetite for risk on the bank's funding base by:

- Providing funding rates which incentivise the BUs to generate profitable assets and liabilities;

- Using the interbank market and liability issuance to source liquidity of a desired term and structure;
- Trading interest rate derivatives to manage the TP book's interest rate risk.

Thus, for instance, given a simple example gap:

	0–90 Days	91–180 Days	181–360 Days	1–2 Years	>2 Years
Gap	(£700M)	£300M	(£400M)	£900M	(£100M)
Cumulative gap		(£400M)	(£800M)	£100M	—

The TP team might conclude on the basis of the bank's risk appetite that this is too risky and so decide to hedge the position. It could, for instance, lengthen the duration of funding, perhaps by issuing 18-month fixed rate term debt, or it could protect the bank against rising rates by buying an interest rate cap.

8.3.1.2 Investing surplus liquidity

After Treasury operations are complete, one common result is for the bank to be slightly long cash overnight: this is safer than having to borrow in a hurry at the end of the trading day. This surplus liquidity is then invested overnight or for a longer term depending on the profile of the cumulative gap.

It may also be decided to keep an investment portfolio of assets available which can be liquidated if the bank needs extra funding quickly. This *liquidity portfolio* typically contains high-quality assets such as AAA ABS which can either be sold or repo'd to raise funds. Moreover, using ABS securities can help to hedge the prepayment risk of the funding base [see section 10.2 for a discussion of the prepayment characteristics of ABS securities].

8.3.1.3 Multi-currency funding

Finally, note that the TP book must operate in all of the currencies the bank has assets or liabilities in. BU assets are funded in local currency, and any funding currency mismatch will be managed by the TP book.

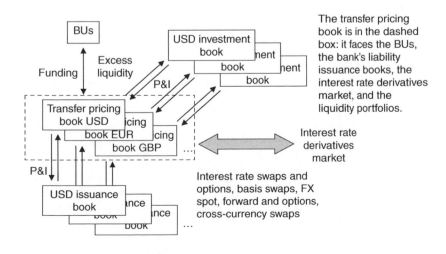

For banks which fund more cheaply in their home currency there will inevitably be a tendency towards running FX risk in the funding base, so the TP book will need limits to reflect the bank's tolerance for this mismatch.

8.3.2 The Market Value of Portfolio Equity

The fact that the balance sheet balances means that we can estimate the value of portfolio equity as

> Sum of PVs of all the expected cashflows from assets − Sum of PVs of all the expected cashflows from liabilities

This expression is known as the market value of portfolio equity (MVPE).*

8.3.2.1 MVPE from gaps

Let us make a few changes and additions to the simple bank's gap report, reflecting a marginally more sophisticated view of the balance sheet.

Assets and Liabilities	0–90 Days	91–180 Days	181–270 Days	271–360 Days	1–2 Years	Perpetual or Not Rate Sensitive
Loans					900	
Loan interest	11.3	11.3	11.3	11.3	11.3	
Securities		306.3				
Fixed assets						60.0
Loan provisions					20.0	
Retail deposits		(100.0)			(400.0)	
Deposit interest	(5.0)	(5.0)	(4.0)	(4.0)	(16.0)	
Issued debt				(400.0)		
Debt interest		(8.4)		(8.4)		
Interbank borrowing	(2.1)	(202.1)		(400.0)	408.2	
Dividends	(2.0)	(2.0)	(2.0)	(2.0)	(9.0)	
Net expenses	(9.0)	(9.0)	(10.0)	(10.0)	(42.0)	

Here the interest items have been given their own line, fixed assets have been added, the release of loan loss provisions against the fall-off of risk loan book is included, dividends have been added and net expenses after non-interest income are there too. Shareholders' funds are omitted. Everything is in million pounds rounded to one decimal place.

---

* It is also sometimes referred to as the *net economic value* or sometimes *economic value of portfolio equity*. This concept is related to the value of the balance sheet if subject to fair value rather than historic cost accounting. See Jean Dermine's 'Asset and Liability Management: The Banker's Guide to Value Creation and Risk Control' (*Financial Times*).

Note that rather than leaving a gap we have included the effect of a *funding strategy* for the bank:

- Retail deposits are used for as long as possible (since they are the cheapest source of funding);
- Interbank borrowing is rolled until the securities mature;
- Once the issued debt expires the bank needs funding again, so we assume that we increase interbank borrowing at this point.

The MVPE calculation therefore estimates the value of the portfolio equity *if this strategy can be executed* and rates do not change. Adding down the columns and picking appropriate discount factors for each bucket:

Sum of column	−6.8	−8.9	−4.7	−13.1	56.1	60.0
Discount factor	0.995	0.985	0.975	0.966	0.943	1

An estimate of the banks MVPE is then just that obtained by multiplying each amount by the discount factor and then taking the sum of these PVs. This is approximately £80M here.

This is a very simple view of the world. Some of the issues here include:

- The funding assumptions;
- The fact that the bank has only looked 2 years ahead, and the time bucketing is still crude;
- The loan loss provisions are released at the end of the loan rather than amortising reflecting the declining risk during the life of the loan;
- There is no amortisation of the fixed assets;
- The non-interest income and operating expenses numbers are necessarily estimates;
- We have not included any off-balance-sheet positions.

8.3.2.2 MVPE duration and leverage

Before bank systems were sufficiently advanced that it was practical to estimate the real interest rate sensitivities of large portfolios, the duration of MVPE was defined as

PV weighted sum of asset durations − PV weighted sum of liabilities durations

This idea was used to estimate the effect on MVPE of a change in rates. However, now many institutions find it feasible to produce a shock to the yield curve and recalculate MVPE directly. Thus, for instance, if interest rates go up 1% and stay at that level, we would find from the table below that the cost of funding increases (since it is floating) but the assets inflows are fixed rate.

Here MVPE declines to approximately £68.6M, indicating that the bank's earnings are fairly sensitive to the level of rates. From an NII perspective, things are dire: at 4% 1-month Libor, NII is £10M, but this falls to a loss of £1.6M if rates rise 1%.

8.3.2.3 Non-parallel rate moves and scenario analysis

Once the basic technology to produce MVPE for a given evolution of the yield curve exists, there is no need to stick to the simple scenario of an immediate parallel rate rise. The bank can run a variety of scenarios involving slowly rising rates, falling rates, flattening or steepening of the curve, faster changes and so on. A thorough approach would involve recalculating the expected cashflow map for the change in rates using prepayment and deposit migration modelling where necessary.

Assets and Liabilities: +1%	0–90 Days	91–180 Days	181–270 Days	271–364 Days	1–2 Years	Perpetual or Not Rate Sensitive
Loans					900	
Loan interest	11.3	11.3	11.3	11.3	11.3	
Securities		306.3				
Fixed assets						60.0
Loan provisions					20.0	
Retail deposits		(100.0)			(400.0)	
Deposit interest	(6.3)	(6.3)	(5.0)	(5.0)	(20.0)	
Issued debt				(400.0)		
Debt interest		(8.4)		(8.4)		
Interbank borrowing	(2.6)	(202.6)		(−400.0)	(410.2)	
Dividends	(2.0)	(2.0)	(2.0)	(2.0)	(9.0)	
Net expenses	(9.0)	(9.0)	(10.0)	(10.0)	(42.0)	

8.3.2.4 Leverage

This analysis also gives a slightly different view of leverage in that it allows us to look at things from a PV perspective:

$$\text{PV of portfolio leverage} = \text{MVPE/PV(assets)}$$

> *Exercise.* How different is this from the usual definition of leverage for a typical bank?

8.3.3 Strategic Risk and Real Options

The analysis of the previous section relied on assumptions about the evolution of the bank's funding. This has not been locked in, so the bank has both liquidity risk [discussed in the next chapter] and flexibility. For instance, rather than rely on interbank borrowing at Libor + 10 bps from the expiry of the debt to the maturity of the loans, it could choose to offer a new fixed rate term deposit account in the hope that this would gather the desired £400M funding. This is a *real* option as opposed to a financial one.

Real options management is really strategic risk management: the bank's management must decide which business choices to make on the basis of reputational considerations, likely income and risk profile. Thus, the new deposit account suggested in the previous paragraph does not just involve a choice about funding: it is also a choice about the kinds of account the bank wishes to be seen to be offering. Moreover, successful funding will depend on the bank actually being able to open the accounts efficiently, to service them and so on.*

8.3.4 ALM Risk Reporting

As with any other risk reporting, the key to effective ALM reporting is simple, relevant, accurate reporting. Typically, this is scenario-based, focussing on how key variables change with various interest rate and FX scenarios:

- Current estimates of NII, gap by bucket and cumulative gap by bucket;
- Estimate of the effect of the chosen movements on these parameters and on MVPE;
- Investment portfolio scenario analysis, as with any investment portfolio, but perhaps with more emphasis on the prepayment behaviour to the extent that this is intended to hedge prepayment risk elsewhere in the banking book.

Limits may well be set on the basis of these scenarios. For instance:

Interest Rate Scenario	NII	Change	Limit	Breach?	MVPE	Change	Limit	Breach?
Immediate move								
+2%	−13.5	−23.4	−5	Yes	59.2	20.8	20	Yes
+1%	−1.6	−11.5	−3	Yes	68.6	11.1	12	No
0%	9.9		N/A	N/A	80.0		N/A	N/A
−1%	21.4	11.5	−3	No	91.2	11.2	12	No
Gradual moves								
+2%					62.1			
⋮					⋮			

The Treasurer and his team clearly have some work to do here to get back within limits.

> *Exercise.* What disclosures do banks provide on the interest rate sensitivity of NII and MVPE?

---

* While this might not sound difficult, a number of banks have suffered reputational damage when it was perceived that their new Internet-based current accounts were not being administered with the efficiency the customer expected.

8.3.5 P/L Translation and Hedging

International banks typically find mismatched assets and liabilities in a number of currencies. In addition to the problems of assets in one currency and liabilities in another, such as the sensitivity of NII and MVPE to FX rates, there are also several further issues.

8.3.5.1 Capital hedging

The potential need to hedge capital comes due to investments in foreign subsidiaries which must be capitalised in local currency:

- If the sub is wound down, capital will have to be repatriated to the parent;
- But this happens at the then current FX rate.

Since firms typically set up subsidiaries with a business plan which extends several years, this is often long-dated risk. Moreover, it is 'wrong way' risk in the sense that if the country does well, average profits tend to grow and the currency appreciates, whereas if the country falls into a deep recession or suffers a country risk event, average profits are lower and the currency may depreciate.

8.3.5.2 Profit hedging

Subsidiaries often make profits in local currency. FX rates affect what these are worth in the bank's accounting currency. Therefore, hedging could potentially reduce earnings volatility. But how much should the bank hedge and for how long?

- The budget offers one answer, but here there is some delicacy depending on whether this is really an accurate estimate of the expected earnings or a 'stretch goal' intended to motivate managers.
- Suppose a manufacturer can pass on increased costs due to changes in the local currency price of raw materials to customers at the end of an account period. Then it makes sense to hedge for that period since that is the duration of the exposure.
- Financial services firms, however, often cannot easily increase their local currency profitability to reflect adverse FX movements. Therefore, it is less obvious how long to hedge expected local currency profits for.

Exercise. Devise a local currency profit hedging strategy for an international bank. You should consider: how much to hedge, how long to hedge for, whether to lock in the current rate via forwards or whether to buy options to hedge downside.

Suppose the FX rate you are hedging moves 20% and stays at the new level. What is the impact at the end of your hedge programme, and what would you do next? Does your answer here affect whether it would make sense for the bank to hedge capital or not?

8.3.6 The Role of the ALCO

The 'ALCO' was historically one of the most important committees in the bank since it was responsible for setting policies and risk appetite for the ALM process, monitoring the implementation of funding and ALM strategies by Treasury, and often other areas of bank risk management and external relationship management too.

The ALCO is usually chaired by a board member, and staffed by senior risk management, Treasury and finance personnel. It may have a number of sub-committees focussing on particular areas: indeed, in some banks, both market and credit risk management committees report into the ALCO, whereas in others, the ALCO focuses on Treasury matters and other risks are dealt with in a separate committee structure.

8.3.6.1 Process and organisation

The ALCO is given a remit from the board.

- In turn it delegates authority either to functional groups or to sub-committees.
- That authority will include the reporting the ALCO desires, details of when an ALCO must be convened and representation in other parts of the bank's organisation.
- The ALCO is also sometimes given a role in investor relations such as managing communications with ratings agencies and responsibility for certain aspects of the annual report.
- The main board of the firm will receive regular reporting from the ALCO and will monitor its performance.

8.3.6.2 Capital and liabilities

The ALCO typically has the following functions relating to issued securities:

- Defining economic capital allocation policy;
- Reviewing capital allocations;
- Either taking action on the basis of risk-adjusted performance measures or recommending such actions to the main board;
- Setting policy for capital security and other liability issuance;
- Reviewing the current state of the bank's capital and liability base and any proposals for significant changes to it;
- Monitoring the bank's liability diversification and making changes where necessary;
- Taking responsibility for the management of regulatory reporting and strategies relating to the maintenance of capital adequacy in some banks.

8.3.6.3 Liquidity and funding

The ALCO's role here relates to activities undertaken in the TP book:

- Definition of the bank's liquidity policy and liquidity risk appetite [as discussed in Chapter 9];

- Definition or approval of the bank's TP policy, setting of high-level limits for the TP book and regular review of its risk levels;
- Responsibility for the management of the sensitivity of key financial variables such as NII and MVPE to interest rates and FX movements;
- Review of consolidated liquidity risk reporting and management of this risk where it is inconsistent with the risk appetite;
- Review of the bank's liquidity stress plan and the results of liquidity stress testing;
- Definition of the policy for the FX hedging of capital and income, and management of the implementation of this process.

8.3.6.4 Investment

The ALCO also oversees the investments made by Treasury:

- Definition of the policy for the investment of surplus liquidity and setting limits for it;
- Review of risk reporting relating to the liquidity portfolio.

Exercise. See if you can get the reporting pack and minutes for a real bank's ALCO.

8.4 TRADING BOOK ALM

The funding of trading book assets offers the choice between a number of alternatives. In this section we discuss the forms of funding available and the basis on which the choice between them is made.

8.4.1 Repo and Other Forms of Secured Funding

A bond repo [discussed in section 1.1.5] is a form of collateralised lending: the bond is collateral against a loan at the repo rate. The repo markets are enormous, and they usually offer considerable liquidity to banks either looking to place surplus funds (by repoing in bonds) or to fund their securities portfolios.

Another trading book funding structure already discussed is the total return swap. Here again an institution with low funding costs or surplus liquidity can earn a spread by financing another firm's asset.

8.4.1.1 Funding bond positions

Thus, a securities trading book has a number of routes available for funding a bond:

- Borrow from the TP book at the rate offered by Treasury;
- Repo the asset overnight or to term;
- Total return swap the asset.

8.4.1.2 Funding equity positions

A similar range of options is present for equities, although matters are slightly more complex as the financing counterparty might have different motivations:

- Some players in the *equity repo* or *equity swap* markets provide financing for equities just as in a bond repo.

- Others want stock to be able to sell it short, possibly with the expectation of being able to buy it back more cheaply later. Thus, in a *stock borrowing* or *stock lending* transaction, one counterparty lends an equity position to another in exchange for collateral (which can in turn be repo'd) and a fee. The borrower thus has stock to deliver into a short sale, and the lender has collateral which might be easier to finance than the original stock.

Another extra complication for equity comes from dividends. In a total return swap on a bond, all bond coupons have to be passed on to the return receiver. Similarly, in an equity swap, dividends payments are part of the total return. However, there is often a different tax treatment on a *manufactured dividend* provided synthetically under an equity swap to a real one paid to the holder of record of an equity.

8.4.2 Practical Issues in the Funding of Trading Books

Given a book including securities, the bank has to decide on the right form of funding on the basis of cost, and on the term of the funding available. This process is complicated by several factors.

8.4.2.1 Funding at Libor flat

Recall that the Black–Scholes formula for pricing an equity or commodity option includes an interest rate input. This is because the hedging argument which gives the Black–Scholes price relies on being able to borrow at this rate to support a long or receive it against a short. Many trading systems use the Libor curve as this input. However, this may not give the correct rate:

- Unsecured Treasury funding via TP is not usually at Libor flat, and may often be at a spread above it.

- Shorts are effectively financed at the stock borrow cost, which is usually above Libor, and may for some stocks be significantly above it.

- In contrast, repo markets often offer funding cheaper than Libor flat for GC bonds.

Some trading systems allow for this effect by including a financing spread input: where this is not present, traders need to be aware that their real financing cost will introduce an extra P/L not predicted by the mark-to-model valuation.

8.4.2.2 Ownership of trading book risks

Who is responsible for funding the securities in the trading book? This needs to be clear so that someone is optimising the bank's cost of funds under the constraint of its risk tolerance. For instance, if traders are responsible, they can chose between taking the transfer price rate or funding their position elsewhere via total return swaps or repo. This has the advantage of making the trader completely responsible for all aspects of their book, but the disadvantage that valuable repo lines may be used without Treasury being aware that this is happening, so reducing the sources of funds available in a crisis. Moreover, traders will tend to take the cheapest source of funds, resulting in many bonds being funded using rolling repos; this in turn means that if these repos are terminated, the bank has to find alternative financing in hurry. In effect this position is short a liquidity option that Treasury may not be aware of.

On the other hand, if Treasury is responsible, they will tend to prefer term funding in some cases to reduce the bank's gap position. If funding then becomes unnecessary, for instance because the delta of the option the securities are hedging approaches zero, there may be a cost in terminating the unwanted financing transaction.

It is good practice then for there to be a close dialogue between the traders of a book with large securities positions and Treasury. Treasury needs to know what the securities are for and how long they might have to be financed for; trading needs to know the range of financing alternatives available given the bank's tolerance for mismatch in the funding base.

CHAPTER 9

Liquidity Risk Management

INTRODUCTION

The last chapter considered the interest rate and FX risks an institution might take as part of its funding strategy. One of the central questions was how much the bank's cost of funds might change if interest rates moved given how it plans to borrow. Here we look at another aspect of funding: the risk that the firm might not be able to borrow *at all*, or only at prohibitive cost, perhaps due to a fall in confidence or to a market-wide crisis. This is *liquidity risk*.[*]

Liquidity risk in a firm occurs due to the mix of assets and liabilities, so we look at both sides. Asset liquidity concerns the ability to turn an asset into an amount of cash close to where it is marked; liability liquidity concerns the behaviour of a firm's liability base in various conditions. Once we have seen how some of the different parts of the balance sheet behave, liquidity risk management is introduced; this process is designed to keep liquidity risk within bounds without subjecting the firm to too high a cost of funds.

Contingent liquidity instruments are the ones which supply or demand liquidity under certain conditions such as uncommitted LOCs. We touch on these to highlight the liquidity options a firm might have positions in.

Finally, liquidity in a crisis is discussed and the techniques firms use for dealing with these events are touched upon.

9.1 LIQUIDITY OF SECURITIES AND DEPOSITS

Liquidity is the ability to meet demands for cash. These demands might be either *expected*, as in a coupon that we know we have to pay on an issued security, or *unexpected*, as in the early exercise of an option.

---

[*] For a wider and more comprehensive discussion of liquidity risk, see Avinash Persuad's *Liquidity Black Holes: Understanding, Managing and Quantifying Liquidity Risk* or Erik Bank's *Liquidity Risk*.

9.1.1 What Is Liquidity Risk?

Liquidity risk, therefore, is the risk that a firm may not be able to meet its commitments when they become due at a reasonable cost. Consider the bank from the last chapter funding its loan portfolio with a mixture of deposits and interbank borrowing. The bank funds itself this way to enhance its NII but, in doing so, it takes interest rate risk and liquidity risk.

To see this, suppose that one of the bank's largest counterparties defaults and, as a result, rumours arise questioning the solvency of the bank. It might then have difficulty in rolling its interbank borrowing: professional counterparties will not want to lend to it any more. There are then a few alternatives:

- The bank can raise funds in the secured market, perhaps by repoing its securities.
- It could try to attract more retail deposits.
- Alternatively, if both of these fail, it might have to sell its securities to raise enough funds to meet its expenses. At this point, it is a forced seller, and the liquidation prices obtained for the assets could be fairly far from the bank's mark-to-market prices for them.

Notice that the bank is *not* insolvent or even necessarily anywhere close to insolvency; it is just that the rumour that it *might be* is interfering with its ability to borrow in the interbank market. If the rumours become widespread, they might affect retail deposits, and the bank's problems worsen.

Liquidity risk arises in any situation where assets and liabilities are not completely matched. In financial institutions, average liability duration is often shorter than average asset duration and some forms of the funding used, such as deposits and CP, can be withdrawn at short notice. This gives rise to risk as a decline in confidence makes funding expensive or even impossible to obtain. On the contrary, if assets mature first, there is reinvestment risk on the cash. This is usually not as serious.

9.1.1.1 Sources and sinks of cash

The major drivers of planned payments have already been discussed [in Chapter 8]. Some of the causes of an unexpected need for cash are:

- Disappearance of expected cash inflows due to a counterparty suffering a credit event;
- Operational risk;
- Early exercises of derivatives or drawdowns on LOCs;
- Puts of putable debt;
- Withdrawal of funding due to a counterparty withdrawing from a rolling repo or a bank cancelling an LOC.

9.1.1.2 Bank failures

Liquidity risk is independent of other risks in that it can cause a bank to fail even though it has not lost money. For market risk to cause a bank failure, the bank must lose a large amount of money. But with liquidity risk, all that is needed is for the bank to have mismatched funding and be unable to borrow. Then a failure to meet any senior obligation—including principal or coupons on issued debt, repaying called deposits or payments on derivatives—causes the bank to default, even though it has more assets than liabilities. Indeed, some estimates suggest that roughly half of historical bank failures were in institutions that would have been adequately capitalised under the Basel I rules. Their problem was that they could not borrow to meet a senior obligation, so they defaulted. Presumably in this situation, the senior debt recovery would have been close to 100%, but that is no consolation to the bank's staff or subordinated creditors.

Exercise. Investigate the timeline of a bank failure. When did the critical events that led to the failure happen, and until when could depositors withdraw money? Was deposit insurance involved and, if so, when did it pay out?

9.1.1.3 Endogenous and exogenous liquidity crises

A liquidity crisis caused by the bank's actions or strategy without a broader market crisis is sometimes called an *endogenous liquidity event*. In contrast, sometimes there is a broader rise in the cost of funds for banks caused by a market crisis. The Southeast Asian, Russian and Brazilian crises all resulted in a flight to quality with funding becoming more expensive. This kind of event is known as an *exogenous liquidity crisis*.

9.1.1.4 Central bank liquidity provision

In many countries, the central bank often has a role in mitigating an exogenous liquidity crisis in that it can open the window. The *central bank window* or *discount window* is a borrowing facility whereby the central bank offers loans to banks secured by collateral. Think of an old-fashioned office where you approach a counter, hand over a bond and walk out with cash. In that sense, the discount window offers similar facilities to the repo market, but since the central bank controls both the rate at which banks can borrow—known as the *discount rate*—and which assets qualify as collateral, it give authorities flexibility in the provision of liquidity.

In ordinary conditions, the central bank uses the discount window as an instrument of monetary policy,* but in emergencies by offering a lower rate or increasing the range of acceptable collateral or both, the central bank can mitigate the effect of a liquidity crisis. This is sometimes known as the central bank acting as a *lender of last resort*.

---

* Opening the discount window wide—lending more—increases the money supply. Closing it contracts the money supply. The central bank's money supply objectives can be achieved either via changing the rates offered or (less commonly) via changing the range of collateral accepted.

9.1.1.5 The nature of financial firms' liquidity

Most financials are subject to liquidity risk because they have liquid liabilities expiring before illiquid assets:

- Banks use short-term paper or demand deposits to fund long-term loans.
- Broker/dealers use short-term paper or rolling repo to fund long-term securities.
- Insurance companies invest premiums in long-term assets which have to be liquidated if a claim is made.

This situation is sometimes called *maturity transformation*. It means that the inability to roll liabilities or liquidate assets gives rise to a risk of default. Therefore, to understand liquidity risk, we need to review both sides of the problem: liability liquidity and asset liquidity.

> *Exercise.*   Examine the annual report of a large broker/dealer. What can you deduce about its use of the repo market for funding?

9.1.2 Liability Liquidity

The following table summarises the various sources of liquidity and their characteristics.

Liability	Contractual Term	Liquidity Risk
Retail deposits	Often demand	Medium to low (if insured)
Commercial deposits	Demand or short term	High: confidence sensitive
Interbank borrowing	Short term	High: very confidence sensitive
Commercial or other short-term paper	Short term	High: very confidence sensitive
Purchased LOC	Short term, may be cancellable	Medium if not cancellable, high if cancellable
Term debt without puts or calls	Long term	Low: term funding
Capital security	Long term, may be perpetual or extendable	Very low: especially for perpetual securities

Liquidity is a two-sided phenomenon in that institutions that require access to liquidity place demands on institutions or individuals that supply it: a loan is a liquidity source for the borrower, and an asset that has to be funded for the lender. Thus, the best liabilities for the issuer are also often the worst assets from a liquidity perspective for the holder.

9.1.2.1 Positive liquidity

There are a few firms that have the opposite problem to the usual one: excess positive liquidity. This typically happens either in banks specialising in deposit taking, such as some Internet banks, or in corporates which manage to persuade their clients to pay for their products before they have to pay their suppliers for the materials used to produce them. In this case, liquidity risk management focuses on the short-term investment of this positive cash position. This shows that there is a spectrum of liquidity risk with sovereigns (that can print their own money), cash generative corporations such as utilities and deposit-rich banks with few long illiquid assets at one end of the spectrum and those broker/dealers or hedge funds that are highly reliant on one source of funds at the other, as in the illustration below.

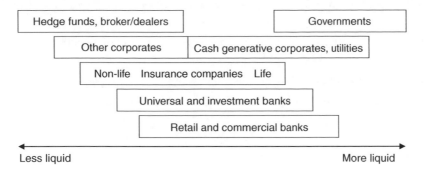

9.1.3 Asset Liquidity under Ordinary Conditions

Asset liquidity matters because if a firm tries to raise liabilities to finance assets and fails, it has to try to liquidate assets. There are two aspects to asset liquidity:

- Dealing with those assets that we *know* might take some time to sell, such as commercial property, bonds or equities which do not trade regularly, large positions or highly structured positions;

- Managing the situation where an asset we think is liquid turns out not to be when we try to sell it.

The first situation will be the topic of this section.

9.1.3.1 Securities

Earlier [in section 1.5.2], the need for marking to a liquidation price was discussed in the context of valuation policy and accounting standards. This practice—perhaps achieved using mark adjustments where data feeds of bid or offer prices* do not provide an accurate assessment of the realisable value of a position—is also an important control for liquidity risk management purposes.

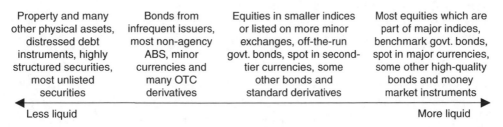

A firm's valuation policy should therefore ensure that position valuations really do reflect the worth of its security and derivatives positions. The less liquid something is, the more conservative it is necessary to be here.

---

* It is prudent to be skeptical of some screen prices for securities. Although the well-known data vendors can probably be relied on for major market equities and the best-known bonds, away from that screen prices can be out-of-date, unreliable, only representative of a single dealer (perhaps in your institution) or all three.

Typically, security liquidity is good for ordinary size position in major equity markets and in some bonds from a few thousand major issuers; away from that, or for larger positions, liquidity can fall off fast. Thus, there is a spectrum of liquidity as shown in the illustration above.

The liquidity of a security has an impact on several aspects of a security:

- The *time* potentially required to sell the position;
- The *transactions costs* incurred on selling the position and most importantly;
- The magnitude of the *uncertainty in the liquidation value* of the position.

It also has the following effects:

- Liquid assets trade at a premium to illiquid ones with similar risk profiles (which is why some financial institutions prefer holding illiquid instruments). Part of the reason for this *liquidity premium* is the fact that
- Liquidity affects funding: the repo haircut, repo level or total return swap spread for an illiquid asset is likely to be less advantageous than for a liquid asset, increasing funding costs.

9.1.3.2 Block trading and position liquidity measures

The *block trading* desk in an equity trading firm is one area with expertise in the management of liquidity risk. This group bids for large blocks of stock from clients—typically pension funds or investment managers who build up substantial positions—and then liquidates them in the market. The block trading desk usually concentrates on positions that are a substantial fraction of the total daily market volume in a stock, perhaps 20% or more.

Two forms of block trade are common:

- In a *risk bid*, the trader provides a cash bid to buy the block of stock, and then takes the risk that it will not be able to sell the equity position at a higher level. Typically, this bid will be significantly lower than the current market price to reflect the likely fall in the price during the liquidation of the position;
- In a *VWAP bid*, the bid is a spread to the *volume-weighted average price* over the liquidation period. Thus, a client might offer a number of brokers a chance to bid on a position of 300M shares of Diageo, a FTSE 100 component. Diageo's average daily market volume is 250M shares, so this is a sufficiently large position to be considered a block trade. The winning bid is VWAP − 20 bps over 5 days, meaning that the client will receive the VWAP over a 5-day period times 99.8%.

A VWAP bid means that the broker then takes the risk that they cannot execute at the average price, but they do not take the risk that the stock value falls significantly as the position is being liquidated. The spread represents the broker's dealing costs and profit on the trade.

The difference between where a broker would quote a risk bid on a stock and the spread to VWAP it would quote is a measure of the expected *price impact* of the trade. For the Diageo position, if the risk bid is 6% below market, this means that the broker expects the position to fall by no more than 5% or so during its liquidation on the basis of their experience of selling large positions.

Ultimately, fair value accounting depends on being able to execute trades at the mark. Therefore, a good test of market besides any academic debates about valuation is to force a sale: if a trader can sell 10% of his position quickly and in one go at better than the marked level, then you can have a measure of confidence in the mark. If not, a mark adjustment is called for.

Note that block trading desks tend to be a large user of risk limits during the liquidation of positions but to have *de minimis* risk otherwise. Backtesting an SR VAR model on such a desk would provide an interesting test of the model's accuracy.

Exercise. Examine the trade reporting for a major stock: this is often available from the exchange's website.* Try to get a sense as to where large trades (10% of daily volume in one trade or more) are done compared with small trades.

Why might a broker/dealer bid particularly aggressively on a block trade?

9.1.3.3 Security-based liquidity measures

There are four common measures of the liquidity of a security:

- *Bid/offer spread.* This measures how tight the market is.
- *Volume.* The larger the daily volume, the larger an order can be executed without moving the market.
- *Price impact of a trade.* How much the market moves if a single trade of, say, 10% of daily market volume is executed is a good indication of the extent to which larger blocks can be absorbed.
- *Recovery time.* The period of time it takes the market to return to equilibrium after a large trade has been executed is another measure of liquidity.

It is important to note that these measures give different perspectives; for instance, the volumes traded in most securities markets have been steadily rising, but that does not imply a falling price impact of a trade if there is less risk capital present.

Twenty years ago, some market makers would absorb big sales onto their balance sheets and warehouse them for days or even weeks, allowing the market to recover quickly, and the market maker to take a profit as they dribbled the position back into the market.

Currently, many firms seem less willing to commit their own balance sheets, so although flow is larger, the market may be less resilient. The availability of risk capital to provide liquidity for large positions in turbulent markets may even have fallen.

---

* See, for instance, www.londonstockexchange.com.

9.1.3.4 Pools of liquidity

Liquidity requires willing buyers and willing sellers. That in turn implies standardisation: agreement on using a particular instrument. Thus, for instance, there is no particular reason that the future on the 10-year bond has become a standard long-dated interest rate hedging instrument rather than, say, a future on an 8- or a 15-year bond. But now that this contract is liquid, there is considerable incentive for someone wishing to hedge 12-year rates to use the 10-year future: prices are easily observable, trade execution in size is (almost) certain and bid/offer spreads are low. A 12-year OTC future, in contrast, might not offer any of these advantages despite being a nominally more accurate hedge.

> *Exercise.* Examine some successful (high open interest) and some failing or failed exchange contracts. Why were users of the successful ones prepared to sacrifice precision of their hedge for liquidity, and why did the unsuccessful ones not attract interest in a similar compromise?

The nature of market participants' commitment can also be important for the liquidity of an instrument. In the equity markets, there is a long tradition of *market making*, so even where participants have no legal obligation to make continuous prices, there may be a reputational incentive to do so. In the OTC markets this is much less common, so these pools of liquidity are potentially more volatile.

9.1.3.5 On- and off-the-run bonds

Treasury bonds give a good example of *liquidity premiums*. The *on-the-run* Treasury is the most recently issued bond of a given maturity (and hence probably the one whose coupon is closest to the current yield); older bonds are known as *off-the-run* and these are less liquid. There is typically a spread between the on-the-run and off-the-run treasuries despite having identical maturity and credit risk: for the 30-year Treasury bond, this varies between 2 and around 20 bps. Tighter liquidity premiums imply more risk capital in the market and a lower level of risk aversion; larger ones imply that investors want more compensation for taking liquidity risk, and hence typically that the market is in a more stressed condition.

> *Exercise.* What are the correlations between the on-the-run/off-the-run spread, the average spread of long-dated AA bonds, and the swap spread?

9.1.3.6 Corporate loans

At least some securities can be readily liquidated. The situation is more bleak for corporate loans in that—away from a number of names that are relatively liquid in the syndicated loan or loan trading markets—finding a buyer for a single loan is not easy, and may anyway be impossible due to restrictions on transferability. Securitisation is one alternative, but given the relatively extended period needed to arrange these transactions, this is unlikely to be a useful way of raising funds in a crisis. It does, however, provide some banks with an effective source of funds in ordinary markets: most lower-credit-quality banks—those rated A+ or below—would have a cost of term funds some tens of basis points above Libor, yet the supersenior tranche of a CLO might trade as tight as Libor plus 5 bps, thus providing an attractive source of term secured funds for an illiquid loan portfolio.

9.1.3.7 Retail assets: mortgages and unsecured loans

A variety of retail banking assets including mortgages and unsecured loans can also form the collateral for securitisations [and we discuss mortgage-backed securities in more detail in section 10.2.1]. Another possibility is to issue a *covered bond*: this is a debt security issued by the originator but also backed by mortgage collateral. Thus, unlike a securitisation it is an obligation of the issuer; for a securitisation, the tranches are usually issued by an SPV, so it is the collateral quality alone that backs the bonds. This form of asset finance for banks is particularly important in Germany, but it is also used elsewhere.

9.1.3.8 Receivables financing

The flipside of banks raising liquidity for themselves by securitisation is banks providing liquidity for others. A good example is receivables financing [discussed in section 10.3].

9.2 LIQUIDITY MANAGEMENT

If a financial is in the happy position of:

- Being able to hold all assets and liabilities until maturity (so that in particular market events and the presence of risk limits do not ever require the early liquidation of assets);
- Having duration-matched funding (and in particular no liabilities which can be presented for repayment earlier than the matching asset);
- Having an immaterial risk or size of unexpected payments;
- Accounting in such a way that unrealised losses do not cause a material risk of recapitalisation;
- And having positive NII

then liquidity risk is likely to be very low. However, this situation is unusual and expensive to achieve: firms usually prefer to have some liquidity risk and a lower cost of funds. But it means that liquidity risk management is necessary. In this section, some of the tools involved in this process are discussed.

9.2.1 Measures of Liquidity Risk and the Firm's Liquidity Profile

Some simple measures of liquidity risk are based on the gap analysis we discussed in the previous chapter.

9.2.1.1 Daily cash management

A more detailed version of the gap report is a good starting point for daily management. A typical short-term report might look like

Expected	Overnight	1 Day	2 Days	3–5 Days	Next Week	3 Weeks	<1 Month
Gap	101.2	−50.4	−82.1	−34.2	10.3	−51.6	35.3
Limit	−5	−10	−50	−50	−100	−150	−200

The overnight position is long cash, which is safe. It is better to be investing excess funds at the end of the day, and run the risk of getting a slightly lower rate for them than expected, instead of being short cash and being at the mercy of the interbank market. In this case, half the excess will be used up tomorrow, so we only have to place some of the excess overnight: the rest can be placed for longer. Later in the week, the funding profile goes negative, indicating a need for funds: the firm has a few days to source this cash. The limit expresses the ALCO's tolerance for gaps at various horizons.

Some firms—especially those with volatile cash needs—also calculate an expected cash position on the basis of a probabilistic model of cashflows. For instance, deposit migration is modelled and the funding outflow at 99.9% confidence is estimated; loan prepayments are estimated; and so on. This gives a cashflow report at a 99.9% confidence interval with correspondingly larger limits:

99.9%	Overnight	1 Day	2 Days	3–5 Days	Next Week	3 Weeks	<1 Month
Gap	−9.6	−92.0	−104.5	−96.1	−20.4	−141.2	−63.1
Limit	−5	−100	−200	−300	−400	−500	−500

This shows that the current position is above the firm's stress liquidity limit in several buckets, so term borrowing would be increased slightly to generate funds. This probably will not be used, but it gives the firm a margin of safety if unexpected cash outflows have to be made.

9.2.1.2 The cost of borrowing

The liquidity management process aims to keep the net funding basis within reasonable bounds. Part of the reason for this is that as more funding is needed, it gets more expensive; and as surplus cash increases, the rate at which it can be placed declines. The situation is complicated by the use of LOCs, the repo market and so on but it does emphasise the importance of controlling liquidity.

Note that the market often *rations* liquidity rather than pricing it: a firm might be able to borrow up to £20B at Libor plus 10 bps in the interbank market, but if it tries to borrow £21B in the same place, the funding cost *on the whole notional* can go up to Libor plus 20 bps.

9.2.1.3 Liquidity scenarios and liquidity limits

The unexpected cashflow report captures a bank's estimate of cashflows on the basis of probabilistic modelling; this is an ordinary conditions report, so it is reasonable to assume that some of these could be usually covered by borrowing, for instance in the interbank market. Different considerations apply to markets in stress [and we leave that to section 9.4.1].

9.2.1.4 Medium-term liquidity management

Medium- and long-term liquidity management usually concentrates on liabilities. Thus, a large gap in the 6-month bucket might be met by a planned MTN issuance, with the firm relying on purchased liquidity via LOCs to cover any shortfall if this cannot be executed.

9.2.1.5 Liquidity ratios

A variety of financial ratios are sometimes used to give a quick insight into liquidity risk [as mentioned in section 5.4.5]. For traditional banks, deposit and borrowing ratios are popular:

$$\text{Deposit ratio} = \frac{\text{Cash} + \text{Highly liquid marketable securities}}{\text{Total deposits}}$$

$$\text{Borrowing ratio} = \frac{\text{Total deposits}}{\text{Borrowed funds}} \quad \text{and} \quad \text{Volatile borrowing ratio} = \frac{\text{Total deposits}}{\text{Volatile borrowed funds}}$$

More sophisticated views would include contingent assets and liabilities as discussed below.

9.2.2 Policies, Procedures and the Regulatory Perspective

Since a run on a bank is a classic way for financial institutions to fail,* regulators have a range of requirements concerning the management of liquidity risk.† These include the following areas:

9.2.2.1 Policies

Firms should have documented liquidity and funding strategy and policies approved by the board of directors. There should be a management structure in place to execute the liquidity strategy: typically, this might involve a subcommittee of the ALCO, the *liquidity committee*.

9.2.2.2 Risk measurement

Firms must have adequate information systems for measuring, monitoring, controlling and reporting liquidity risk including daily processes, scenario analysis and longer-term planning. So far, this is motherhood and apple pie, but interestingly the requirement continues:

> Firms should review the assumptions that underlie their liquidity strategy and ensure that they continue to hold. They should consider testing their name in the market on a regular basis even if they have no need for funds.

---

* Runs can happen to insurance companies too where policies can be presented early for cash: a good example of this is the failure of General American Insurance.
† See, for instance, the Basel Committee on Banking Supervision's *The Management of Liquidity Risk in Financial Groups*.

In other words, firms should have a realistic assessment of their ability to use the interbank market, and they should not assume that it will always be available in the desired size. This hints at the importance of signalling in the liquidity market: if a firm is not seen borrowing from one year to the next, it should be no surprise if the market concludes that it is in dire need of liquidity when it does try to borrow and charges accordingly. A regular borrower is more likely to be able to access the market without inadvertently raising concerns.

Further regulatory requirements relate to contingency planning, disclosure and supervisory action.*

9.2.3 Upstreaming, Downstreaming and Corporate Structure

The obvious approach to liquidity planning is to have a centralised Treasury function and to manage liquidity on a group-wide basis (perhaps using local entities for security issuance if needed). However, this can give rise to a number of issues, particularly for financial conglomerates:

- There may be regulatory requirements which prohibit upstreaming or downstreaming of funds within the group. For instance, local regulators may be unwilling to permit deposit funding to leave their jurisdiction or they may not wish to rely on an off-shore parent's ability to provide liquidity in a crisis. Streaming funds from a bank within a broader non-bank group can be particularly problematic.

- Banks can organise themselves using *branches*. This permits bank-wide liquidity management as there are no constraints on inter-branch funds transfer. In contrast, if a non-bank wishes to set up an entity in a new country, it must use a subsidiary company and potentially suffer more constraining rules on intergroup exposures.

- Local bank branches can also be useful in that they can take deposits. A bank that can fund in local currency via deposits may have both access to a larger funding base and an all-in lower cost of funds than a non-bank which is forced to fund in dollars and swap to local.

- Life insurance subsidiaries within banking groups can pose particular problems: if their solvency falls, the parent has to inject extra cash to retain their capital adequacy, but extracting that capital again when the situation improves can be difficult. Decreasing solvency often happens when equity markets are falling, so the connection between market risk and recapitalisation risk can be troublesome.

9.2.4 The Implications of Illiquidity for Pricing and Risk Measurement

The standard analysis of security prices is based on a number of assumptions:

- Security returns are based on random variables following a random walk at fixed volatility.

- In particular, this walk follows a continuous path.

---

* See the BCBS publication *Sound Practices for Managing Liquidity in Banking Organisations* available on www. bis.org for more details.

- So though we cannot predict a future price or return, we can engage in continuous hedging.

This continuity is important for two reasons:

- Black–Scholes option pricing depends on it.
- Some risk measures such as VAR depend on it.

Both are potential problems for illiquid assets.

9.2.4.1 Derivatives pricing under illiquidity

The Black–Scholes formula [as we discussed in section 2.3.1] relies on a replicating portfolio. We construct an instantaneously risk-free portfolio consisting of delta of the underlying security, the derivative and a position in cash. This portfolio requires continuous rebalancing to remain hedged, and the Black–Scholes option price is an estimate of the cost of this hedging process. Therefore, the correctness of Black–Scholes (or indeed any other approach based on a replicating portfolio) depends on the ability to perform this replication. For illiquid assets, perfect replication may not be possible and therefore Black–Scholes prices are questionable.

There are numerous models of derivatives pricing under illiquidity, but most of them display the following effects:

- In general, hedging is costlier: selling an option at the Black–Scholes fair value often results in a loss.
- High gamma causes the need for more rebalances and hence makes the problem worse.

One easy technique for estimating the impact of this effect is to use a practical re-hedge analysis [as discussed in section 2.3.2].

> *Exercise.* Assume a 2% bid/offer spread and the ability to trade twice a week. Price a 1-year at the money call on a 40% volatility asset using Black–Scholes and using a hedge simulation incorporating illiquidity and transaction costs. How different are the two prices?

9.2.4.2 Liquidity-adjusted VAR

The calculation of 99% n-day VAR is based on the following paradigm:

- We have a position in a security with returns distributed in some known way (typically either normally or according to some historical data).
- We hold the position for n days.
- We calculate how far its value could have moved during that period at a 99% confidence interval.
- Because we assume that the position can then *be sold*.

This is clearly a liquidity assumption. A simple approach to fixing it is to adjust the holding period for illiquid underlyings. Consider the following Monte Carlo or historical simulation:

- The whole portfolio is simulated for n days.
- Then we remove the liquid part of the portfolio and continue the simulation for the illiquid positions, reducing their size as we think they could be sold.
- We continue until the last position has been liquidated.
- This gives rise to a P/L for the portfolio that reflects a realistic liquidation horizon.
- This is repeated many times to obtain the P/L distribution and hence an estimate of its 99th percentile.

This does not quite capture the full phenomenon, though, as large falls in an asset are likely to be accompanied by increased illiquidity. A more sophisticated approach would attempt to estimate the likely liquidation horizon contingent on being in the 1% tail of the distribution.

9.3 OFF-BALANCE-SHEET LIQUIDITY AND CONTINGENT FUNDING

Contingent liquidity occurs when a firm can access liquidity when it is needed. The ideal arrangement will give certainty of rapid access to cash: the obvious example is a purchased committed LOC. This and other instruments used by financial institutions to provide backup liquidity are discussed in this section.

9.3.1 Positive Contingent Liquidity

Purchased LOCs form an important source of contingent liquidity. However, here it is important to be aware of the precise nature of a line: cancellable lines, or short-term rolling lines, are much less useful than long-term irrevocable lines. Some previous flights to quality such as the LTCM event [discussed in section 6.2.2] have lasted 6 months or more, so arguably firms should expect that if they have to draw on emergency liquidity support, it might be for at least this long. Moreover, it is prudent for a large firm to have a diversified collection of LOCs purchased from a range of institutions in different countries, providing at least some hedge against a country-wide banking crisis.

9.3.1.1 Contingent issuance facilities
Another form of positive contingent liquidity is provided by a *contingent issuance facility*. Here a financial institution buys the right under certain conditions to issue paper—either senior term debt or a capital security—from a counterparty. Thus, for instance, a bank might identify a downgrade as a potential liquidity risk event. The downgrade is viewed as unlikely, but if it happens, the impact on its ability to raise short-term funds will be considerable. Therefore, the bank buys a downgrade contingent issuance facility: it pays a fee of 50 bps on $500M every year for 5 years to a counterparty. During that period, it has

the right if it is downgraded by one or more of the major ratings agencies to issue $500M of 10-year subordinated debt securities paying 3m USD Libor + 100 bps to the counterparty and receive their face value.

9.3.1.2 Total return swaptions

What is the right but not the obligation to enter into a total return swap worth? One answer would be to look at the volatility of the total return swap spread and use this to derive a price. But that might not capture the real issue: in ordinary conditions, total return swap spreads are fairly stable. They only move significantly when liquidity becomes tighter, that is, when some players are in funding stress. So a total return swaption is really a form of liquidity risk hedge in that if you have the right to put an asset to a counterparty and force them to fund it to term at a fixed spread, you own a potentially valuable instrument in the event of a liquidity crisis.

9.3.2 Conduits

An ordinary securitisation involves the transfer of a pool of assets to an SPV with the SPV issuing tranched securities. These are term debt: the contractual maturity of the issued securities is usually slightly longer than the average expected life of the collateral pool. Another approach is possible, though: the SPV could issue shorter-term securities with the intention of rolling them.

9.3.2.1 Conduit structure

This gives rise to the idea of a *conduit*:

- An SPV is set up and capitalised. This initial capital is similar to the equity tranche of a conventional securitisation.
- The sponsor transfers a range of assets into the SPV. Unlike a traditional securitisation where typically one class of assets is used (just mortgages or just credit cards say), a blend of assets is used with a weighted average life (WAL) typically in the 3- to 5-year range.
- The SPV issues short-term paper, often in the CP market. This is known as asset-backed commercial paper (ABCP) and typically has a WAL of 6 to 9 months.
- Interest on the CP will be paid from the cashflow generated by the collateral pool, just as in a conventional securitisation.
- The SPV purchases liquidity support in the form of a line which can be drawn down in the event of a dislocation in the ABCP market which prevents the paper being rolled. This protects the SPV against the risk that an exogenous liquidity crisis might cause it to have to liquidate its assets.
- To get the highest A-1/P-1 rating for its CP, the conduit will use one or more forms of *credit support* including *overcollateralisation* (it has more assets than issued CP, the excess funding coming from its initial capitalisation and retained spread); *collateral support* (where, for instance, the sponsor agrees to substitute collateral which is

downgraded); or *credit support* (where the sponsor or a third party agrees to provide protection against some level of losses).

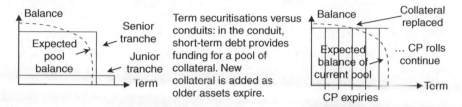

9.3.2.2 Conduit assets and their sellers

A wide range of assets can be funded in conduits including:

- Short-term receivables such as trade or credit card receivables;
- Any asset suitable for a term securitisation including mortgages, loans and some leases;
- Rated ABS including tranches from other securitisations.

The diversity of assets in a large conduit means that it is in effect similar to a traditional 4-6-4 bank: assets are originated for it; they are funded using the (short term) CP market; and the vehicle acts as a maturity transformer taking liquidity risk.

> *Exercise.* Are there any situations under which most or all of that collateral could decline in value at once?

Two forms of conduit are commonplace:

- If a single firm sponsors the conduit and contributes assets to it, we have a *single-seller conduit*.
- Whereas a *multi-seller conduit* takes collateral from more than one originator. Multiseller conduits often have two layers of credit support: pool-specific support often provided by each seller to their assets and programme-wide enhancement supporting losses above that across the whole conduit.

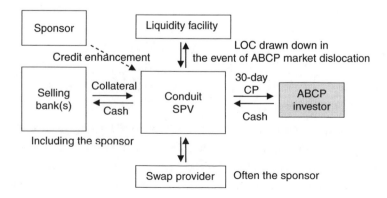

9.3.2.3 The advantages of conduits

The use of conduits allows banks to significantly reduce the liquidity risk of their balance sheets while retaining many of the advantages of the risk position: the assets go off balance sheet; the majority of the funding is done by the SPV, reducing the bank's funding requirement; yet the bank retains much of the upside of both the yield curve position on the funding and the ES of the collateral via its equity stake in the SPV, through retaining the first loss tranches of the assets and only selling the higher tranches into the conduit, or both.

One potentially considerable advantage of conduits is that they are often *blind pools*: subject to credit quality and diversification criteria, the sponsor can substitute collateral at will. Therefore, in a liquidity crisis, the sponsor can put assets which cannot be repo'd at a reasonable level into the conduit, and take out more liquid assets to use in the secured funding markets. Conduits can therefore be thought of as *liquidity arbitrage* vehicles.

Another advantage of sponsoring a conduit is that it is often a firm buyer of securitisations: if you have a client who wants to fund an asset, but the resulting securitisation will be too small, obscure or difficult to understand to sell the paper in the tranched ABS market, that is fine: provided the paper meets the required rating, you can put it into your conduit.

Some conduits issue both CP and, as a form of liquidity enhancement, term paper. These are known as structured investment vehicles (SIVs). The term SIV also tends to refer to vehicles which are more highly leveraged than standard ABCP conduits. They retain their rating provided that they adhere to a set of rules—diversification and concentration criteria, overcollateralisation and interest coverage tests, and so on—and compliance with these rules is regularly monitored by the ratings agencies. They may also issue liabilities such as CLNs to hedge specific assets, or subordinated term debt to further enhance leverage.

9.3.3 Negative Contingent Liquidity

One firm's positive liquidity is another's negative liquidity. Since banks provide contingent liquidity to their clients, they need to manage not only their current liquidity needs but also the portfolio of liquidity options they are short. These options arise from:

- Liquidity facilities granted to corporates, conduits and other SPVs;
- Standby LOCs, guarantees and written bond wraps;
- Other credit support provided to its own or others' SPVs;
- Market making in repo or other forms of secured funding, where the bank has a reputational obligation to provide liquidity to clients;
- Off-balance-sheet liquidity options embedded in derivatives, for instance via early termination agreements.

9.4 STRESSES OF LIQUIDITY

One risk of writing contingent liquidity is that many of these options could be correlated: a crisis in the ABS market affecting the ability of conduits to roll ABCP, the credit quality of ABS securities and the liquidity of clients engaged in structured finance activities could result in multiple liquidity options being exercised against the firm at once. A good liquidity stress test would capture this kind of risk, so, with that in mind, we turn to stresses of liquidity.

9.4.1 Liquidity in a Crisis

The late 1990s were interesting times for market risk managers. The 1997 Southeast Asian crisis was rapidly succeeded by Russia's default, the bailout of LTCM and the Brazilian crisis. These events were stressful for some people and some firms, but they also gave some insight into market crisis dynamics. The common pattern is shown in the illustration.

Financial liberalization and the
expansion of credit

Speculative investment raises
prices causing an asset
price bubble to form

A shock (real or finanical)
triggers a fall in asset prices

Defaults by investors, a flight to
quality a loss of confidence
and/or causes a crisis

9.4.1.1 Gestation of a market shock

Many market shocks form in the same way. In the late 1990s, the asset price bubble was in emerging market assets specifically, but to a lesser extent in equity prices and credit risky instruments too. External shocks then came repeatedly, causing a reassessment of the risks in these markets.*

9.4.1.2 Risk propagation

After a shock, asset prices fall. Some leveraged players cannot meet margin calls and are thus forced out of the market, resulting in further asset prices falls as their positions are

---

* See, for instance, E. Philip Davis' *A Reappraisal of Market Liquidity Risk in the Light of the Russia/LTCM Global Securities Market Crisis*. It is worth comparing this analysis with descriptions of earlier crises. For instance, the account of the 1826 London Financial Crisis reprinted recently in the *Guardian* (see www.guardian.co.uk) shows a familiar pattern of an asset bubble, a failure of confidence and a rescue based on opening the discount window wider.

liquidated. Volatilities rise, causing VAR and other risk measures to increase. Mark-to-market losses and risk limit breaches cause other players to cut their positions, intensifying the fall of the market.

Risk then propagates from one market to another. After the Russian default, for instance, there were few buyers of the recently defaulted debt: it had gone from being a fairly liquid instrument to a more or less totally illiquid one. Therefore, to cut positions and reduce risk, traders had to sell something else, causing the crisis to spread from one market to others and increasing the flight to quality. Finally, the hedging of short call positions again exacerbates the effect: deltas fall, hedges are sold and asset prices fall further.

The following general effects are therefore often observed in a market crisis:

- A flight to quality.
- Risky instruments become more illiquid, and liquidity premiums rise.
- Volatilities increase.
- Many return correlations tend towards either $+1$ or -1.

9.4.1.3 Liquidity black holes

The term *liquidity black hole** refers to the situation immediately after an asset price bubble bursts. Suddenly, there are many more sellers than buyers and prices plummet with very little trading. The possibility of this situation demonstrates the risk of taking volume as a proxy for liquidity: in ordinary conditions, there might be a lot of liquidity, but if many or all the buyers can withdraw from a market simultaneously, liquidity risk is high.

A particular issue here is diversity of views: liquidity black holes are less likely and last for a shorter period if speculative investors step in to 'bottom fish', buying the fallen asset and ending the period of illiquidity. Proprietary traders and hedge fund managers used to take that role, but are increasingly unable to do so. The consolidation of financial services into fewer, larger firms, the rise of VAR models as a ubiquitous risk control technique [as discussed in the introduction to Chapter 7] and the use of VAR-based collateral models to control hedge fund leverage [as discussed in section 7.1.4] all mean that the financial system may actually be more susceptible to liquidity black holes now than hitherto.

> *Exercise.* How could a central bank best manage the dramatic rise of liquidity premiums during a liquidity black hole? In answering this question, consider the balance between the cost of funds and their availability: does it help more to be able to repo a liquid asset cheaply through the discount window, or to be able to repo an illiquid asset at GC?

---

* This term has been popularised by a number of authors: see, for instance, Avinash Persuad's *Liquidity Black Holes: Understanding, Managing and Quantifying Liquidity Risk.*

9.4.2 Liquidity Stress Testing

Just as market stress testing estimates the effect of extreme market moves on a portfolio, liquidity stress testing estimates the effect of exogenous or endogenous liquidity crises. Typically, this will include a market or credit risk event, so in fact liquidity considerations should be incorporated into other stress tests.

9.4.2.1 Common liquidity scenarios

Typical liquidity stress tests considered by firms include the following (with more extreme versions of the same scenario in parentheses):

- Country risk events such as the suspension of convertibility of the largest emerging market the firm does business in (The same event but all on-shore assets and no liabilities are sequestered and funding is impossible in local.)
- A flight to quality like the one which occurred around the Russian/LTCM crisis (This event combined with the closure of the CP market for 6 months.)
- A major operational risk event such as 9/11 (And all major market counterparties are affected by the same event.)
- Default of the counterparty which owes the largest amount to the firm in the near term (The same, with a knock-on default of a major liquidity provider.)
- A ratings downgrade (The downgrade is caused by a $2B rogue trader loss.)

The important point here is to estimate not only the immediate quantitative impact of the event but also the knock-on effects to customers—including draws on lines—and on the market perception of the firm. This last may in turn cause an increased rate of termination of derivatives, a run on deposits or much tighter rationing of credit in the interbank market.

9.4.2.2 Contingency planning

A *contingency* is anything that can disrupt operations. Examples include:

- Natural disasters including severe weather, fire or earthquake;
- Local emergencies including transport disruption, demonstrations or power outages;
- Key personnel risks.

Contingency planning is part of operational risk management, but it also has liquidity risk characteristics since the firm has to be able to manage payments to retain confidence and avoid a default. Therefore, Treasury needs to be actively involved in the disaster recovery process, and it must have a robust strategy for reacting to contingencies. Note that this is not necessarily just a short-term problem: after 9/11, some firms could not return to their offices for 6 months or more.

9.4.3 The Liquidity Plan

The liquidity plan is a collection of strategies for dealing with a liquidity crisis. Liquidity plans are developed by institutions running significant liquidity risk as part of their planning of a crisis. There are several elements to most plans:

9.4.3.1 Trigger

There are a number of general indicators that indicate that a firm might be drifting towards a liquidity crisis:

- Declining liquidity ratios;
- Growth in highly illiquid assets (especially if this is funded from purchased liquidity rather than deposits);
- An increase in the bank's short-term cost of funds, senior debt credit spread/CDS spread or secured funding spreads;
- An increase in the rate of withdrawal of short-term funds, cancellation of LOCs or rate of exercise of early termination agreements;
- A ratings downgrade.

Any or all of these events could be defined as the trigger for the firm to enter into the liquidity crisis management programme. This trigger should be objectively defined, so, for instance, the firm will automatically start the plan if it is downgraded below AA− or if more than one relationship bank terminates a strategic LOC.

9.4.3.2 Elements of the plan

The essential elements of the plan include:

- Increased frequency of review of predicted cashflows;
- Much smaller gap limits;
- Modification of the firm's liability profile with duration lengthening where possible;
- The suspension of discretionary cashflows;
- Increased use of secured borrowing to raise term funding;
- Decreased market and credit risk limits to reduce funding needs caused by P/L volatility;
- Sale or repo of assets in the liquidity portfolio;
- Drawdowns on LOCs and use of other contingent sources of liquidity;
- Opportunistic use of off-balance-sheet instruments, for instance, by terminating in the money swaps to raise cash;
- Increased high-level management communication with investors, the media, regulators and ratings agencies to restore confidence;

- If necessary, a gradual liquidation of the firm's assets starting with the most liquid securities;
- And finally, if absolutely necessary, application to the lender of last resort.

In summary, firms take liquidity risk to enhance income: it means that they need strategies for managing everyday liquidity and for dealing with endogenous and exogenous liquidity events. Liquidity is a lightening conductor allowing firms to survive when a crisis strikes.

Part Four

Some Trading Businesses and Their Challenges

CHAPTER 10

An Introduction to Structured Finance

INTRODUCTION

Structured finance is a somewhat vague term: at its broadest it covers any activity involving a bond, credit derivative or loan where the credit risk transferred is not just that of a single corporate name.* Typical structured finance activities separate into several strands:

- The structuring, underwriting and trading of various forms of asset-backed security, i.e., *funded structured risk transfer*;

- The transfer of the risk of a single or multiple financing activities using credit derivatives or similar transactions, i.e., *unfunded structured risk transfer*.

Structured finance transactions may involve the *pooling* of risk and possibly its tranching—so securitisation technology is important—so too is the *credit enhancement* of some otherwise undesirable or risky debt.

In many structured finance transactions some element of risk transfer is achieved either through a derivative transaction or though a financial insurance policy. Therefore, we begin by looking at these contractual relationships. The subsequent section discusses ABS and their analysis. Finally, various structured finance transactions involving securitisation structures are reviewed.

10.1 CONTRACTUAL RELATIONS

There is sometimes a temptation to assume that just because a trade is booked in a system, that booking represents the economic reality of the firm's position. Leaving aside mis-bookings, there is still the issue of whether the trade is *enforceable*, and, if it is, how to control the extent of the firm's exposure to the counterparty.

---

* Structured finance is perhaps a more secretive activity than some parts of trading, and it is certainly fast moving. This means that there are few comprehensive texts: some discussion can be found in Sanjiv Das' *Credit Derivatives: CDOs and Structured Credit Products* and *Euromoney* magazine is occasionally useful.

This section reviews the standard mechanism for documenting OTC derivatives transactions—ISDAs—and discusses insurance as another mechanism sometimes used in structured finance transactions. Finally, we touch on whether the contract is effective, and whether it is worth enforcing it.

10.1.1 The Documentation of Derivatives and Credit Risk Mitigation

Most OTC derivatives transactions between two parties are defined by:

- An ISDA *master agreement* between the two parties. This is an overarching arrangement available to govern all derivatives transactions between the two parties.
- The *schedule* to the master and any credit support annex (CSA).
- *Definitions* for each class of derivative.
- And a *confirmation* for each transaction which references the definitions.

10.1.1.1 Master agreement

The master agreement* covers terms which are common to all or most transactions between counterparties. For instance, it typically includes clauses relating to:

- The *netting* of ongoing payments, so that if A owes B a payment of £10.3M on one swap today and B owes A £3.2M on another, only the net amount of £7.1M is paid.[†]
- *Close out netting* is an agreement to settle all contracted but not yet due liabilities to and claims on an institution if a defined event such as bankruptcy occurs. Close out netting thus accelerates all payments under the master, and allows one counterparty to crystallise a total amount owing or owed as soon as a close out event occurs.
- Tax withholding representations, so that each counterparty has certainty on whether it needs to withhold part of any payment to the other under tax regulations.
- Other representations including matters each counterparty might wish to know about the other, including their current solvency, absence of litigation, their legal capability to engage in derivatives transactions and so on.
- Agreements to comply with laws, maintain authorisations, etc.
- Definition of default events usually including failure to pay, failure to perform under the terms of the CSA, misrepresentation and breach of the terms of the master as well as bankruptcy, repudiation or merger without assumption of obligations and cross-default.
- Other termination events such as changes in the law or in the tax treatment of transactions.

---

* The discussion here is very much a sketch: see Paul Harding's 'Mastering the ISDA Master Agreement' for more details.
† Netting is a delicate topic especially when multiple currencies and jurisdictions are involved, and one of the advantages of using ISDA documentation is that ISDA has coordinated an effort to obtain a measure of legal certainty on various netting arrangements: see their website for more details.

- Timelines specifying notice periods, grace periods for curing breaches and so on.
- Finally, of course, the master needs to be *signed*.

One of the purposes of the master is to define exactly when a close out event has occurred and what is owing in this situation. As such it is both a credit risk mitigant—in that netting reduces credit risk—and an operational risk mitigant in that properly drafted masters reduce legal risk.

10.1.1.2 The schedule to the master

The schedule allows the parties to a master to select, de-select or otherwise customise various provisions of the master agreement. Typically, it might include:

- The precise names and addresses of the entities concerned, and perhaps where one or both entities is a subsidiary, details relating to the credit support it receives from its parent;
- A threshold amount needed for a failure to pay close out event to be triggered;
- Details of any additional termination events;
- A list of any documentation each party agrees to provide to the other.

10.1.1.3 The credit support annex and collateral

The CSA is an addendum to the master which determines the nature of the credit mitigation provided by each party to the other. For instance, it may contain conditions defining:

- The requirement for the posting of *collateral* if the mark-to-market of the portfolio of transactions between the parties exceeds some threshold amount;
- What collateral is acceptable, how often the collateral requirement is calculated and how long after a collateral call is made the counterparty has to meet it;
- Additional collateral or other credit support required if one party is downgraded, Acceptable substitutes for collateral such as letters of credit or guarantees.

The CSA therefore determines the conditions under which credit is extended by each party to the other and, as such, it is critical for limiting the PFCE in derivatives transactions.

Amounts owing under an ISDA master usually rank *pari passu* with senior debt as a legal matter. However, the combination of credit support and the ability to assign transactions [discussed below] mean that ISDA claims often have effective recoveries higher than that of senior debt.

10.1.1.4 ISDA definitions

Derivatives transactions rely on a web of nomenclature. To take a simple example, an IRS of 4% versus 3-month Libor flat in EUR with quarterly resets on the 2nd of January, April and so on defines a set of cashflows if we agree:

- Exactly how the payments on each leg are calculated, including the details of the calculation of interest including the day count;

- Exactly when Libor is fixed for a given period;
- Exactly when the payments are made, including details of what to do if the relevant days are not business days.

We also might need to deal with the possibility that Libor is not available, that the market is disrupted and so on. These issues are dealt with in a common set of *definitions* which are available for any swap transaction; similarly, the equity derivatives definitions deal with a range of issues including the treatment of special dividends, corporate actions such as a merger or spin-off and so on.

10.1.1.5 The confirmation

Each transaction is documented using a confirm, which in turn refers back to the relevant definitions. Thus, a large dealer might have 10,000 or more equity derivatives confirms, most of which will use the latest set of ISDA equity derivatives definitions. If this dealer has 1,000 counterparties, it would typically have signed masters, schedules and CSAs with 95% or more of them, with a handful of stand-alone transactions not documented under a master agreement and a few transactions where the master is still being negotiated.

The reason that netting and collateral calling on the net exposure to a counterparty work is that the master and all the confirms entered into under it form a single unit: signing a master amongst other things amounts to an agreement by both parties to consider all transactions under it *together* for the purpose of determining who owes what to whom.

Exercise. Try to gather all the documentation supporting a single transaction: master, schedule, CSA, definitions and confirm. Review how the various parts of the documentation relate to each other and how the trade is handled by various functions within the firm. How are the terms of the CSA reflected in the systems for credit risk monitoring and collateral management? How does the position valuation from the trading system net with other transactions to the same counterparty to drive the collateral process?

10.1.1.6 Assignment

One particular feature of many contractual arrangements is worth noting: some derivatives done under a master are *assignable*. That is, one party can pass their rights and obligations under the derivative on to a third party. Typically, permission from the counterparty is required to do this, but often that cannot be unreasonably withheld.* Suppose that we have two investment banks, A and B, doing business with a corporate C. C defaults, leaving A and B with the exposures in the first column.

---

* One reason that permission could reasonably be withheld might be that the new counterparty was of a low credit quality.

At this point, C owes A \$4M: that is the mark-to-market of the portfolio transactions between them. If C's senior recovery is 50%, A will lose \$2M on its exposure. Meanwhile, B will pay the liquidators of C \$5M, as that is the amount it owes on its portfolio.

	Mark-to-Market of Portfolio with C	
	Before Assignment	**After Assignment**
Bank A	+\$4M	0
Bank B	−\$5M	−\$1M

But suppose A offers to assign B its portfolio in exchange for \$3M. This is good for A as it makes \$1M more from this than from going through recovery.

After assignment, B can *net* its negative exposure of \$5M with the positive \$4M from A's portfolio under its master with C. It pays the liquidator the result, \$1M, and its total cost is \$4M, a saving of \$1M, so it makes money too.

This looks like magic: both banks book a \$1M profit. But consider the liquidator: before assignment, they were expecting to receive \$5M and to pay out \$2M, giving a net available for other creditors of \$3M. Now there is just \$1M available: the million dollar profit for each bank came at the expense of other creditors of C.

> *Exercise.* Assume this is legal: is it ethical? Is there significant reputational risk in doing this?

10.1.2 Credit Derivatives in the Form of Insurance

Reviewing a different approach to transacting derivatives will give some insight into the advantages and disadvantages of using the ISDA master/schedule/CSA/definitions/confirm approach. Therefore, consider a transaction that has a similar effect to writing a CDS, but in the form of insurance.

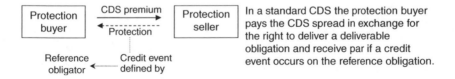

In a standard CDS the protection buyer pays the CDS spread in exchange for the right to deliver a deliverable obligation and receive par if a credit event occurs on the reference obligation.

This would typically be documented under a master between the two parties and using a confirmation that refers to the latest ISDA credit derivatives definitions. In contrast, in the insurance version of the structure we would have:

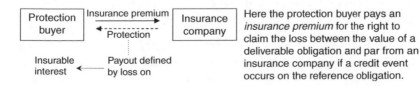

Here the protection buyer pays an *insurance premium* for the right to claim the loss between the value of a deliverable obligation and par from an insurance company if a credit event occurs on the reference obligation.

10.1.2.1 Some features of insurance

This seems rather similar; however, there are certain important differences:

- First, the purchaser of insurance must usually have an *insurable interest*: I can buy fire insurance on my house because if it burns down, I will lose something. But I cannot profit from buying fire insurance on your house because I have no interest in it. With a CDS, the protection buyer can go short by buying the CDS without owning the underlying: if there is a credit event, they make money. With a credit insurance policy, in contrast, the protection buyer needs to have an insurable interest to claim, so they have to own a deliverable obligation or otherwise have exposure to the credit at the time of the credit event.

- A claim on an insurance policy is usually subject to *proof of loss*—the insured can only make a claim to the extent of their actual loss, and before a payment is made, this must be demonstrated to the satisfaction of the insurer. In contrast, a payout on a cash-settled CDS is usually fixed and automatic after a credit event.

- There are typically regulatory requirements in order to be an insurance company and, in many jurisdictions, only regulated insurance companies can write insurance. In contrast, most corporate and many other bodies can enter into ISDA transactions.

- In English insurance law, there is a *duty of good faith* by the insured to disclose to the insurer any fact that might be material to their underwriting decision.

- Because insurance policies are not executed under a master, netting does not apply, and if the protection buyer requires collateral in that the event that the credit spread of the underlying increased, this would have to be agreed by the protection seller and documented as part of the policy.

- The *accounting status* of an insurance policy may be different to that of a credit derivative. The CDS is clearly a fair value instrument, but deciding whether accounting policy permits the use of fair value for a purchased or sold insurance policy would require detailed accounting advice. Similarly, if the purchaser is regulated, the policy might not count as a hedge for the purpose of calculating regulatory capital on the net SR position in the underlying.

Exercise. Why is a standard CDS not insurance? What would be the consequences for the market if it were deemed to be legally insurance?

Crude though this summary is,* it indicates some of the issues that arise as soon as we go beyond standard documentation: not only do legal questions arise, so do ones of accounting, regulation, credit support and so on. None of these are insurmountable, but clearly an

---

* These notes are merely a high-level sketch of a legally complex situation. As with any other area, only an experienced lawyer can provide advice relevant to a particular transaction or issue.

extra return will be required to pay for the extra costs of doing business in a non-standard way and the extra operational, legal, documentation and credit risks run.

> *Exercise.* Research the financial guarantee insurers: find out what they do, what their business model is, how they account for their transactions, what their regulatory and ratings agency capital regime is and how they are similar to and different from a conventional insurance company. Would this business make as much sense if it were conducted in the trading book of a regulated bank or broker/dealer?

10.1.3 Enforceability and the Pros and Cons of Enforcement

If a firm buys an interest rate cap from a market counterparty—an investment bank, say, or a large corporate—and rates move so that this cap is worth far more than the premium, the firm can be reasonably confident that money has really been made: caps are standard transactions; they are suitable for a wide range of counterparties; and the legal framework for transacting them is well developed.

Matters may not be so clear-cut in other areas. For instance, suppose a firm engages in a prepaid commodity swap transaction [similar to the prepaid IRSs discussed in section 2.4.9], hedges all the commodity risk, and then buys credit insurance on the performance of the swap counterparty from a number of insurance companies. This transaction might be profitable if the cost of the purchased insurance is significantly lower than the implied loan spread on the prepaid swap.

However, it may not be prudent to recognise all of this P/L,* since there are a number of unusual risks in the structure including:

- Lack of legal certainty on purchased protection. The combination of the duty of good faith and a relative paucity of case law means that legal risk is elevated.
- Reputational risk if it is necessary to go to law to enforce a claim on the policy.

This highlights a feature of some structured finance transactions, namely, their potential for reputational damage if a hitherto private arrangement becomes public in negative circumstances surrounding legal action. The possibility of this—particularly for transactions involving a significant arbitrage which could be portrayed as 'taking advantage' by a hostile counterparty or press—should be considered when authorising these transactions.

> *Exercise.* What criteria would a firm use in making the decision on whether to sue a counterparty for payment or not? Who would be involved in this decision?
>
> Review the SEC document referred to in the footnote and research this case further.

---

* See http://www.sec.gov/litigation/litreleases/lr18252.htm for details of a situation somewhat like (although rather more complex than) the one described.

10.2 ASSET-BACKED SECURITIES

An asset-backed security is one whose principal and interest payments are primarily serviced by the cashflows generated by an identified pool of collateral, together with other rights or assets which aid the servicing of the pool or the distribution of scheduled cashflows to security holders. This pool can be either fixed, revolving or managed, and the rights or assets can include interest rate derivatives, liquidity facilities, servicing arrangements or spread accounts and other forms of credit enhancement.

10.2.1 Mortgage-Backed Securities

Securities backed by the cashflows from pools of residential mortgages were one of the first types of ABS. Their genus was the establishment of the U.S. agencies.

10.2.1.1 The agencies

The term *agency* in the MBS market refers to one of three corporations sponsored by the U.S. government: the Federal National Mortgage Association or *Fannie Mae*, the Federal Home Loan Mortgage Corporation or *Freddie Mac* and the Government National Mortgage Association or *Ginnie Mae*. All of the agencies are mandated to have a role in promoting the U.S. residential mortgage market, and as part of this activity they buy mortgages from originators. Thus, a small regional bank can offer residential mortgages to its clients and, providing the mortgages *conform* to agency standards, these can then be sold to the agencies, reducing the bank's funding costs and allowing it to continue to provide finance to its clients without ballooning its balance sheet. This also extends to non-bank providers of mortgage finance.

The agencies are viewed as essentially U.S. government credit risk, so they can fund themselves cheaply by issuing term debt. However, just doing this is potentially problematic for the agencies due to the nature of most U.S. mortgages.

10.2.1.2 Mortgages and prepayment

The standard U.S. residential mortgage is a 30-year, fixed rate, level pay structure: the loan is for 30 years, it is secured by a first claim (known as a *first lien*) on a residential property, and the borrower pays a fixed coupon for the entire life of the loan.* This coupon includes both interest payments and a principal *amortisation*, so early payments are mostly interest and later ones are mostly repayments of principal. Crucially, though, these loans are *prepayable*: the borrower can prepay the mortgage balance early, often with little notice and with only a modest fee.

---

* There are a huge variety of other mortgage types including *ARMs* (adjustable rate mortgages), *Alt-As* (mortgages from good credit quality borrowers but where the standards of documentation provided by the borrower are not sufficient for the mortgage to conform), *jumbos* (where the mortgage is too big to conform) and *balloons* (where the interest payments on the mortgage are lower due to an additional requirement to pay a large sum, the balloon, at a point perhaps 5 or 7 years into the mortgage). A separate mortgage sector is *sub-prime* where the borrower is not sufficiently good credit to conform.

Prepayments are caused by:

- Disaster. If a home is destroyed, for instance by fire, the mortgage is often paid off by home insurance.

- Cash-out refi. If house prices rise significantly, home owners sometimes monetise their gain by refinancing using a mortgage with a higher notional value (but roughly the same loan-to-value ratio).

- Moves. If a home owner switches properties, they typically use the sale price of the old property to pay off their old mortgage, and take out a new mortgage on their new home. A few mortgages are portable from one home to another, but this is fairly rare.

- Payment-based refinancing. Home owners that can get a mortgage with lower payments than their current one sometimes do so. This typically happens either because the borrower's credit has improved—their FICO score has increased so the spread required from them decreases—or because interest rates have fallen.

The first three causes give a predictable level of refinancing for a given level of housing market activity. The last means that a mortgage is like a bond with an embedded interest rate derivative: if rates rise, there is no incentive to refinance, so the fixed rate payments are likely to continue; but if rates fall, there is an incentive for the mortgage holder to refi, so the expected duration of the mortgage falls.

10.2.1.3 Pass throughs

Suppose we have a pool of 30-year fixed rate prepayable mortgages.* The right approach to ALM is not obvious [as we discussed in section 8.2.1] as the expected cashflow profile depends on interest rates. One answer to this is simply to make it someone else's problem: we can issue a security that simply passes on the principal and interest payments of the underlying mortgage pool to an investor. This is known as a *pass through* (PT) certificate for obvious reasons.

The first PTs were issued by the U.S. agencies. They do not bear default risk, in that the issuing agency guarantees that if any mortgage in the pool defaults, the agency will make good the payments, but they do bear prepayment risk in that whatever prepayment behaviour the mortgages display is passed on to the investor, so the PT is a fully amortising security. Note also that default is a prepayment event: once the mortgage is recognised as delinquent, it is removed from the pool, and the agency makes good its principal balance.

The construction of a PT certificate is shown below. The originator or some other party is paid a fee for *servicing* the mortgage—collecting payments, following up arrears and so on—and the agency also takes a spread for guaranteeing the payment of timely interest and ultimate principal.

---

* Pools of different underlying mortgage types have different behaviour: see Lakhbir Hayre's *Salomon Smith Barney Guide to Mortgage-Backed and Asset-Backed Securities* for more details.

Pass through investor	Cash → ← P&I minus fees on pool	U.S. agency issuer	Cash → ← P&I minus fees	Servicer Originating mortgage bank	Loan → ← Level pays

Purchases of mortgages from other originators   Other mortgage loans

10.2.1.4 The agency MBS market

The agencies are major issuers of RMBS, between them issuing around $1T of MBS in recent years. Many of these are PTs: others pass on risk in more structured ways.

First we review the risks of residential mortgage pools, then turn to the various types of securities available which are backed by those pools.

10.2.1.5 Prepayment behaviour

Each mortgage in the pool backing a PT is prepayable so the PT itself does what it says on the label, and passes that risk on to the investor. Typically, data on the cashflows from collateral pools are available monthly from the servicer, and many ABS pay monthly, so prepayment analysis is usually phrased in terms of monthly repayments. For a constant level of rates and a large, diversified pool, the level of prepayments tends to display the following behaviour:

- Prepayments depend on housing market activity: there tend to be more house moves and significantly more cash out refis if the market is rising.

- The amount of general economic activity is important too, in that it drives job mobility and hence the number of house moves.

- The ages of mortgages in the pool are important. For instance, new mortgages are much less likely to prepay as the borrower may recently have moved or refinanced, and they are unlikely to want to do so again soon. Very old mortgages are also less likely to prepay, in that if the borrower was going to move or refi, they would probably have done so already.

For most pools, interest rates are the most important factor in determining the level of prepayments. The tendency of borrowers to refinance more cheaply depends on how much they have to gain from this and on their understanding of the issue: a high-value borrower is unlikely to go to the trouble of refinancing to save $20 a month on their mortgage repayments, but they might well do it to save $200. Moreover, a financially sophisticated borrower is probably more interest rate sensitive than someone who does not read any financial journalism.

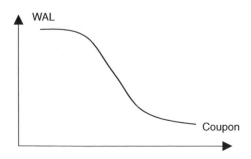

The coupon on the mortgage determines the extent of the refi incentive: low coupon mortgages—which were probably originated when rates were lower—produce losses on refinance, whereas higher ones are ripe for refi. Therefore, if we plot the weighted average life of a pool versus the average coupon in it, we typically find a picture like the one above. Note the convexity whereby very large coupons do not produce significantly different WALs from moderately large ones (and similarly for small coupons).

10.2.1.6 Measuring prepayment: CPR and speed

The *single monthly mortality* of a collateral pool is the fraction of the beginning month balance that prepays during the month without being scheduled. Prepayment rates are usually quoted in terms of conditional prepayment rate (CPR): this is the cumulative prepayment rate over 12 months which gives the same single monthly mortality per month. This is sometimes referred to as the *speed* of the pool: faster pools prepay more quickly. The WAL versus coupon graph translates into CPR in an obvious way: the higher the refi incentive—the higher the average coupon on the pool—the quicker the CPR as in the illustration on the left below.

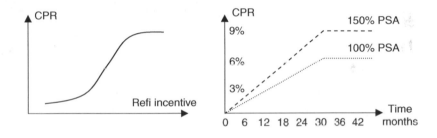

Even when rates are constant, mortgage pools do not display constant CPR due to the tendency of new mortgages to be slow. A simple approximation to this effect was introduced by the Public Securities Association (PSA). This metric—the PSA—assumes that the pool's CPR starts at 0 and increases by 0.2% every month until month 30.

At this point it remains constant at 6%, as in the illustration on the right above. This is known as 100% PSA: a faster pool might be referred to as 150% PSA. The illustration on the left below shows the prepayments on a pool at 100% assuming a 50 bps servicing fee, whereas the one on the right shows the timing of the payments from the same pool at 150% PSA: the duration is shorter as principal repayments come in faster.

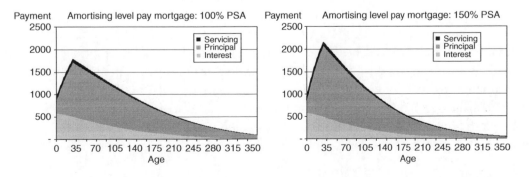

10.2.1.7 Prepayment modelling, interest rate optionality and OAS

Prepayment modelling is the process of estimating the likely prepayment behaviour of a pool of collateral for a given interest rate scenario. This is a complex process as there are many variables:

- The overall levels of housing market and general economic activity including *seasonality* (the tendency of the housing market to be slow early in the year and during the summer);
- The age, geographical distribution, loan-to-value ratio and type of mortgages in the pool;
- The interest rate scenario, including whether rates change slowly or quickly.

The output of the model would include a projected CPR over time as shown in the illustration below. Typically, we find:

- As rates drop, prepays accelerate until there is sufficient incentive to encourage all home owners who are interested in and capable of refinancing to do so.
- This takes some time, as there is a lag between a rate change occurring and a new mortgage being granted to a refinancer.
- If rates do not change again, CPR then slows reflecting a pool where all or most of the potential refinancers have exited.
- If rates increase, speeds slow and the pool becomes less sensitive to further rate changes as most of the mortgages in the pool now have below market rates.

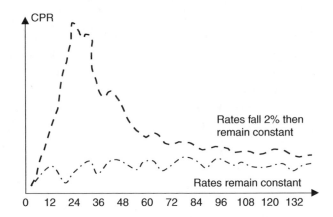

These phenomena mean that PTs are negative convexity securities: they prepay faster at lower rates. This effect could be hedged using a portfolio of interest rate options that compensate us for the loss of duration as rates fall. Therefore one aim of a prepayment model is to estimate the sensitivities of the reserved portfolio of derivatives allowing some players to try to arbitrage between the MBS and interest rate derivatives markets.

Since the holder of a PT is long a fixed rate bond and short a portfolio of interest rate options, the spread of the MBS includes compensation for prepayment risk. The pure spread is sometimes known as the *yield curve spread* in contrast to the *option adjusted spread*, which subtracts the value of the embedded interest rate options.*

> *Exercise.* Historical prepayments from pools backing MBS are often available from their servicers. Download a few and review the PSAs versus the level of interest rates.

10.2.1.8 Tranched agency MBS

PT certificates pass on prepayment risk equally to all holders. Instead of a PT consider *tranching* the cashflows from a pool of mortgages. The resulting securities are a form of RMBS known as CMOs. The waterfall is often arranged so that:

- Bonds at the top of the waterfall—the senior tranches—are paid first;
- Bonds at the bottom get paid last, so the cash flows down.

This means that bonds at the top have a shorter WAL than the underlying pool: those at the bottom have a longer one. Tranching thus has the effect of concentrating prepayment risk towards the senior tranches: unlike a CDO, where the default risk is often concentrated in the junior tranche, in an agency RMBS CLO, prepayment risk affects the top end of the waterfall, i.e., the senior securities as these are prepaid first.[†]

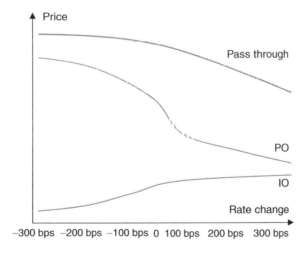

---

* See Lakhbir Hayre's book (*op. cit.*) for more details of prepayment modelling, OAS and the MBS markets.
[†] The precise allocation of prepayment risk depends amongst other things on the tranching. For instance, a thin top tranche will amortise fast in most conditions due to the ambient level of prepays caused by ordinary house moves and defaults, so it will have a short WAL in most environments. In this situation, the next few tranches down are likely to shorten significantly if rates fall and lengthen if rates rise: they have more of the convexity created by prepayment risk.

10.2.1.9 IOs and POs

A different approach to constructing MBS with altered prepayment risk is to separate out the interest payments and the principal repayments on a mortgage.

Thus, the cashflows from a pool can support the issuance of interest only certificates and principal only certificates, sometimes known as *IO-strips* and *PO-strips*:

$$PT = IO\text{-strip} + PO\text{-strip}$$

The holder of an agency PO is guaranteed to receive the face value: the issue is when. As prepayments accelerate, the PO holder is paid quicker, and so the price of the PO increases with falling rates. The IO holder, on the other hand, only gets paid for mortgages that remain current, so the IO falls in value fast as prepays accelerate, and rises in value as the WAL of the collateral pool extends.

For large moves the behaviour levels off so the convexity profiles are complex: a single greek or two does not give the whole picture, and analysis of a variety of interest rate scenarios is needed to understand the full range of behaviour.

10.2.1.10 Floaters and inverse floaters

Another structuring choice is to split the fixed coupon of a pool into:

- An FRN paying Libor + spread;
- An inverse FRN paying strike − Libor.

Typically, this is done in such a way that the WAC on the pair equals the coupon on the original pool. Here the FRN investor is not too concerned about rising rates as they are receiving a floating rate, but the inverse FRN investor has a leveraged exposure via both decreased coupon and extending life.

> *Exercise.* Plot the duration and convexity of a typical IO-strip, PO-strip, floater and inverse floater as a function of interest rates.

10.2.1.11 The O.C. event

The dangers of negative interest rate convexity in structured securities are illustrated by the events in Orange County (O.C.), California, in 1994. O.C. had an investment fund which supported the county's liabilities including some pensions to county employees. This was managed by the county treasurer, Robert Citron, under supervision from a five-person board.

The investment strategy taken for the fund involved significant positions in structured bonds. It was also leveraged: those bonds were repo'd out and the cash raised was used to finance further bond purchases, turning an initial investment of roughly \$7.5B into a portfolio with exposure roughly three times the size. This portfolio had two major types of risk exposure:

- Term structure risk where the bet was that medium-term rates would stay above the short rate;
- Convexity exposure via positions in structured notes such as inverse FRNs which had enhanced coupons due to their negative convexity.

While rates stayed low, the portfolio return was good. However, during 1994 the Federal Reserve raised rates on a number of occasions, triggering a bond market crisis as fixed rate bonds fell in value.

The portfolio managed by Citron suffered very heavy losses due to both types of risk exposure, and the county was forced to declare bankruptcy in late 1994.*

> *Exercise.* Research the O.C. event and try to find more details of what was in Citron's portfolio. What kind of risk reporting would have been needed to reveal its sensitivities?

10.2.2 Other ABS and Pool Modelling

Agency MBS, although they can have highly complex prepayment behaviour, are at least simple in one sense: they have a guarantee of timely interest and ultimate principal. Other ABS do not necessarily share this feature, and so defaults in the collateral pool can lead to lower cashflows on securities backed by the pool. This section examines the consequences of this.

10.2.2.1 Non-agency MBS

The U.S. agencies are not the only issuers of mortgage-backed paper. Some originators issue MBS backed by their own mortgages rather than selling them into an agency deal. There are several reasons for this including:

- The loans may not *conform* to agency standards, for instance, due to reasons of size, structure, or underwriting standards.
- The originator may take the view that the cost of the agency guarantee is not worth-while given where unenhanced MBS would trade.
- The originator may wish to tranche or otherwise structure the deal in ways that are not available in a pooled agency transaction.

The originator can then either issue non-agency PTs backed purely by the pool, or structure a more complex deal, perhaps with extra credit enhancement.

10.2.2.2 Default and delinquency measures

Away from the U.S. MBS market,[†] there are no institutions with the role of the agencies, so issuers are on their own. This brings the issue of credit enhancement to the fore, since few collateral pools can support the issuance of a highly-rated ABS backed by the entire pool using a PT structure. In these securities too, then, collateral default rates and credit enhancement are key issues.

---

* See Philippe Jorion's *Big Bets Gone Bad: Derivatives and Bankruptcy in Orange County. The Largest Municipal Failure in U.S. History* for more details.

† See Lakhbir Hayre's book (*op. cit.*) or John Deacon's *Global Securitisation and CDOs* for more details of MBS and ABS markets.

CPR measures the speed of prepayment: the analogous measure of default is *conditional default rate* (CDR). This is the proportion of the outstanding collateral pool which defaults during a period, usually quoted on an annualised basis as for CPR.

For some collateral pools such as retail mortgages, a more sophisticated analysis than simply default/not default may be necessary. The following categories are common:

- Current. The borrower has made the monthly payment on time;
- 30 days delinquent. A scheduled payment has not been paid for 1–30 days;
- 60 or 90 days delinquent;
- In foreclosure and owned real estate. The collateral is being seized and liquidated, but the proceeds of this are not available yet;
- Liquidated.

Delinquencies are insightful in that they show borrowers who may be in some distress: not all of them will turn into defaults, as some borrowers are in temporary cashflow difficulties (or have simply forgotten to make their scheduled payment), but they are a useful early warning signal. The table below shows a sample delinquency analysis for a non-agency MBS pool.

Region	30 Days (%)	60 Days (%)	90 Days (%)	Foreclosure (%)	Liquidated (%)
New England	5.5	1.4	0.3	1.2	1.1
Other East Coast	7.1	2.0	0.4	3.7	3.1
Mid West	6.3	1.7	0.4	3.1	2.9
Pacific	3.7	1.0	0.2	0.7	0.8

> *Exercise.* Review the delinquency statistics for the pools you used earlier to investigate prepayments, then extend your search to non-MBS ABS such as credit card securitisations, sub-prime manufactured housing and future flow deals. How do the CPRs and CDRs compare?

10.2.2.3 Collateral pool analysis

Collateral pool analysis aims to understand what the CPR and CDR of the pool will be over time, contingent on the level of interest rates and other economic variables. There are several tools available to do this:

- Detailed analysis of the collateral characteristics. For a mortgage pool a large amount of data is usually available including the zip code of the property, loan age and amount, property value at the last valuation, mortgage structure and FICO score of the borrower. All of these are potentially valuable in understanding the potential future CDR: a seasoned pool of low LTV mortgages from high FICO borrowers is much less likely to suffer defaults than a high LTV pool from sub-prime borrowers. Similarly, a pool of adjustable rate mortgage where payments are based on Libor is likely to show more CDR interest rate sensitivity than a fixed rate loan pool.

- Historical behaviour of the pool or similar pools. This will give some insight into the previous CDRs and CPRs. Note though that most pools exhibit a *selection phenomenon*: fast refinancers exit the pool over time, leaving a higher percentage of less rate sensitive borrowers.

It is particularly important to understand if there are any factors which could cause a spike in CDR due to multiple defaults: for instance, a high percentage of Californian mortgages could indicate substantial earthquake risk in the pool.

The same principle of know-your-collateral also applies to non-mortgage pools: for revolving or managed pools, in addition, there is the issue of ensuring that the underwriting criteria which govern the admission of new collateral to the pool are sufficiently robust. The balance between prepayment risk and default risk also differs dramatically, for instance:

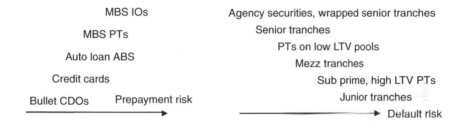

10.2.2.4 Originator default

It is vital that both the legal structure of a securitisation and the practical aspects of it are robust under the default of the originator. The former involves ensuring that the holders of the ABS have a claim on the underlying cashflows of the collateral pool unencumbered by any claims on the originator: the latter is particularly an issue if the originator is also the servicer. In this case, it is prudent to have a backup servicer who can take over the management of the pool if the original servicer fails. If this does not happen, CDRs can rise dramatically on the pool as far more delinquencies will lead to default without active servicing.

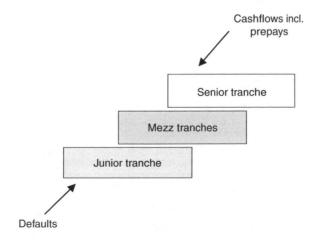

10.2.3 ABS Tranche Analysis

Once the CDR and CPR profiles of the collateral pool are understood, the next step is to understand how that affects the issued securities. This process is known as *waterfall modelling*: the cashflows are introduced into the waterfall, the effect of whatever credit enhancement is available is included and the cashflows on the issued securities and on the credit mitigation are calculated. This gives a benchmark yield and WAL for each tranche.

10.2.3.1 Credit mitigation
Credit mitigations which are sometimes found in ABS waterfalls include:

- Overcollateralisation;
- A bond wrap* or another form of guarantee on one or more tranches;
- ES accounts or interest diversion [as discussed in section 5.3.3].

10.2.3.2 Example
Consider a four-tranche residential mortgage securitisation structure: a senior piece A, two mezz pieces B and C and a junior piece D. The A piece is prepaid first and so on down the structure, so the principal repayments are distributed as in the picture on the left, and these are allocated to the tranches from the top down. (This is known as a *sequential pay* structure: other alternatives are possible.)

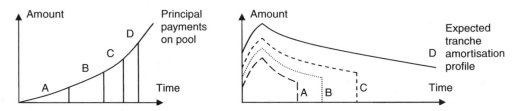

Passing this collateral prepayment through the waterfall gives the amortisation profile of the tranches shown on the right.

10.2.3.3 Sensitivity modelling
This process of modelling is repeated for a range of interest rate scenarios and underlying collateral default rates.

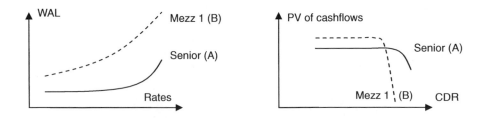

---

* As discussed in section 7.3, there are a number of highly rated specialist insurance companies who provide insurance on ABS. They are paid a fee in exchange for providing a financial guarantee on timely interest and ultimate principal much as the agencies do in the U.S. MBS market.

This type of analysis can be summarised in graphs like those above, but again in a real application more interest rate scenarios and a more extensive set of factors would be reviewed.*

Typically, the risk of extending WAL increases down the structure: our example piece B, for instance, might have the extension profile illustrated below.

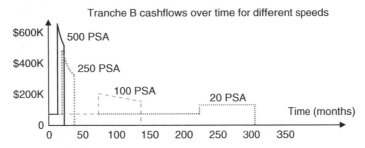

For defaults, the senior piece is often robust even under quite severe CDR shocks, but the more junior C and D pieces tend to suffer badly as CDR increases reflecting their lower subordination.

10.3 SECURITISATION STRUCTURES

This section reviews further aspects of the mechanics and assessment of securitisation structures away from RMBS. In particular, we look at CDO structures, securitisation of the cashflows from whole businesses, and structures where the investor has some degree of ignorance of the collateral pool.

10.3.1 CDO, CLO and Related Structures

Various forms of CDO are common ABS. Here we look at some aspects of how these transactions are done, what features make a good collateral pool and the role of various parties in a securitisation.

10.3.1.1 Securitisation structures

The structure of a typical synthetic securitisation is shown in the illustration.
The various parties involved are:

- The sponsor, who transfers the assets;
- The originator(s), who originate the risk. (In a CBO of public bonds this is irrelevant, but in a CLO or MBS securitisation the investor needs to understand the underwriting standards under which the loans were made.)
- The servicer, who follows up delinquencies on the collateral pool and provides collateral reporting;
- The note trustees and SPV administrators, who collect cash from the collateral, allocate it according to the waterfall, distribute it to investors in the tranche notes and manage the administration of the SPV;

---

* This is especially so since we have skirted around the complexity of the waterfall found in some more structured CMOs. See Chapter 15 of Lakhbir Hayre's book (*op. cit.*) for more details.

- The ratings agencies, who may rate the issued tranches;
- Any credit support or liquidity providers.

10.3.1.2 Collateral transfer

There are several alternative mechanisms for moving risk into a securitisation vehicle:

- The originator can sell assets from their balance sheet into the vehicle;
- The sponsor can buy assets in the market as agent for the vehicle;
- The vehicle can write default swap protection on the desired risks.

The first mechanism is common where the sponsor is also the originator: these are typically deals based on either funding (the originator wants cheap funding via issuing highly rated tranches), risk transfer (the originator wants to reduce their risk to collateral default) or regulatory capital arbitrage, and they are sometimes known as *balance sheet CDOs*. The second approach was typically used for early arbitrage CBOs.

The third mechanism is a *synthetic securitisation*. This can be very convenient: there is no need to acquire a large portfolio of bonds in a CBO for instance—which can take weeks—the terms of the CDS on each bond can be identical, removing the need for a liquidity facility to manage cashflow mismatches, and there is no need for IRSs to manage the cashflows from fixed rate collateral. Therefore, this has become an important mechanism for arbitrage transactions.

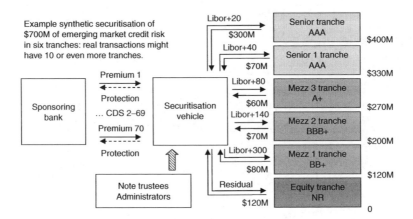

10.3.1.3 Collateral transfer and management

Some CDOs are structured just like CMOs: cashflows from the collateral are used to pay principal and interest on the tranches, possibly with the help of liquidity facilities or other collateral. These are known as *cashflow* deals. In contrast, *market value* transactions rely on the mark-to-market value of the collateral to support tranche payments (and so overcollateralisation tests are based on market value rather than par value here). Again there may be borrowing against the collateral via a liquidity facility or the collateral may be traded to raise cash.

In particular, there may be a *collateral manager* in a market value deal. Their role ranges from highly active, where they have the discretion to trade collateral at will within certain diversification and concentration limits, to purely passive management where collateral is only traded if an overcollateralisation or interest coverage test is breached.

10.3.1.4 What makes a good securitisation pool?

Like any investor, the buyers of securitisation tranches want to be well compensated for the risks they are taking. With CDOs the risks concern:

- The behaviour of the collateral pool;
- How that is transformed by the waterfall into behaviour of the tranches.

The first concern highlights the importance of *predictable* cashflows from the collateral pool: a lot is known about the behaviour of a diversified pool of conforming mortgages, so investors are comfortable that they can price the risks involved. A different collateral pool, even if on average it has less prepayment and default risk, may nevertheless be a less suitable asset for securitisation if investors are less comfortable assessing the risk. A good example here is the securitisation of *catastrophe risk*: insurance companies have had only limited success at selling the earthquake risk they originate to investors, partly because it can be highly idiosyncratic. Assessing a particular pool of earthquake policies is difficult for many ABS investors, so they often prefer risks they think they understand better.

Predictable pool cashflows can usually be transformed into attractive cashflows for tranched securities via a judicious design of the waterfall and credit enhancement. For instance, if the prepayment risk on a particular mortgage pool is too high but otherwise the pool behaves well, the sponsor can simply buy one or more interest rate floors to modify the risk profile of the issued securities and make them more attractive. If the pool itself has unknown or highly volatile behaviour, though, there is less that can be done. A key issue is *common risk drivers*: any circumstances which can cause a spike in security defaults or prepayments is potentially troubling, and most serious problems for investment grade ABS have had this cause. Examples include:

- *Servicer risk*. If only the originator has the skills to service the collateral, then the ABS may have a similar risk profile to unsecured debt.
- *Credit support risk*. When the credit quality of the securities relies on credit support such as a bond wrap rather than the behaviour of the collateral pool, the ABS buyer is highly exposed to the performance of the support provider.
- *Collateral value correlation*. Here the problem is a common factor causing a rapid decline in collateral value. For instance, in a housing market downturn, high LTV MBS can turn into unsecured debt as collateral value decreases.
- *Systematic collateral risk*. This situation is a disguised version of the one above: the collateral appears to be uncorrelated, but there is a common risk factor which may not be evident in normal market conditions such as country risk.

> *Exercise.* Review the reasons for downgrades/spread widening of securities associated with the Boxclever, Greentree and Hollywood Funding securitisations.

CBOs involving a fixed pool of liquid public securities offer investors a clear view of the collateral pool so these concerns are often minor. CLOs may be a different matter particularly if the collateral has been chosen by the originator and complete information on the obligators is not available. The pool of small business loans a national champion bank does not want in their home country and is not willing to take a first loss piece on may not be good risk.

> *Exercise.* Review the term sheet (or ideally the prospectus) for a range of balance sheet and arbitrage CDO transactions. Try to find both cash and synthetic deals.

10.3.2 Banking Using Securitisation

Securitisation is important as an enabler for a key trend in financial services: *specialisation*. Before securitisation, banks perforce kept most of the loans they originated. They were too illiquid to sell, sale might anyway be contractually or reputationally difficult, and most buyers would demand too high a premium due to the asymmetry of information present: the bank knows the obligator far better than the loan buyer does. Securitisation allows a bank to transfer some of its risk privately—a synthetic securitisation would not even require the loan counterparties to be notified—and to transfer a diversified portfolio with a relatively predictable loss distribution. In turn this allows banks to concentrate on *origination*—seeking out well-priced risks—rather than long-term risk management. The attributes needed to originate and administer risk, such as branding, customer contact and efficient administration, are rather different from those needed to assess and manage portfolios of credit risk.

A traditional bank transformed illiquid assets—loans—into liquid assets: equity and debt in the bank. In this sense, a bank is just a large securitisation: and a conduit or a SIV is a purer version of the same thing. Securitisation can then be seen as enabling a similar transformation: it can turn illiquid assets into liquid (or at least semi-liquid) tranche notes without the same degree of pooling involved in a large bank's loan bank.

The scale of the risk transfer out of the banking sector via securitisation has been considerable: over 20% of the total notional of credit risk on some estimates (although not 20% of the risk given that many banks retain junior pieces in their securitisations). This is changing the character of banks and the banking system: there is increasing attention on originating risk that is easy to securitise, on active credit portfolio management, and on using banks' preferential access to liquidity where it can earn the highest return. Meanwhile it will be interesting to see the extent to which the traditional skills involved in risk managing a buy-and-hold portfolio of credit risk migrate from banks these to the eventual holders of securitisation tranches.

10.3.2.1 The Securitisation Treadmill

Suppose a bank sells a pool of mortgages to an SPV as part of a securitisation. Before the sale, the bank would record the net spread of the mortgage assets as net interest income, and this would trickle into its earnings over the life of the mortgages. On sale, it records a gain of the entire difference between the purchase price and the value of the pool on the balance sheet. This has the effect of enhancing earnings as most or all of the excess spread is PV'd up front. However this is a one off gain: to keep earnings at the new enhanced level, the bank has to keep originating and securitising mortgages. This gives rise to what is known as the *securitisation treadmill*: banks must keep originating securitisable assets and they must actually be able to securitise them at a gain in order not to suffer an earnings shock. If good assets become difficult to find, that may then give rise to a temptation to originate less good assets simply to assure the gain on sales needed to meet the bank's earnings targets. This may have led banks towards lending to worse and worse credit quality borrowers during the U.S. house price bubble of 2001–2006.

10.3.3 ABS in Principal Finance and Whole Business Securitisation

The more predictable the cashflows on a collateral pool, the easier it is to get a large AAA tranche. This insight suggests that it is not just pools of assets like mortgages or bond cashflows that can be securitised: there are some whole businesses which have predicable cashflows too such as utilities or certain service businesses. This gives rise to whole business securitisation (WBS), a technique that is becoming increasingly popular in PE transactions.

10.3.3.1 Perfecting a claim

In a perfect securitisation, the tranche security holders have 100% legal certainty that they own a right to the cashflows as transformed by the waterfall. There are two principal ways of achieving a measure of legal certainty:

- True sale either of the collateral pool to the SPV or of a collection of derivatives that pays off on the basis of its performance;
- Granting a *first priority security interest* over the collateral to an SPV. This is a feature which is particularly well developed in English law: it allows a borrower to grant a lender the right to a first charge not just over current assets, but also over future assets such as as-yet-unearned profits.

Most WBSs involve the latter concept (and so WBS is usually only possible where the law permits this particularly wide ranging form of security interest). Suppose a company identifies a business group which has a predictable cashflow profile. An SPV is set up to issue notes, as in a standard securitisation.

The business to be securitised is isolated under a *securitisation group* and the SPV lends the funds raised from issuing the notes to the group and in exchange receives a promise to repay backed by a security interest in the assets of the securitisation group. The scope of this charge can be very wide ranging, including current and future assets, tangible and intangible. The terms of the loan are arranged to match precisely the issued notes, and the security interest expires after they mature.

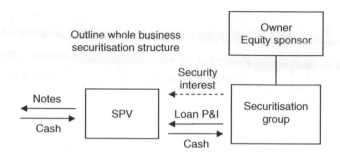

10.3.3.2 Features of the structure

The advantage of this structure is that the ownership of the securitisation group remains with the equity sponsor company—so there are no tax concerns involving crystallisation of gains on a sale—and the original owner keeps any profits after those necessary to service the securitisation (although there may be a requirement to fund a reserve account in the SPV, and this has to be met before the securitisation group can dividend profits back to the parent).

The note holders have no legal recourse towards the equity sponsor, so the financing structure and ability of timely and full repayment hinges on the stand-alone future cashflows generated by the operating companies in the securitisation group and not the credit quality of the equity sponsor.

10.3.3.3 Advantages of WBS

The conventional capital model of a public company is to have a large piece of listed equity which acts as credit support for unsecured debt either in the form of bank loans or debt securities (or both). More innovative companies might use assets such as property or receivables as collateral for secured borrowing, but the scope for extensive use of the secured borrowing market is limited by the ratings agencies: a company which places a charge on too many assets is liable to find itself downgraded.

WBS, in contrast, offers an alternative paradigm: fund the majority of the business using securitisation, and keep a smaller, more highly geared piece of PE as credit enhancement. This technology has transformed PE from a seeker after high-growth companies early in their life cycles to a much broader investor. In particular, the kinds of companies which work well in WBS—cash cows—tend to be mature firms in relatively unglamorous sectors: exactly the opposite end of the corporate spectrum to the traditional PE-financed company.

The use of WBS has allowed PE firms to identify likely targets amongst public companies, bid on them, then finance much of the bid using the issuance of WBS notes. This works well where there is a substantial difference between the companies' credit spread for unsecured senior debt and the weighted average cost of capital in a PE and WBS funding model.*

10.3.3.4 Candidate businesses

The lower the volatility of the expected cashflows from the securitisation group and the longer those cashflows are stable for, the better a WBS deal is available. By better we mean:

- Longer duration;
- Higher leverage;
- Lower requirements for external liquidity or credit support;
- Higher credit rating available for the WBS notes.

Good candidate businesses can be found in various sectors including:

- Healthcare, with nursing homes being particularly attractive due to the combination of property collateral and ongoing fees;
- Transport, including toll roads, airports (with their lucrative retails assets and landing fees) and sea ports;
- Utilities, especially where the delivery infrastructure has a long life and a high initial cost, as in water, gas or electricity;
- Service businesses such as hotels.

10.3.3.5 Further credit considerations

Another issue in WBS beyond the cashflow profile of the securitisation group is the likely scenario if the business has to be liquidated. Note buyers will value the ability to replace the businesses' management without severe cashflow interruptions, to sell the business to a trade buyer or to liquidate physical assets such as a property. Easy to run businesses which do not depend on a particular management team are therefore to be preferred. Moreover, WBS transactions often include a purchased liquidity facility to support some or all of the tranches during a period where the tranche holders are replacing management or selling the securitisation group.

---

* Arguably, if ratings agencies and debt analysts were doing their job properly, this gap would usually be small: after all, the cashflows of the business are available to support the senior debt before a WBS; the security interest just makes this support explicit, focussing attention on the stability of the cashflows that were there all along. One explanation for the value created by WBS is that typically such deals involve very tight covenants on the management of the securitisation group which require it to focus on generating cash from its core business: such tough restrictions are not present in typical senior bond offerings. Perhaps investors value the ability to force management not to innovate or diversify rather highly.

10.3.3.6 Example

Several early UK WBS transactions were done on public house businesses which owned and operated portfolios of pubs. This sector had ideal characteristics for a WBS including:

- Tied arrangements under which the pubs were required to buy a large proportion of their beer, wine and spirits, and in some cases soft drinks, from the business generating a stable income stream;
- Long-term leases giving rental income on the pubs themselves, secured by property.

The first transactions also benefited from good timing: thanks to changes in the law concerning tied arrangements there was a good opportunity for a pub operator to acquire more pubs and significantly increase its scale and profitability. However, bank finance would have been difficult and costly to acquire, and the equity market did not seem to value the stable earnings of these businesses.

An initial WBS pub transaction was done, a number of tranches were issued and the securitisation group invested the funds raised partly in buying further pubs. Under the terms of this transaction the issuance SPV had a security interest on the current and future assets of the business, and the business agreed to a number of covenants designed to preserve the stability of the cashflows including:

- An undertaking to adhere to a minimum EBITDA to debt service ratio;
- A minimum capital spending per year on each pub (so that the value of the pubs does not fall too far, and they retain their appeal as drinking venues);
- Acquisition and disposal criteria governing when pubs can be bought and sold by the business.

The covenants allowed note holders to seize the business in the event of under performance. Clearly, the assets would be valuable to other potential buyers given the barriers to entry for a national pub operator, so even if the current management had difficulty in extracting sufficient value from the pubs to service the notes, there was a fall back plan. The transaction also included liquidity facility to support cashflows on the more senior notes in the event of a business interruption during sale or a temporary fall off in cashflows. Finally, of course, the group management were highly incentivised to generate cash as they did not receive a dividend until the note holders had been paid and then only if the EBITDA to debt service ratios were sufficiently high.

The combination of these features made the transaction a success, and the same model was subsequently followed in a number of other pub-related deals.

> *Exercise.* Research the kind of transactions described above and the subsequent performance of the tranche securities. Why is this kind of WBS deal not done in the United States?

10.3.4 Some Revolving and Blind Securitisations

There are several reasons why a securitisation sponsor might wish to buy protection on a pool without revealing the contents of it to the tranche buyers:

- The pool changes so fast that details of it would be out of date by the time they were communicated to tranche investors.
- They cannot reveal the details of the pool due to the nature of the originator's contract with the underlying borrowers.
- The originator does not want to reveal the details for business reasons.

The first of these is typical of revolving transactions such as credit card securitisations: we deal with these first. The last two give rise to trades involving blind pools: an example here is also discussed.

10.3.4.1 Revolving securitisations

Revolving securitisations are transactions where for some period principal repayments are reinvested in the purchase of new collateral rather than being used to repay principal on the tranches. In contrast to term securitisations such as MBS or most CDOs, then, new collateral regularly enters the pool. Typically, this structure is used when the collateral pool has a relatively short life—as for credit card or trade receivables—and so replacement is needed to keep the collateral balance roughly constant for an extended period. This revolving stage typically lasts from 2 to 4 years, after which the pool is allowed to amortise, and principal is repaid on the tranches.

The tranche securities may be either

- *Bullet*, in which case sufficient cash needs to be present in the securitisation SPV to repay principal on time, either through collateral prepayment or through the use of a liquidity facility;
- *Scheduled amortising*, with the SPV being committed to paying out fixed amounts per month to some or all tranches on a schedule;
- Or *uncontrolled amortising*, in which case the tranche holders receive some share of the principal repayments actually received by the SPV from the collateral.

It is common for revolving securitisations to include clauses which permit the suspension of the revolving period and an early amortisation of securities. Typically, these *early amortisation* events may include:

- Operational risk–related events such as inability to transfer receivables into the SPV, changes in the law, in taxation and so on;
- The monthly prepayments on the pool falling beneath some *trigger* level. This trigger is usually a multiple of the amount needed to ensure that scheduled amortisation or bullet payments can be made;

- ES is not sufficient to fund a designated spread account;
- Defaults or delinquencies eroding the principal balance sufficiently to endanger principal repayment of the mezz tranches. (Typically the originator retains a seller's interest equivalent to a junior tranche.)

If there is an early amortisation event, prepayments are directed towards the securities according to rules in the waterfall. The spread account or ES from the pool may also be used to accelerate prepayment.*

> *Exercise.*   Review the prospectus for a typical credit card securitisation with particular attention to the default and amortisation risks of each tranche and of that retained by the seller. What compensation are investors receiving for their share of these risks?

10.3.4.2 Blind managed pool

One good reason for accepting a blind pool is that we think the collateral manager is better at selecting collateral than we are. Thus, if:

- The manager has a good reputation for managing the collateral asset class;
- Their compensation depends partly or completely on good collateral management;
- The manager has no incentive to substitute bad collateral for good;
- Anyway we are protected by strict diversification, concentration and overcollateralisation requirements.

Then investors can perhaps be reasonably comfortable in purchasing tranche securities where the underlying pool is blind and there is collateral substitution. If not, then some investors will demand an excess yield for buying the tranches and others will not invest at all.

10.3.4.3 Derivatives receivables securitisation

A good example of a situation where a blind securitisation with collateral substitution makes sense is *derivatives receivables securitisation*. Recall that regulated institutions have a capital requirement for current future credit exposure and PFCE on derivatives transactions. Perhaps this capital requirement can be reduced by buying protection on counterparty performance? The outline structure is as shown in the illustration.

---

* Basel II distinguishes between *controlled* and uncontrolled early amortisation in revolving securitisations. The former requires that the prepayments are shared strictly *pro rata* between the tranches and the seller's interest during an early amortisation event, and that the SPV has 'sufficient' liquidity in the event of early amortisation. The CCFs for banks' exposures to transactions with uncontrolled early amortisation are typically much higher than for controlled amortisation deals.

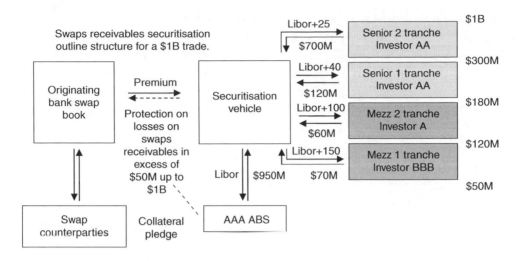

The collateral pool is the collection of net receivables on all swap transactions the originating bank has with investment grade corporates under a signed master agreement, subject to certain limits:

- Country limits;
- Diversification limits on the sectors of each counterparty;
- Concentration limits which require no net receivable from a single A+ or better-rated counterparty forms more than 3% of the pool and no net receivable from any counterparty rated BBB+ to A forms more than 2%.

If a net receivable partially or fully included in the pool is not paid, the originating bank has the right to add the amount not received to the cumulative losses of the pool under a portfolio default swap written by the securitisation vehicle. The swap allows the bank to substitute new exposures into the pool up to the maximum AID covered and subject to the diversification limits, so as the mark-to-markets of the swaps in the portfolio move around, the bank retains cover on whatever exposures it has, rather than on a fixed portfolio of obligators who may or may not owe it money at any given point.

The securitisation vehicle uses the premium received on this swap to pay interest on a number of tranched securities. Once cumulative losses reach $50M, claims can be made on the securitisation vehicle and so the tranches start to lose principal starting as usual at the bottom.

The cash raised from issuing the securities is invested in high-quality securities, and these are pledged as collateral against the portfolio default swap.

The trade caps the bank's losses on qualifying receivables to $50M for the term of the transaction. If it can persuade its regulator that this is sufficiently effective, then it may also be possible to get capital relief on the counterparty credit risk capital required for the transactions in the pool.

This type of transaction is a good example of a trade that makes sense from both parties' perspective: the bank gains not just from risk reduction but also potentially has a regulatory capital saving. This saving would justify paying a bigger premium on the portfolio default swap. Investors can understand why the pool must be variable—to give the bank cover on its current swap receivables portfolio whatever that is—but they are protected by the country, diversification and concentration limits. Moreover, it makes sense for the pool to be blind: the bank does not want to disclose who it has a big portfolio of swaps with. Investors might be concerned that the bank could be incentivised to originate swaps with more risky counterparties, but the $50M retention is some protection here, and in addition there may be covenants which require the bank to use the same credit approval criteria for exposures to be included in the pool as for other transactions.

Notice that under a master agreement, default of the bank causes all of the swaps to close out and the bank will no longer be able to pay premium on the portfolio default swap, so the tranche rating is capped at the credit rating of the originating bank.

Exercise. How much might the capital charge on a $1B portfolio of current swaps receivables be? Given this, how much can the bank afford to pay over the value of the protection provided by the swap?

Now suppose you are a regulator looking at a proposed transaction of this form. Is it reasonable to grant capital relief on the charge on the current value of swaps receivables included in the pool? What about the PFCE charge?

CHAPTER 11

Novel Asset Classes, Basket Products, and Cross-Asset Trading

INTRODUCTION

This chapter examines several relatively novel and innovative areas of trading, beginning with inflation-linked instruments. Equity basket trading comes next. CBs are then discussed, and a need for modelling equity and credit instruments in a uniform framework is encountered. The subsequent section introduces those models and shows how they can be applied to arbitrage trading, CB modelling and the analysis of loan portfolios. Finally, we discuss new products in general and how their introduction can be managed.

11.1 INFLATION-LINKED PRODUCTS

Consider a basket of goods and services that represents a typical consumer's costs. Over time, the cost of this basket will change: some things will become more expensive, some will become cheaper and (mostly) the price of the basket will drift slowly up. This is *inflation*.*

Inflation affects the purchasing power of money: the higher it is, the more quickly purchasing power is eroded. If inflation is higher than government bond rates, then investors have to take risk to retain constant purchasing power. Whereas if it is lower, then investors can make a *real return*: their investments will be able to buy more of the basket in the future than they can today. To provide investors with a certain real return, governments (and others) issue *inflation-linked bonds*. These provide investors such as pension funds with an instrument which guarantees that the purchasing power of an investment will increase.

---

* See Mark Deacon, Andrew Derry and Dariush Mirfendereski's *Inflation-Indexed Securities: Bonds, Swaps and Other Derivatives* for a more extensive discussion of inflation and inflation-linked products.

11.1.1 Inflation Indices

Consider a basket representing the typical costs of an investment banker during the average month:

- Twenty assorted take-away meals;
- Twenty sandwiches;
- Ten bottles of 10-year-old Chateau Latour;
- A weekend's skiing in Val d'Isère;
- The current monthly rent of a luxury flat in Docklands.

Suppose we reprice this basket at the start of every month by checking the prices of each item in a uniform way.

Over 6 months, we observe that the price of the basket increases by 17%. This change represents the increase in the cost of living for a typical investment banker: to have the same standard of living, as measured by the basket, their nominal income must increase by 17% in the same period.

In general, an inflation index measures the cost of living for some consumer. Our basket has been skewed to a (caricature of a) typical banker: other consumers will need a different index to represent their costs.

11.1.1.1 Inflation indices

The inflation index is usually calculated in each country by a government agency using a suitable local basket. The design and calculation of the basket is key to the utility of an index: typically, the baskets used to calculate inflation are intended to reflect the costs of a 'typical' retail consumer, so they contain a wide range of goods and services. Equally, the composition varies from country to country: food is a bigger percentage of the Eurozone basket than the U.S. one, for instance.

Notice that the basket needs to be rebalanced regularly to remain representative: video cassettes would have had a place in the basket in the 1980s, but now DVDs are much more common, and soon we will probably have high-resolution DVDs.

Exercise.   Review the composition of the basket used to calculate inflation in a country of interest. How often is the basket composition revised and by whom?

Some of the major indices are as follows:

- In the United States, the consumer price index (CPI) for all urban consumers, or CPI-U;
- In Eurozone, the harmonised index of consumer prices (HICP) excluding tobacco;

- In the United Kingdom., the retail price index (RPI) and the CPI. The main differences between them are that the CPI excludes certain housing costs such as average mortgage repayments but the RPI includes them.*

Inflation is usually quoted on an annualised basis. Note that some countries have experienced deflation, so the index can be negative: a good example of this occurred in Japan for much of the 1990s.

11.1.1.2 Seasonality and lag

The calculation of the basket price is often not a simple matter: tens of thousands of individual prices from different parts of the country may go into the calculation. Therefore, there is often a *lag* between the publication of inflation data and the date they refer to. In the United Kingdom, for instance, this is approximately 6 weeks, so the data for December are available in mid-January.

As an inflation index is a more complicated thing to calculate than, say, an equity index level, there is the risk of *restatement*. This can happen when errors are made in the calculation of the basket price, or later data invalidate earlier assumptions. Obviously, it is important to understand the impact that this might have on any financial asset linked to inflation. This is especially true in emerging markets where inflation is potentially higher and operational risk in its calculation may be greater.

Inflation usually displays *seasonality* due to sales periods. Winter and summer sales tend to depress prices in December and January versus November and in July and August versus May or June, so the month-on-month data can be negative for these periods even if the overall trend is up.

11.1.1.3 Inflation targeting

The control of inflation has become macro economic orthodoxy: many governments take the view that inflation should be kept within limits, and interest rates are used as a control: if inflation is too high, rates are raised to slow down the economy and lower inflation. Although there is broad agreement about this, the details vary considerably from country to country:[†]

- Some countries have a target *band*, indicating that a particular measure of inflation should be not only lower than *x* but also higher than *y*. This is often because a small amount of inflation is often considered good for an economy, deflation being essentially a tax on the holders of assets versus the holders of cash.

- Others simply have a general notion of *price stability* rather than an explicit inflation target, with the central bank or Treasury being given some discretion as to the interpretation of this.

---

* See www.bls.gov/cpi/home.htm, epp.eurostat.ec.europa.eu/pls/portal/url/page/PGP_DS_HICP and www.statistics.gov.uk for more details of these indices. The decision as to whether to include housing costs can have a major impact on the inflation level: for instance, the RPI in December 2006 was 4.4%, but the CPI was only 3.0%.

† This is a very short summary of a complicated macro-economic puzzle. See Ben Bernanke, Thomas Laubach, Frederic Miskin and Adam Posen's *Inflation Targeting: Lessons from the International Experience* for a more comprehensive discussion.

This is important from a trading perspective as if inflation targeting is successful, it obviously places limits on the price changes experienced by inflation-linked securities.[*]

> *Exercise.* Download a long history of inflation data from a range of sources. How does earnings inflation relate to wage inflation over the long term?

11.1.2 Inflation-Linked Bond Design

There is one common form of inflation-linked security—the capital-indexed bond (CIB)—and a variety of other structures. We begin the discussion of security design by looking at the idea behind the CIB.

11.1.2.1 Real and nominal yield

In the design of inflation-linked products, the term *real* refers to something after the effect of inflation, so that if you have a 4% real yield, it implies that the buying power of your asset versus the basket increases by 4% annually. In contrast, the term *nominal* refers to an ordinary, non-inflation-adjusted yield.

11.1.2.2 Capital-indexed bonds

A CIB provides investors with a fixed real yield if held to term just as an ordinary fixed rate bond provides investors with a fixed nominal yield. To do this, all of the cashflows of the bond accrete at the rate of inflation.

The typical structure for a semi-annual pay CIB is therefore as follows:

- Maturity: N years
- Initial notional at issue: 100
- Notional(n): $100 \times$ index(n) for $n = 1 \ldots 2N$
- Index(n): basket price at time(n)/basket price at time(0)
- Real coupon: c% for some fixed c
- Cashflow at time n: $c \times$ Notional(n)/2
- Principal repayment at time N: Notional(N)

The following example shows the cashflows for 100 face of a 2-year, 2% real-yield bond, given a certain level of inflation:

n	Real Coupon (%)	Index (%)	Compounded Inflation (%)	Coupon	Principal
1	1	1.5	1.5	1.0150	
2	1	1.4	2.92	1.0292	
3	1	1.2	4.16	1.0416	
4	1	2.1	6.35	1.0635	106.35

---

[*] In the United Kingdom the situation is even more complex as the CPI is the targeted measure of inflation, but most securities pay on the basis of the RPI.

The CIB pays the stated real yield as compensation for the risks the holder is taking such as liquidity risk. This is achieved via the indexation of all its cashflows.

11.1.2.3 Interest-indexed bonds

The interest-indexed bond (IIB) structure is less common than the CIB. Here the rate of inflation is simply added to the stated coupon. Principal is not adjusted. The cashflows are therefore: .

n	Real Coupon (%)	Index (%)	Compounded Inflation (%)	Coupon	Principal
1	1	1.5	1.5	2.5	
2	1	1.4	2.92	2.4	
3	1	1.2	4.16	2.2	
4	1	2.1	6.35	3.1	100

The IIB is therefore somewhat like an FRN, but where the floating rate is inflation.

11.1.2.4 Zero coupon inflation-linked bonds

Zero coupon bonds are useful in thinking about discount factors: for inflation, the analogue is the inflation-linked zero. This simply adjusts a single terminal cashflow for the effect of inflation.

11.1.2.5 Government inflation-linked bonds

Many governments issue inflation-linked bonds including those of the United States; various Eurozone countries including France and Germany; Japan; and the United Kingdom. Inflation-linked issuance ranges from single-digit percentages of total government debt to (in the case of the United Kingdom) over 20% of the total. At first, the signalling aspects of inflation-linked issuance may have been an important motivation for issuance: the interest costs on inflation-linked bonds are higher when inflation in high, an governments have a direct incentive to control inflation if they have a significant portion of their debt in inflation-linked instruments.* Now, though, the provision of real yield for retirement savings is often viewed as an aim in itself. This is particularly the case in jurisdictions such as the United Kingdom where some pensions are linked to inflation: good ALM would suggest that a pension fund with an inflation-linked liability needs an inflation-linked asset with a matching duration. Without inflation-linked government bonds such assets would be hard to find.

> *Exercise.* If inflation-linked bonds were not available, what investments would you use to try to construct a portfolio that was highly likely to have a positive real yield?

---

* Another government motivation is that market expectations of future inflation have tended to be higher than realised, so debt service costs for inflation-linked bonds have been lower than borrowing with conventional nominal instruments.

Most government inflation-linked bonds use the CIB structure. For instance the U.S. government's inflation-linked bonds, or Treasury inflation-protected security (TIPS), are CIB bonds where the principal value is adjusted for inflation by multiplying the value at issuance by an index ratio as described above. The inflation accrual on principal is paid at maturity, and the coupon payments are a fixed percentage of the inflation-adjusted value of the principal.

One issue where government bonds differ is their treatment of deflation. For the TIPS, the principal repayment at maturity is protected against deflation, so the principal repayments are never less than the initial face. The coupon payments are, however, not protected against deflation. This structure is therefore known as a 0% floored principal bond or, less precisely, a bond with a *deflation floor*.

U.K. government inflation-linked bonds or *gilt linkers* also use CIB structure, but they do not include a deflation floor on either coupons or principal.

> *Exercise.* Review the prices of inflation-linked gilts on the U.K. debt management office website www.dmo.gov.uk. What do the quoted yields mean?

11.1.2.6 Inflation-linked bonds through the cycle
The ubiquitous use of short-term rates as the only inflation control mechanism suggests that we can modify the table of section 1.2.3 to include the inflation-linked bond market:

Point in the Economic Cycle	Inflation Expectations	Central Bank Action	Real Bond Market	Nominal Bond Market
Overheating	Low, increasing	Increased rates	Good	Bad
Cooling	Overshooting	Rate increases slowing	Declining	Improving
Recession	High, decreasing	Rate cuts	Bad	Good
Recovery	Undershooting	Rate cuts slowing	Improving	Declining

11.1.2.7 Corporate inflation-linked bonds
A range of corporates naturally have inflation-linked assets including:

- Utilities, especially where regulation limits price rises to some spread to inflation;
- Flow businesses such as supermarkets whose profits depend on average prices.

These are natural candidates to issue inflation-linked corporate bonds, and most corporate issuance has been either in these sectors or by financials shadowing government issuance.

11.1.3 Retail Inflation-Linked Products

Inflation-linked products have a natural attraction to anyone saving to defer consumption thanks to the certainty of a positive real yield they provide. This has resulted in an increasingly large array of retail products being structured with some inflation exposure. Examples include:

- Inflation-linked deposits;
- Mixed savings products which offer some combination of an inflation-linked and a nominal yield;

- Structured products combining inflation linking with other risk. A simple example here would be a variant of the GEB [discussed in section 2.3.3] but where instead of having a zero coupon nominal bond plus an equity option, a zero coupon *real* bond plus an equity option is used. This offers a return of constant real principal plus equity upside participation.

> *Exercise.* Find out what inflation-linked retail products are available in your jurisdiction and try to break them down into simpler products as in the inflation-linked GEB example above.

11.1.4 Lags and the Inflation-Linked Curve

The payments on bond coupons are not based on the current month's inflation because this is not known due to the publication lag. Instead, there is an interval between the date a CPI figure refers to and the coupons it affects. This is often 3 months, so that, for instance, December's inflation data are known in mid-January, and this determines coupons due in March.

11.1.4.1 Accrued inflation

The idea of accrued interest gives fair compensation to both the buyer and the seller of a nominal bond: the bond trades on the dirty price which assigns some of the current period coupon to the buyer and some to the seller. For an inflation-linked bond, the idea is similar, but we need to assign a real rather than a nominal coupon so we need to know the price of the basket at trade date. At any given date during the month, this could, for instance, be interpolated between the index value for the start of the previous month and that for the start of the next, as in the illustration.

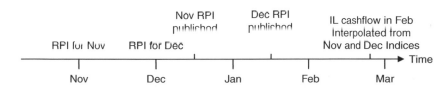

This is the method used for TIPS, for instance.*

11.1.4.2 Coupon calculation

Accrued inflation is not just relevant to trading inflation-linked instruments away from coupon dates: we also need a method of calculating the level of RPI for any date to be able to calculate what coupon payments are due on a CIB with arbitrary payment dates.

This is also done by interpolation. Consider a coupon due on 15 February 2007 for a 2% semi-annual pay CIB issued on 4 August 2006. We will use straight-line interpolation between the RPIs as this is the convention for many linkers.

---

* Other inflation-linked bonds have slightly different methods: see Deacon et al. (*op. cit.*) for more details and detailed calculations, and note in particular that some U.K. linkers have an 8-month lag rather than the more usual 3.

- The coupon payment is 1% times the reference (ref) RPI for 15 February 2007 divided by the ref RPI for the CIB issue date, 4 August 2006.

- The ref RPI(15 February 2007) is equal to the ref RPI(1 February 2007) plus 14/28 of [ref RPI(1 March 2007) − ref RPI(1 February 2007)].

- Similarly, ref RPI(4 August 2006) = ref RPI(1 August 2006) + 3 × [ref RPI(1 September 2006) − ref RPI(1 August 2006)]/31.

The ref RPI for the 1st of the month is the RPI for 3 calendar months earlier, so the ref RPI(1 March 2007) = RPI for December 2006.

Consider the following RPI data:

May 2006	June 2006	July 2006	August 2006	September 2006	October 2006	November 2006	December 2006	January 2007
143.4	145.0	146.5	147.7	148.5	149.2	150.1	151.2	150.4

This gives ref RPI(15 February 2007) = RPI(November 2006) + 14 × [RPI(December 2006) − RPI(November 2006)]/31 = 150.1 + 14 × 1.1/28 = 150.65. Similarly ref RPI(4 August 2006) = RPI(May 2006) + 3 × [RPI(June 2006) − RPI(May 2006)]/31 = 143.4 + 3 × 1.6/31 = 143.55.

Therefore, the inflation uplift is roughly 150.65/143.55 = 1.05, so the real 1% cashflow on 15 February is actually a nominal cashflow of approximately 1.05 in this case. This amount would have been known when the RPI for December 2006 was published in mid-January 2007.

11.1.4.3 Break-even inflation

If inflation-linked government bonds are fairly priced compared with nominal bonds, the rights to the same PVs should have the same prices. The *break-even inflation* for a bond is the level of inflation that would need to be experienced in order for a real bond to generate the same yield to maturity as a nominal bond with the same maturity and coupon frequency. Roughly* if both 2% 5-year real and 5% 5-year nominal bonds trade at par, then 5-year break-even inflation is 3%.

11.1.4.4 The inflation curve

If a range of real and nominal instruments trade from the same issuer, the break-even inflation curve can be calculated. Thus, for instance, knowing the prices of gilts and gilt linkers allows the sterling inflation curve to be derived.

11.1.4.5 Seasonality

Note the seasonality in the data above: the December-into-January inflation is negative as the basket was cheaper in January than in December. This means that the later

---

* The situation is slightly more complicated for several reasons: first, the governing equation—known as the Fisher equation—is sometimes taken to include extra compensation for the volatility of future inflation, adding an extra risk factor which requires compensation; and second, there are various small convexity adjustments. See Deacon et al. (*op cit.*) for more details.

in March a coupon is due, the *lower* it is. The inflation curve is therefore typically not smooth.

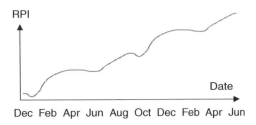

11.1.4.6 Lack of protection

Note that in a very high inflation environment, a CIB does not give complete protection against inflation due to the lag. If inflation is 10% per month, for instance, by the time a real coupon is paid, it will have lost roughly 27% of its buying power since $90\%^3 = 73\%$.

> *Exercise.* Examine the connection between the price of Brent crude and inflation levels. Given that inflation volatility tends to be rather low, and Brent options trade on higher vols, is a trade suggested by your analysis?

11.1.5 Inflation Swaps

IRSs allow interest rate risk to be managed: a fixed rate bond plus a matching pay fixed receive interest rate swap, for instance, is equivalent to a floating rate instrument. Similarly, *inflation swaps* allow inflation risk to be managed. The simplest version index links a single cashflow.

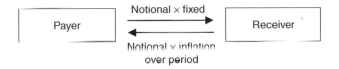

11.1.5.1 Zero coupon inflation swaps

Above we saw that a cashflow that was uplifted by inflation from 4 August 2006 to 15 February 2007 increased in nominal terms by about 5% on the basis of the RPI data given. The zero coupon swap uses the same idea for inflation linking one cashflow: this trades versus fixed. A typical structure fixing at some fixing date with a fixed rate of $r\%$ would then have the outline* form:

- Maturity date: fixing date $+ n$ years
- Fixed rate payer pays: notional $\times (1 + r)^n$
- Fixed rate payer receives: notional $\times$ ref RPI(fixing date $+ n$)/ref RPI(fixing date)

---

* A full term sheet would include precise dates, day count conventions, business day conventions and so on, all of which are omitted here for simplicity.

11.1.5.2 Bond hedging

A *revenue swap* is a hedge for the inflation risk in a CIB. It simply packages up a number of zero coupon swaps, one for each coupon paid by the bond before maturity, and one larger one for the final coupon and return of principal. Thus, if we had £10M face of a 2-year semi-annual pay 2% CIB, we would need three zero coupon swaps of 6-, 12- and 18-month maturity on £100,000 notional, and one £10.1M notional swap of 2-year maturity.

Most financial issuers of inflation-linked bonds hedge their issuance using either matching revenue swaps or broad portfolio hedges with few taking significant inflation risk.

11.1.5.3 Asset-swapping inflation-linked bonds

An inflation-linked bond plus a matching revenue swap generates a pure inflation leg versus fixed, and these fixed cashflows can be swapped for floating in the usual way. Another way to do this is to *asset swap* an inflation-linked bond. Here a linker is sold and simultaneously a pay floating receive inflation swap is entered into. Revenue swaps on the one hand and asset-swapped inflation-linked bonds on the other are therefore basic sources of inflation legs for derivatives structuring and risk management.

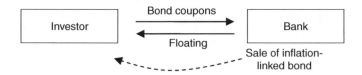

11.1.5.4 Year-on-year inflation swaps

Another variety of inflation derivative is the *year-on-year* swap: this pays the annual change in inflation versus fixed. The outline form is:

- Maturity date: fixing date + n years
- Fixed rate payer pays annually: notional $\times r$
- Fixed rate payer receives in year i: notional $\times$ ref RPI(fixing date + i)/ref RPI (fixing date)

The year-on-year swap hedges the inflation realised year by year, and hence has no consideration of cumulative inflation. This makes sense for cashflow-oriented users such as banks trying to hedge typical inflation-linked deposit accounts.

11.1.5.5 Other inflation derivatives

The inflation derivatives market away from swaps is still in its infancy. Caps and floors on inflation do trade, but the market here is sometimes distorted by an oversupply of 0% floors stripped from deflation-protected bonds. Pricing is usually based on the foreign currency analogy where the inflation index is viewed as an exchange rate (of cash versus the basket) and inflation as the interest rate in the foreign currency. This allows the standard technology of pricing cross-currency interest rate derivatives to be applied to inflation derivatives.

11.1.6 Pension Fund Risk Management

There are two broad classes of pension:

- *Defined benefit* (DB), where the pension provider has an obligation to pay a defined amount (which may, for instance, be some percentage of final salary adjusted for inflation);
- *Defined contribution*, where the beneficiary receives a pension on the basis of performance of the funds invested on their behalf.

Pension funds are typically the collateral for employers' pension obligations. The details differ from country to country, but often both employers and employees pay regular amounts into a pension fund. The employer has to pay promised pensions, and the pension fund is available to assist in meeting those payments.

Defined contribution funds are simply asset managers: they typically offer a range of investments to beneficiaries, and they have no obligation beyond following the directions they receive. Pensioners have no credit risk to their employer—unless they invest in their firm's stock or bonds—but equally they have no recourse if their pension fund investment performance is poor.

Exercise. Review the documentation for your pension. What type is it? How often do the trustees report to you? What would be the consequences if your fund were unable to meet its obligations?

11.1.6.1 Judging the funding level and solvency of a pension scheme

In many jurisdictions, firms are obliged to fund DB schemes to an adequate level. This adequacy is regularly reviewed by an independent actuary for instance every 3 years in the United Kingdom. This functionary makes a judgement on the level of contributions needed over the long term to allow the fund to pay the pensions promised. The actuary has to make judgements about future investment returns in this process, so they rely on modelling the future behaviour of both the fund's assets and its liabilities.

Note that if we were simply to compare the mark-to-market value of a pension scheme's assets with the PV of its liabilities, it might well not be solvent: most schemes rely on continuing contributions and excess investment returns over the risk free rate to provide the promised benefits.

11.1.6.2 Defined benefit funds

The obligation to pay DB pensions creates a much more significant risk management problem than defined contribution benefits:

- Pensions are very-long-term liabilities. Many of today's 20-year-old workers will probably live to be 90 or more, and they and their employers may make pension contributions over much of that time.
- Longevity is increasing. In some countries at the moment, for instance, the average age of death is increasing by around 4 months per year.

- A scheme offering DB pensions must invest the available contributions to provide the promised benefit. Pensioners living longer make their problem worse, since a pension has to be provided for longer: they are said to have *longevity risk* on the liability side.[*]

- If a fund is not judged to be adequate to meet the future claims on it (in many countries at least), employers have a legal obligation to make good the shortfall.[†]

These risks have caused DB pensions to be much less attractive to employers than defined contribution ones, and it is now fairly rare in some countries to find a private company offering a DB pension to new employees. However, there are a huge number of employees who are contractually entitled to DB pensions, so the risk management issues discussed above will continue for many years.

> *Exercise.* How fast is longevity increasing in your country? How different is it between men and women? Is there data available to allow you to deduce the effect of income or education level on longevity?

11.1.6.3 The DB liability profile

One of the roles of a pension scheme actuary is to estimate the liability profile. For a typical scheme, the actuary has:

- The current scheme membership, their ages, sexes and current contributions;

- Information from which estimates of the retirement and death profiles of scheme members can be made.

Using this information, they estimate the payout profile of the scheme. The illustration shows a typical result of this process for a fairly young scheme: payments go up as more and more beneficiaries retire, and then slowly decrease over 70 years or so as beneficiaries die.

Note that this profile is not fixed for several reasons:

- Increased longevity extends it.

- Many DB pensions are inflation linked, so a pensioner might, for instance, have the right to receive an annual pension of 75% of their final salary adjusted for RPI. Higher inflation expectations increase the scheme's liabilities.

- Similarly, wage inflation increases likely final salaries and hence also negatively impacts the scheme's funding levels.

---

[*] Life insurers have the opposite problem: the sooner people die, the quicker a life insurer has to pay out, and hence the larger the PV of their liability. Therefore, combining life insurance with pensions can sometimes provide a rough and ready longevity risk hedge.

[†] The precise calculations of how adequacy is judged and what must be done if a fund is inadequate vary from country to country; see, for instance, David Blake's *Pension Finance* for a discussion of the U.K. situation.

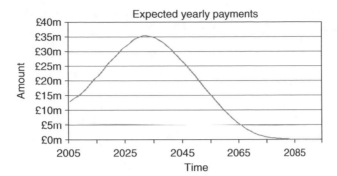

The traditional approach to meeting these liabilities was to invest in long-term assets—typically equities and long-term corporate bonds—and reassess the balance between assets and liabilities occasionally.

Actuaries would estimate long-term growth rates for the fund's assets, and hence provide an opinion on the ability of the investment strategy to meet the liability profile as part of their funding adequacy assessment.

Unfortunately, it turned out in some cases that the actuary's predictions of fund asset performance were optimistic, at least in the short term. Equities do not grow at 9% every year. This resulted in some schemes showing significant under-funding as the equities markets fell, for instance, during 2000–2001. In some cases, employers (or even employees) had to increase their contributions to the fund.

> *Exercise.* What growth rates do the actuaries in your scheme assume? Given that you might well be fired for modelling an equity index derivative assuming that the broad market grows at 9% per annum, are you comfortable with the actuarial assumptions?

11.1.6.4 Liability driven investment

Liability-driven investment (LDI) is the process of designing and implementing an investment strategy to meet a liability profile. It came to the fore as pension scheme trustees realised the extent of the market risks taken in the traditional approach to pension fund investment:

- Equity risk, thanks to the investment of fund assets in equities;
- Interest rate risk from two sources: the investment of fund assets in fixed rate bonds, and the use of a long-term interest rate to discount fund liabilities back to the present so that if rates fall, liabilities increase, and so solvency falls;
- Inflation risk due to the indexation of liabilities.

LDI is just good ALM, controlling the mismatch between assets and liabilities. Unlike banking book ALM, however, where we can control asset/liability mismatch risk to a large extent, LDI for DB funds is less straightforward since:

- The problem is very long term—at least 75 years in the example above.

- There is a considerable shortage of very-long-dated assets, especially long-dated inflation-linked ones.

- Most schemes are not so well funded that they can afford not to take any risk, so they need the yield available from some degree of credit spread or equity risk taking.

> *Exercise.*   Research the long-dated inflation-linked bond market. What issues are available? Where do they trade versus short-dated bonds from the same issuer?

11.1.6.5 LDI-using swaps

The considerations above suggest a strategy where the fund buys medium and long-dated corporate bonds for yield enhancement, and then lengthens and inflation links the portfolio returns synthetically via inflatio swaps.

The aim of this process is to produce a portfolio with lower sensitivities to the market risk factors, a higher probability of meeting the scheme's liabilities and hopefully a low probability of significant under-funding under a broad range of market conditions. The scheme will still have longevity and wage inflation risk.

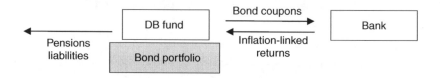

IRSs can also be included to reduce the interest rate risk on the discounting of liabilities, and the fund will probably also have other sources of return such as an equity or alternative investments portfolio.

> *Exercise.*   Investigate the investment strategy of a well-funded DB scheme which uses LDI.

Notice that the fund is taking considerable credit risk on the inflation swap provider in this structure—the PFCE on a revenue swap is usually much higher than the matching IRS—so some form of credit intermediation may be necessary.

11.1.6.6 Fund demand in the inflation-linked market

The success of the LDI paradigm has resulted in a skewed market for long-dated inflation-linked assets in some countries. In the United Kingdom, for instance, the 50-year gilt linker has traded recently on a real yield of around ½%. This is very little compensation for such long-dated risk: the explanation is not huge confidence in U.K. monetary stability, but massive pension fund buying of the bond. This level of demand makes long-dated inflation-linked bond issuance particularly attractive at the moment.

11.2 EQUITY BASKET PRODUCTS

Volatility trading was one of the dominant themes in equity derivatives during the 1990s, thanks in part to the demand for vol via GEB structures. This resulted in products with increasingly sophisticated volatility exposures being created such as *volatility* and *variance swaps*. More recently, equity derivatives dealers have expanded into the trading of *covariance* via products that depend on the correlations and volatilities of multiple underlyings.

11.2.1 Basket Options and Rainbow Products

Basket options are structures whose payout depends on the behaviour of more than one underlying. The simplest forms of basket option are reviewed first, and then we introduce some more sophisticated products which have recently become popular.

11.2.1.1 Basic baskets

Suppose we take $1000 and invest it as equally as possible in three stocks with prices as below. We end up with the following positions:

	Stock 1	Stock 2	Stock 3
Stock price	$42.72	$20.21	$3.70
Position	8 shares	16 shares	90 shares

The total cost of the portfolio is $998.28. This investment is held for a year, and at the end of this period, the share prices are as follows:

	Stock 1	Stock 2	Stock 3
Stock price	$44.14	$18.50	$4.66
Percentage return	3.3%	−8.5%	26%

The portfolio is worth $1068.52 and we have made money. The simplest form of basket option is just a put or a call on the value of the basket: thus instead of investing our $1000, we could instead buy a $1000 strike call on the basket, and benefit if the portfolio of 8 shares of the first stock, 16 of the second and 90 of the third rises in value over $1000 during the life of the option.

Simple puts and calls on baskets can be priced using the idea of a *composite underlying*: the volatility of the basket is derived using the volatilities of the components and their correlation. Technically, combining implied volatilities with historical correlations is inconsistent, but at least this approach gives a quick estimate of the basket option price.

11.2.1.2 Correlation sensitivity and hedging

The higher the average correlation between components of a basket is, the higher the basket volatility is, and hence the more valuable options on the basket are. Therefore, the business of designing attractive basket products depends to some extent on identifying collections of single stocks which make sense from a buyer's perspective but which happen to have low average correlations: this explains some of the popularity of global baskets, since diversifying across many markets tends to reduce correlation.

Typically, the demand in basket trading is for options on baskets, so selling them leaves the book short correlation (since if correlation rises and stock vols are constant, the basket vol rises and the bank is short vega on it) and short vol on most of the components (since a rise in the vol of any component usually increases the basket vol).

The basket book usually buys back single-stock options either from the street or from an internal single-stock book leaving the trader free to manage the pure covariance risk. This is sometimes hedged with a *correlation swap*: an analogue of the volatility swap whose underlying is realised correlation.

Exercise. How would you decide on the limits for correlation risk in the equity baskets book?

11.2.1.3 Best of baskets, worst of baskets

The term *rainbow option* is used to refer to various derivatives on baskets whose payouts depend on the relative performance of the components. Some of the simpler kinds of rainbow option include:*

- A *best of* option, which pays out on the basis of the performance of the best return in the basket;

- A *worst of* option, which pays out on the basis of the performance of the worst return in the basket.

	Payout Based on	Payout for 100% Strike
Best of call	Stock 3	26%
Best of put	Stock 3	0
Worst of call	Stock 2	0
Worst of put	Stock 2	8.5%

The table above shows the payouts of the best of and the worst of the example basket. Notice that if there are only two assets in the basket, having the best of call and the worst of call is just equivalent to having calls on both assets.

Thus in the limit of correlations going to -1, the best of call on two assets with the same volatility tends towards twice the price of a vanilla call on one of them. The worst of put tends to twice the price of a vanilla put, as shown in Figure 8.

The best of call becomes less valuable with increasing correlation, as the stocks tend to move more together decreasing the potential for one to outperform. Therefore, it is said to be *short correlation*. Similarly, the best of put is *long correlation*: it rises in value if the two components of the basket tend to move in the same direction for constant volatility of the components.

---

* Also included within rainbow options are spread options [as discussed in section 2.2.2] and a variety of other structures. See Hull, Wilmott or Jarrow for more details (*op. cit.*).

11.2.1.4 Baskets with recomposition

More complex basket structures can be created by allowing the basket composition to change over time. A *multi-period rainbow option* will provide a total return on the basis of a series of period returns, with the basket composition recalculated at the start of each period. Examples of basket composition rules include:

- Picking some collection of stocks from a defined universe such as the 10 highest dividend stocks in the FTSE 100 each year in a 5-year structure;
- Ejecting stocks from a fixed basket on the basis of performance, so that, for instance, we might start with 20 stocks, but after each 6-month period, we throw out the best-performing stock, ending up with just one stock at the end of 10 years.

> *Exercise.* Research the 'mountain range' rainbow structures popular with some dealers.

11.2.2 Copulas and the Problem with Gaussian Correlation

The usual correlation coefficient is more properly named the *Pearson linear correlation coefficient* since it measures the comovement of two variables which are linearly related. It would not be sensible, then, to talk about the correlation between CPR on a mortgage backed security (MBS) and interest rates because we do not expect them to be linearly related.

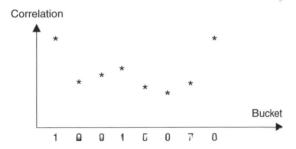

Even when two variables might be conjectured to be linearly related, their comovement can be too complicated to capture using a single correlation coefficient. A good example of this occurs with single stocks. Suppose we take an equity and divide its returns into eight buckets: those days when it had a negative return bigger than 3 S.D.s; those days when it returned between −2 and −3 S.D.s; and so on up to more than 3 S.D.s. Now let us compute the correlation coefficient between returns of our equity and those of another stock for each bucket. Typically, the structure of correlation shown in the illustration is observed: the stocks tend to move together more for large negative returns than for ordinary conditions. This effect has been christened the *correlation smile* by some authors, in analogy with the volatility smile (although note that other authors use this term for a phenomenon in the credit markets: the variation of implied correlation with tranche attachment point in pricing CDOs).

> *Exercise.* Study the actual tail association of an equity basket or index of your choice.

11.2.2.1 Copulas

One fundamental question when dealing with baskets of assets is, given the univariate return distributions of each basket component, what is the right multivariate distribution for modelling the basket? The discussion above shows that even if the asset return is individually normal, the true multivariate distribution might not be multivariate normal.

A copula is a function which amalgamates univariate distributions into multivariate ones. Technically, a two-asset copula on u and v is a function $C : [0,1]^2 \to [0,1]$ obeying:

- Where either variable has zero probability, then so does the copula, that is, $C(u, 0) = C(0, v) = 0$
- Where either variable has probability 1, then the copula has probability given by the other variable, that is, $C(u, 1) = u$ and $C(1, v) = v$
- If in each univariate distribution probabilities increase, then that is true in the copula too.[*]

The simplest copula of all just enforces independence of the variables:

$$C_{\text{Independent}}(u, v, \ldots) = uv$$

The multivariate normal distribution is built by using the *Gaussian* or *normal copula* and a correlation matrix. If u, v, ... are univariate normal and ρ is a correlation matrix, the Gaussian copula is given by

$$C_{\text{Gaussian}}(u, v, \ldots) = \Phi[\Phi^{-1}(u), \Phi^{-1}(v), \ldots, \rho]$$

Here Φ is the cumulative multivariate normal distribution function and Φ^{-1} is the inverse cumulative normal distribution function. [This is the same expression that occurred in the discussion of maximally likely normal variables in section 4.3.2.] Thus, the specification of the multivariate distribution just involves giving the univariate distributions—u, v and so on—and then specifying the comovement by giving the correlations ρ.

11.2.2.2 Tail association

There are a variety of copulas that allow the comovement in the tails of the multivariate return distribution to be different from that in the centre.[†] One of the simplest examples is the Student's t copula:

$$C_{\text{Student's } t}(u, v, \ldots) = T[T^{-1}(u), T^{-1}(v), \ldots, \rho]$$

Here T is the multivariate Student's t distribution with n degrees of freedom and correlation matrix ρ, and T^{-1} is the inverse cumulative univariate Student's t distribution. For small n, this gives some measure of correlation smile.

---

[*] See Umberto Cherubini's *Copula Methods in Finance* for the detailed definition and a much more extensive discussion.

[†] Much longer accounts can be found in Cherubini's book or Alexander McNeil, Rüdiger Frey and Paul Embrechts' *Quantitative Risk Management: Concepts, Techniques, and Tools*.

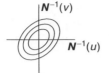

Rough sketches of the contours of constant probability for the bivariate normal distribution with $\rho = +0.5$ (on the left) and the Student's t distribution for $n = 3$ and $\rho = +0.5$ (on the right).

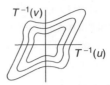

11.2.2.3 The copula problem

Modelling multivariate return distributions requires us to select a suitable copula. If the correlation smile is not significant and fat tails are not an issue, then the multivariate normal distribution is an easy choice. However, if we do have to model these effects, then there is no standard choice of copula. Moreover, even if there were, calibration is not straightforward as there is rather little information available on implied comovements. At least with univariate problems, the implied volatility smile can be used: for multivariate problems, the same help is not available since there are few liquidly traded options on baskets.

The exception to this, of course, is index options. An option on the CAC 40 is after all an option on the 40-stock basket that comprises the index. Therefore, we do have at least that data available, and a successful basket model should be able to infer the observed index volatility surface from the surfaces of the components and their assumed comovement structure. However, there are reasons to be careful:

- First, there may be convenience or liquidity yield effects associated with index products which do not carry over to custom baskets: for instance, the correlation smile might be less pronounced on an index simply because an index is likely to be more liquid in the event of a market crash, and hence not to suffer the same measure of liquidity risk as an arbitrary basket.

- Second, at least for many of the common copulas, the implied correlation necessary to recover even something as simple as the at-the-money index volatility from the component volatilities is highly unstable over time. This suggests either that the right copula is not a common one, or that the right calibration process has not yet been found.

The use of copula methods in pricing basket products is currently an area of active research in the quantitative finance community, with the use of a wide variety of different copulas being investigated. This work has highlighted some of the shortcomings of the multivariate normal distribution with its simplistic notion of linear correlation, and work is continuing on finding more satisfactory replacements.

Exercise. Calculate the average implied correlation under the Gaussian copula for an index of your choice based on the implied volatilities of the components. How does it move over time?

11.2.2.4 Dispersion trading

Index arbitrage trading used to be a term which was used for fairly simple strategies such as trading stock baskets versus index futures or even *ins and outs*—trading around an index recomposition date, usually shorting stocks which are about to leave the index and buying those that are about to enter. More recently, however, basket trading technology has been used for indices, leading to a form of index arbitrage known as *dispersion trading*. Here the trader takes a position on index volatility versus volatility positions on the stocks in the index, or a representative selection of them.

Typically, dispersion traders would sell index vol and buy single-stock vol in the expectation that profits on the single-stock vols will be greater than the index vega losses, i.e., the stocks will disperse more around the index return than they are paying for. Thus, days with returns like the one illustrated on the left are good, whereas ones like that on the right are bad.

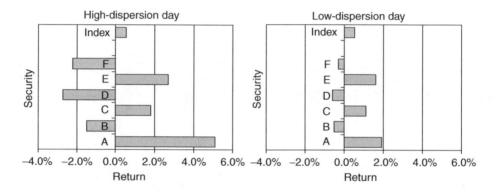

In both cases, the index return is the same—0.53%—but on the high-dispersion day, stocks move considerably versus the index, whereas on the lower dispersion day, they are more tightly grouped around it.

Another way of thinking about this style of trading is as implied versus realised correlation trading: the returns of this strategy depend on the actual realised correlation versus the correlation the trader was paid for in selling an index option.

> *Exercise.* Compare the balance between buyers and sellers of vol on indices with that on single stocks. Does this suggest that the index vol surface should be higher or lower than the fair value implied by the index components?

Finally, note that, at least theoretically, if most people are trading on a normal copula and the market actually behaves more like the *t* copula, then it makes sense to be long-index vol versus single-stock vol on the far downside* and the other way round in the centre of the return distribution.

---

* A short dispersion on the downside position would however be rather exposed to a jump to default event such as Parmalat or Enron where an index component goes bad very quickly.

11.2.2.5 Risk reporting for equity basket books

An equity basket book will need the usual kind of derivatives risk reporting:

- Delta, gamma and vega by underlying, strike and maturity;
- Rho and theta by underlying;
- Concentration reporting.

In addition, though, we will need to understand the impact of the book's sensitivity to changes in stock/stock or stock/index comovements. For instance, a firm might examine:

- The P/L caused by a 1% move up or down in all correlations;
- The biggest individual correlations by sensitivity;
- The impact on total book valuation and the biggest instrument mark-to-market changes caused by a change in copula assumptions;
- The P/L caused by an increase and decrease in stock versus index dispersion.

A key issue here is *index decomposition*: if we do not decompose the index into positions on each component, then the greeks will be large and it will be hard to understand the net position. If we do decompose, though, it is important to have additional controls which monitor:

- The risk of index composition changes and;
- The risk that the comovement structure of the index basket might change and hence that a tracking basket will no longer track as well. The correlation and copula sensitivities suggested above go some way to capturing this issue, but other risk measures may be necessary too.

> *Exercise.* Suppose a large special dividend is declared on one stock in an index. What is the likely impact on a dispersion trading strategy? How about if an index component is bought at a premium and taken private?

11.3 CONVERTIBLE BONDS

CBs are innovative financial instruments that usually share some of the risk and return characteristics of ordinary corporate bonds on one hand and equity on the other.[*] Like a bond, they have coupon rate and a maturity, but they also have an added feature: they can be converted into equity. This makes them cross-asset products: they have equity-like and debt-like features which interact. In this section, CBs are introduced and a simple model of them, which also gives some insight into other equity-linked products, is reviewed. This model will not capture their cross-asset nature: we move on to some approaches to dealing with that in the next section.

---

[*] See Izzy Nelkin's *Handbook of Hybrid Instruments: Convertible Bonds, Preferred Shares, Lyons, Elks, Decs and Other Mandatory Convertible Notes* for a more detailed discussion of CBs and related instruments.

11.3.1 Convertible Bond Structures

One of the most common CB structures is a fixed rate bond which can be converted by the holder into a set number of shares. This number is known as the *conversion ratio* at any point up to the bond's maturity. Once converted, the bond disappears and the investor simply has the shares; if the CB is not converted, principal is returned as usual.

This simple form of CB is similar to an ordinary corporate bond plus an American style equity call. (It is not quite identical to it since the bond coupons are no longer paid once the call is exercised, so the conversion option has a different interest rate risk profile to a vanilla equity call: this is the cross-asset nature coming out.) Since the CB holder is long an equity option which they would not have with an ordinary bond, they pay for this: CBs typically have lower coupons than straight debt from the same issuer of the same maturity. Note that CBs are often long-dated instruments—10 years is not unusual—so the embedded conversion option is also a long-dated instrument.

11.3.1.1 Example

A corporate, ABC plc, wishes to finance itself more cheaply than it can by issuing straight debt. Therefore, it issues a 5-year CB with a face value of £1000 and a 1% coupon. The holder has the right to exchange each CB for 200 shares of ABC. Since the holder has the right to convert something with a face value of £1000 into 200 shares, they effectively have an equity call on 200 shares of ABC struck at £5 per share.

Investors may be interested in buying this CB for various reasons:

- The investor wants a debt instrument from the issuer, but there are no straight bonds available, or none with the desired maturity.
- The investor wants the long-term equity participation offered by a CB.
- They wish to exploit arbitrage opportunities offered hedging the CB.

11.3.1.2 Issuer calls

CBs can be fairly complex securities as it is commonplace to layer various features into the basic structure. The first of these we review is the *issuer call*: a CB may be callable by the issuer at a date or dates before its stated maturity. This can be either a *hard* call (meaning that the issuer can call the bond regardless of any other circumstances) or a *soft* call where the issuer can only call the bond if some hurdle is met such as the equity price having risen significantly above the strike price of the conversion option.

A typical soft call structure might be a 5-year ABC CB issued at £1000 and convertible in 200 shares but callable by ABC at par (i.e., £1000) any time after the third year, provided the stock has risen above £6. The stock price at issue is £4, so the conversion option initially is 25% out of the money: this is fairly typical for a new CB. The effect of the soft call is to *force conversion* of the bond: if the issuer announces a call of the bond, the investor will convert to get the stock (which is worth at least £1200 since otherwise the call would not be possible) rather than par. There is typically a grace period between the call announcement and the actual call date giving investors an opportunity to convert.

The soft call allows ABC to convert debt into equity financing if it wishes: it no longer has to pay the 1% coupon, and its leverage decreases.

11.3.1.3 Investor puts

Some CBs are putable by the investor back to the issuer. This feature is typically used for weaker credits where the investor requires more protection against the credit exposure than is offered by a relatively long-dated bond.

11.3.1.4 Step-ups and other hybrid features

CBs sometimes incorporate step-ups or other hybrid features [as discussed in section 3.2.1] such as subordination, deferral or alternative coupon satisfaction mechanisms (where a coupon can be paid in stock rather than cash, for instance).

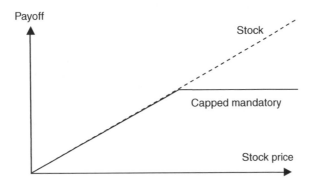

11.3.1.5 Mandatory conversion

A *mandatory* CB is one where instead of having the option to convert, the investor is required to, so that redemption is always in stock rather than cash. The simplest kind of mandatory redeems in a fixed number of shares, so the investor is long a forward rather than a call, and the issuer achieves the same effect as a delayed rights issue

In more complex versions, the number of shares delivered to the investor can vary so that, for instance, in some variants there is a modicum of downside protection via the delivery of larger numbers of shares if the stock price falls.

In others, the upside is capped or otherwise limited allowing the issuer to offer a higher coupon as the holder is short an option.

11.3.1.6 Exchangeable bonds

Most CBs convert into the stock of the issuer: *exchangeables* convert into a different stock. These are sometimes used in situations where the issuer has a large block of stock which it wishes to sell. The exchangeable offers cheap financing for the position and the possibility of disposing of it without having to take it to the stock market.

11.3.2 The Behaviour of Convertible Bonds

The major risk factor in most convertibles is the underlying equity price. The diagram below shows how the price of a simple CB before maturity moves as the underlying stock prices move according to a simple bond plus call model. For low stock prices, the CB price does not move very much. It behaves just like straight debt of the same coupon: CB marketers call this *downside protection* as the CB's downside is limited to the value of this *bond floor*. Once the stock price moves above the strike of the conversion option, however, the CB starts to benefit from *equity upside*.

Fixed-coupon CBs are sensitive to the level of interest rates in a similar fashion to ordinary corporate bonds: as rates go up, they tend to become less valuable, and conversely they tend to become more valuable as rates go down. However, this sensitivity is modified in two ways. First, as the graph shows for high equity prices, the bond is much more like an equity than a debt instrument, and hence has greatly reduced interest rate sensitivity. Second, the interest rate sensitivity of the conversion option is opposite in sign to that of an ordinary bond, so a CB's interest rate sensitivity will generally be somewhat lower than that of straight debt with the same coupon.

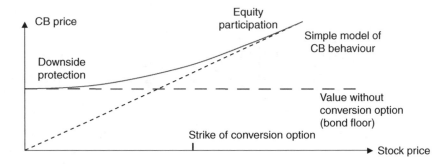

The holder of a CB is thus:

- Long delta because they own the conversion option;
- Long vega for similar reasons;
- Short rho, at least until the conversion option is highly in the money.

11.3.3 Modelling Convertibles

One standard technique for the valuation of equity-linked products is the binomial tree.* This starts from the idea that a stock can either go up or go down each day and we do not know which. However, as an approximation we fix the percentage increase or decrease so that there are only two possible outcomes in each time interval: the stock goes up a fixed amount or the stock goes down.

11.3.3.1 The stock price tree

Suppose that a stock starts at S today and that tomorrow it could be worth either Su, with $u > 1$, or Sd for $d < 1$. The idea is to model the future evolution of a stock price as a *tree* of possible prices.

---

* The binomial technique is well known: a more detailed discussion can be found in most of the standard references such as John Hull's *Options, Futures and Other Derivatives*.

The first few steps in the tree are shown below, with $ud = 1$ so that the tree *recombines*. This has the nice property that once the tree is big, we can arrange things so that the terminal stock price distribution is roughly normal. To see this, just look at the tree below: there is only one path to the largest terminal node *Suuuu*, but four to the next one down *Suu* and six to the middle node. As the tree gets bigger, the probability of getting to a terminal node approaches the familiar bell shape of the normal distribution.

If we assign a probability p to a move-up, so that a move-down has probability $(1 - p)$, we can calibrate p, u and d so that the distribution of final stock prices approximates to a desired return distribution, i.e., has the same forward and volatility.

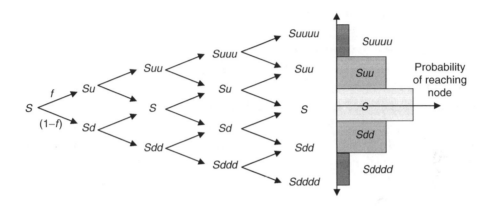

11.3.3.2 Tree building
Let S be the initial stock price, σ the stock volatility and r the risk-free rate. Then without dividends, the tree parameters u, d and p for a tree with a time step of Δt between nodes are given by

$$u = \exp(\sigma\sqrt{\Delta t}) \quad d = 1/u \quad p = (a - d)/(u - d) \quad \text{where } a = \exp(r/\Delta t)$$

11.3.3.3 Example
Suppose ABC plc's equity has a 36% volatility, the initial stock price is £4, and (to make the trees small enough to display all of them) we take monthly time steps and take rates at 5%. The first 4 months of the stock price are then as shown below:

				6.06
			5.46	
		4.92		4.92
	4.44		4.44	
4.00		4.00		4.00
	3.61		3.61	
		3.25		3.25
			2.93	
				2.64

Thus after 4 months, we have five possible stock prices ranging from roughly £6 to £2.64. Calibrating to give a 36% volatility and a 5% forward gives $p = 49\%$, so the probability of getting to £6.06 is p^4 or roughly 6%.

One nice feature of modelling the stock price evolution this way rather than using a diffusion is that we can engage in various pieces of tree surgery to capture features of interest.

For instance, suppose the stock pays dividends every quarter and we use a proportional dividend model so that each dividend is 0.6% of the stock price.* After the dividend is paid, we expect the stock price to fall by the amount of the dividend. Therefore, we can go into the tree and reduce the node on the dividend date by the required amount. For instance, in our example, if the dividend is paid in the third month, the node £5.4634 will fall to £5.4606 and the terminal nodes are now £6.03, £4.89, £3.98, £3.23 and £2.62.

11.3.3.4 The European call price tree

Suppose we have a 4-month £4.50 call on ABC plc stock. The value of the call at each of the terminal nodes is just whatever the call would pay out if the stock were at that level, so, for instance, if the stock level after 4 months was £6.03, the call would pay out max (£6.03 − £4.50, 0) or £1.53. Now consider a node 1 month before expiry, with the stock at £5.46. With probability 49%, the stock goes up to £6.03, and down to £4.89 with probability 51%. Therefore, the weighted average value of the call here is 49% × £1.53 + 51% × £0.39, and this needs to be PV'd back 1 month as a month passes in each branch of the tree.

Stock price Terminal prices of £4.50 call Calculation of call price one node in

Proceeding this way, we can construct a tree of call prices by moving backwards through the tree from the terminal call prices. The value of the root of this tree is just the call price at time zero: the initial value of the option. Even with our crude 4-step tree, this approach is not too bad: the tree price is within 5% of the Black–Scholes price.

11.3.3.5 Backwards induction

The process of moving backwards through the tree to obtain the derivative price is known as *backwards induction*. Notice that we can value any derivative that depends only on the final call price this way: all we need is to be able to evaluate the payoff at the terminal nodes. Of course, in practice a much finer time step would be used, giving a tree with hundreds of terminal nodes rather than our five.

Formally, the backwards induction step to a node $S_i(t)$ if $S_{i+1}(t+1)$ and $S_i(t+1)$ are known is

$$S_i(t) = \exp(-r\Delta t)[p \times S_{i+1}(t+1) + (1-p) \times S_i(t+1)]$$

---

* In the short term, companies tend to keep their dividends fixed regardless of stock price, so this is probably not such a good model for short-dated structures. In the longer term, however, there is evidence to support the use of a proportional dividend approximation.

11.3.3.6 American calls

Is it ever worth exercising an option early? For a non-dividend-paying underlying, the answer is usually no, but dividends make the situation more complex:

- Suppose we are 6 months from maturity, a dividend of 10% is just about to be paid and we hold $100 face of CB which can be converted at any time into 20 shares. The stock price cum div is $7 and volatility is 15%.
- If we convert, we get $140 and lose some time value on early exercise.
- If we do not convert, the stock price will drop to $6.30 after the dividend reducing the intrinsic value of the bond to $126, but we keep the time value.
- Unfortunately, this time value is not worth much: after the dividend, the fair value CB price is roughly $128. Therefore, here we should exercise early.

Another piece of tree surgery allows this effect to be captured. Another tree is constructed where each node $\mathrm{Ex}_i(t)$ is just given by the value of exercising the option now, so $\mathrm{Ex}_i(t) = \max(S_i(t) - K, 0)$.

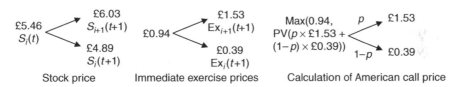

| Stock price | Immediate exercise prices | Calculation of American call price |

The American call price is then obtained by the following backwards induction:

$$S_i(t) - \max(\mathrm{Ex}_i(t), \exp(-r\Delta t)[p \times S_{i+1}(t+1) + (1-p) \times S_i(t+1)])$$

11.3.3.7 CB tree

We now have all the ingredients needed to value CBs: we just add the price of the American call on the underlying to the value of the bond floor. This last is just the PV of the scheduled cashflows on the CB discounted along the issuer's credit curve.

Suppose instead of just adding the bond floor to the option value, we wanted to build a tree of CB prices. At a terminal node $S_i(T)$ of the tree, we know the CB's value: it is just

$$CB_i(T) = \max(\text{Conversion ratio} \times S_i(T), \text{Bond floor})$$

We could then use the same backwards induction as before. But there is one difficulty: what discount factor should we use?

11.3.3.8 Blended discounting

The backwards induction for a call assumed a risk-free counterparty and hence we used the risk-free PV: this was reasonable on the call alone as companies can always print their own shares, so there is little credit risk in buying a call from a counterparty on its own stock. For the bond floor, though, we have a cashflow from a risky counterparty, which suggests using the risky curve. One crude solution to the problem of choosing between a risk-free discount factor for the call and a risky one for the bond is to use a *blended discount factor*: at each point in the tree, we calculate the probability of the conversion option being in the money, $p_i^{conversion}(t)$ say, then use a discount factor that varies between the risk-free DF if conversion is certain and the risky DF if conversion is impossible:

$$DF_i(t) = p_i^{conversion}(t) \times \exp(-r\Delta t) + (1 - p_i^{conversion}(t)) \times \exp(-(r + s)\Delta t)$$

Here s is the credit spread. This gives us a tree of CB prices via a backwards induction

$$CB_i(t) = DF_i(t) \times [p \times CB_{i+1}(t + 1) + (1 - p) \times CB_i(t + 1)]$$

11.3.3.9 Soft calls

If the CB has an issuer call, yet another collection of trees will suffice to include this phenomenon. First, a tree of immediate conversion values Conversion ratio $\times Ex_i(t)$ is built. Next, we work out at which nodes in the stock price tree a call is possible. Finally, a tree CallableCB$_i(t)$ is built where the terminal nodes are the CB values at the terminal stock prices, and where the backwards induction step is

If (CB callable) Then CallableCB$_i(t) = \max($Conversion ratio $\times S_i(t)$, Par)
Else CallableCB$_i(t) = DF_i(t) \times [p \times$ CallableCB$_{i+1}(t + 1) + (1 - p) \times$ CallableCB$_i(t + 1)]$

This assumes that if a CB can be called by the issuer, then it will be, but that is usually conservative: the investor simply gets a windfall gain if there is no call.

11.3.3.10 Equity/credit optionality

Blended discounting is a rough attempt at solving one of the fundamental difficulties with modelling CBs: the equity and credit optionality is intertwined, and in particular, there is information in the equity price about the likely credit spread and *vice versa*. Thus, for instance, if the equity price rises, it is likely that the credit spread will tighten; more saliently, if the equity price falls a long way, the credit spread is likely to increase, and thus the value of the bond floor falls. A more accurate picture of the behaviour of a CB value with equity price is therefore:

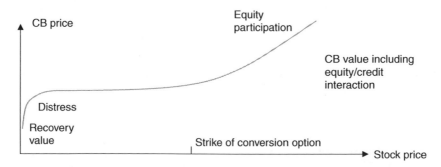

The models discussed above are all *1-factor* models: the only source of uncertainty is the equity price. *2-factor models* with two sources of uncertainty—variable equity prices and interest rates, for instance—or even 3-factor models are sometimes used for CBs. In the next section, models which give some account of the interaction between equity and credit will be discussed. Rather than having both the equity price and credit spreads as separate variables, these *structural models* will concentrate on how both equity and credit spreads change as a result of movement in a firm's *enterprise value*. This will give more insight into a range of situations, including the modelling of *busted* CBs: bonds in the 'distress' area of the curve shown above.

> *Exercise.* Find a non-investment-grade convertible from a distressed issuer with a readily observable price. Why might the delta of the CB be greater than 1? Observe the equity and CB price series and see if this is the case.

11.4 EQUITY/CREDIT TRADING

Intuitively, the connection between equity and credit is obvious: a stock goes up when a company does well, and that implies a tighter credit spread; the more earnings there are to service a given amount of debt, the better the credit quality and the more that will be left after interest costs to pay dividends. This section reviews a number of models that formalise that intuition, and then discusses how they are used to take positions in both equity and credit instruments issued by an obligator.

11.4.1 The Merton Model of Capital Structure

One obvious way of connecting equity prices and credit spreads is to assume they are both driven by the same variable. This is the approach taken by an early model of capital structure: the *Merton model*.

11.4.1.1 Merton's insight

Let us fix a corporation to analyse, and suppose this firm issues both equity (a perpetual instrument) and a single-debt instrument with face value F. The firm has an *enterprise value*, V. Consider the value of the firm at the maturity of the debt:

- If $V < F$, then the firm cannot repay the debt, and it defaults. Debt holders take the firm and liquidate it, receiving a recovery of V/F if we ignore liquidation costs. Equity holders get nothing;
- If $V > F$, then the firm repays the debt, and equity holders are entitled to the value of the firm $V - F$.

At maturity of the debt, therefore, the value of the equity is $\max(V - F, 0)$. This is the pay-off of a call on the firm's enterprise value struck at the face value of the firm's outstanding debt.

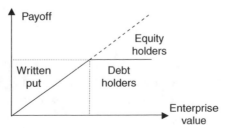

11.4.1.2 A simple capital structure model

Merton then went on to assume that *before* the maturity of the debt, a firm's equity could also be viewed as a call on the enterprise value struck at the debt face. How could this call be valued? If:

- Enterprise value is a stochastic variable with log-normal returns characterised by a volatility σ_V;
- There are no other risk factors;
- The other assumptions of Black–Scholes hold.

Then we could use the Black–Scholes formula. Suppose $E(t)$ is the total value of the firm's equity at some time t, $D(t)$ is the value of the firm's debt and $V(t)$ is the enterprise value so that $V(t) = E(t) + D(t)$. We would then have

$$E(t) = V(t)\Phi(d_1) - e^{-rt}F\Phi(d_2)$$

where

$$d_1 = \frac{\ln(V(t)/F) + (r + \sigma_V^2/2)t}{\sigma_V\sqrt{t}} \quad \text{and} \quad d_2 = d_1 - \sigma_V\sqrt{t}$$

It turns out that in this setting there is a relationship between the volatility of enterprise value, σ_V, and the equity volatility, σ_E:

$$\sigma_E E(t) = \Phi(d_1)\sigma_V V(t)$$

This framework is known as the *Merton model of capital structure*. It identifies a firm's equity as a call on enterprise value and its risky debt as a risk-free instrument plus a written put on the enterprise value.

Note here that the ratio between equity vol and enterprise vol is approximately given by the ratio of enterprise value to equity value since $\Phi(d_1)$ is close to 1: we would expect this, since the higher the leverage, the greater a change in equity value caused by a given movement in enterprise value.

11.4.1.3 Applications

There are three main applications for a capital structure model:

- If there is no traded debt but we do have equity prices and sufficient information to estimate volatility, a credit spread can be inferred. This is the approach used in *structural models of credit risk*.
- If we have a credit spread for a firm but its equity does not trade—for instance, in a PE situation—we can infer an equity price.
- Finally, if both equity and credit trade, the model suggests a relationship between them. If this does not hold, then there may be an arbitrage available by being long the equity and short the debt or *vice versa*. This is *capital structure arbitrage*.

11.4.1.4 Default in the Merton model

The Merton view of default is shown in the illustration: a firm starts off with a positive equity value some time before debt matures, and it defaults if the enterprise value has migrated below the face value of the debt at maturity. At any point, the distance between the enterprise value and the PV of the face is known as the distance to default. In the Merton model, the distance to default can be negative: the firm can have an enterprise value less than the face of the debt now and still not be in default. Here the firm's equity—viewed as an option—is out of the money but it still has value as the firm may recover sufficiently to repay the debt when it becomes due.

The pure Merton model PD is the probability that the enterprise value has migrated so far that it is less than the face of the debt at maturity. Thus, the higher the leverage, the higher the PD since this raises the strike of the option: further, the more volatile the enterprise value, the more likely default is, as this causes the enterprise value to spread out faster.

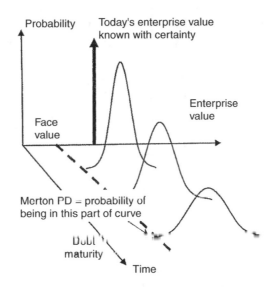

11.4.1.5 Potential issues

Issues may arise in the use of the Merton model if its assumptions break down:

- The notion of default in the Merton model is very simplistic: it only happens at the maturity of the debt, and then debt holders always receive the entire enterprise value. In reality, firms sometimes default for reasons of liquidity or loss of confidence and this can happen at any time. Moreover, in default bankruptcy or other credit event, costs can be large. This problem can be partially addressed by pricing a no touch rather than a vanilla call and using a barrier which represents a higher default point rather the strict face value of the liabilities.

- Enterprise value is not an observable market value: equity and debt values are. Moreover, the assumption that enterprise value follows a random walk is difficult to justify, and inconsistent with standard pricing assumptions.*

- Many firms—especially financial institutions—have a complex capital structure and lots of rolling short-term debt. Application of the Merton model here is complex as we have to make assumptions about how the capital structure rolls forward in time and how to deal with short-term debt in the model.

- Firms sometimes have significant off-balance-sheet liabilities: these need to be included in the model to properly reflect the extra leverage they introduce.

Exercise. Calculate the pure Merton PD for three or four large corporates and compare it with the PD implied by their CDS spreads.

11.4.1.6 A simple structural model of credit risk

The use of the Merton model as a credit risk tool has been pioneered by Vasicek and Kealhofer. The resulting framework is known as a KMV model, and a commercial implementation of it is available from Moody's KMV.[†] The broad approach is a modification of Merton's: first, the Merton model is used to calculate the distance between a firm's current enterprise value and the point at which it will default: this *distance to default* is then scaled to produce an expected default frequency (EDF). The scaling is used to adjust for the effect of the issues discussed above, and is typically based on an empirical approach.

Structural models are useful both for deriving a credit spread when one is not available and for incorporating equity market information into credit analysis. If an obligator's bonds are illiquid, the equity market may price information about changes in credit quality faster or more reliably than the debt market. Therefore, some portfolio managers—while not necessarily trusting the expected default probability produced by a structural model—nevertheless use these models to spot situations where EDFs are changing fast. These might be credits which will see future spread widening, and hence they are candidates for hedging. Portfolio managers certainly need information that changes more quickly than a rating, since by the time an obligator is downgraded it may be too late to hedge.

Exercise. On average, how long before a downgrade is there meaningful information in the equity market?

---

* If returns on enterprise value are log-normal, then equity returns cannot be. Similarly, if debt has a constant hazard rate, then enterprise value returns are not log-normally distributed. The Merton model is therefore inconsistent with standard equity and credit derivatives models.
† See www.moodyskmv.com for more details of the KMV framework. Discussions of structural credit models can also be found in a variety of sources such as Donald Deventer et al.'s *Advanced Financial Risk Management: Tools and Techniques for Integrated Credit Risk and Interest Rate Risk Management.*

11.4.1.7 Structural models of portfolio credit risk

A structural credit risk model may or may not work well for a given obligator: it tends to work better for manufacturing industries than for financials, for instance. But on a portfolio basis, a structural model has one great advantage: since default is driven by equity value and equity volatility, default correlation is driven by these factors too. Therefore, rather than *imposing* a correlation structure on the model as we do in the portfolio credit described earlier [in section 5.4.2], we *discover* them in the comovements of the equity prices of portfolio obligators.

11.4.1.8 The advantages and disadvantages of equity-implied default correlation

In a structural model, if a firm's equity price falls, all other things being constant, then the PD rises. Therefore, if a range of firm's equity falls at the same time, all of their PDs go up together, and so default correlation increases. The model's UL estimate for a credit portfolio will therefore increase.

The advantage of this approach is that a single equity/equity correlation may well reflect information about the comovement of the two stocks if the broad market is flat or up. If the Merton model is right, this information can be transformed into information about the comovement of default probabilities.

The disadvantage is that there are situations when the whole market is down but it is hard to believe that overall credit risk has increased significantly: all that has happened is that the equity market has taken one of its periodic downturns and so equity return correlations here include a systematic risk factor. It may be then that naïve use of structural models for calculating portfolio credit risk capital would produce an unrealistic volatility in capital requirements.

11.4.1.9 Introducing capital structure arbitrage

Suppose a trader wished to take advantage of a possible opportunity suggested by a capital structure model. If the market credit spread is tighter than the model spread, either the debt is too expensive and the equity is the right price or the debt is correctly priced and the equity, which we used to calibrate the model, is too cheap. In either case, the suggested trade is to sell the debt and buy the equity. However, selling corporate debt short is not straightforward as it is hard to borrow. It is easier to use the CDS market, buying protection. On the equity side, we take advantage of the smile, selling downside puts to go long delta.

If the model suggests a trade in the other direction, the play would be to buy bonds (or sell CDS protection) and short the stock (or buy puts).

> *Exercise.* Find any materials you can on hedge funds specialising in capital structure arbitrage. What do they claim to do, and what can you discover about their strategies?

11.4.1.10 Capital structure arbitrage and the big downgrades of 2005

Capital structure arbitrage became a well-known trading strategy in the early 2000s, and by early 2005 there were a number of large players taking 'arbitrage' positions across the capital structure. Their activities and some of the risks of engaging in them can be seen in the General Motors downgrade of 2005.

General Motors was at one point the largest manufacturing company in the world. By 2004, however, it was struggling with high production costs, uncertain and large costs associated with providing pensions and healthcare for its North American workforce, and a declining market share. The problems were by no means idiosyncratic: Ford was in a similar situation.

What did make General Motors unusual was the size of its debt. Both GM and its auto loan financing subsidiary GMAC were large issuers in the bond market. When General Motors was downgraded below investment grade in May 2005, it was the biggest single cut to junk ever.

There was no choice for many fund managers at this point. 'You may only invest in investment-grade bonds' is part of many bond fund mandates, so these investors were forced sellers of GM bonds. Since the downgrade had been widely anticipated, dealers were positioned for this wall of paper hitting the market, and the premium on GM 5-year CDS protection went from around 200 to over 500 bps immediately after the downgrade (although it is not clear whether there was much trading at that level).

The activities of capital structure arbitrage players can be seen in the open interest in far out of the money puts. In the 1990s, there was very little interest in these options—dealers were reluctant to sell them as the premiums were so small and the crash risk was large— and equity downside protection buyers wanted hedges that were closer to the money. Capital structure players use these options against CDS, and hence liquidity here has improved. In GM, for instance, there was significant open interest in the far OTM puts in mid-2005.

The good news for capital structure players the day the downgrade hit was that GM stock fell. However, the fall was not enormous—only around 5%—and implied vols went up, doubling at some points on the vol surface. Moreover, this stock fall was only from levels which had been boosted in previous days by Kirk Kerkorian increasing his stake in GM: overall the stock price was up on the week.

This is a situation where the arbitrage relationship broke down in the days around the downgrade. The equity market was being driven in large part by an assessment of the probability of a Kerkorian takeover. The credit market was digesting the overhang of bonds caused by forced selling from fund managers. Players who were long the credit and short stock could have found themselves losing money on both sides of the 'arbitrage'.

Exercise. Bearing this kind of problem in mind, what kinds of limits would you want to see for a capital structure arbitrage desk? How can the desk be given incentives to ensure reasonable diversification of positions?

How have you accounted for the potential illiquidity or one-way market in an underlying in your answer?

11.4.2 More Sophisticated Capital Structure Models

The fixed default point in the structural models discussed above is a significant issue: in reality, firms tend to take on more liabilities as they slide towards default, so liabilities increase as net enterprise value falls.

11.4.2.1 Moving barriers

One simple modification of the Merton model is therefore to introduce a stochastic default point. In addition to enterprise value, a second variable is introduced which drives the evolution of the default point.

As usual, this is taken to be a random walk characterised by a volatility. Default occurs if the enterprise value hits the default point* as in the illustration.

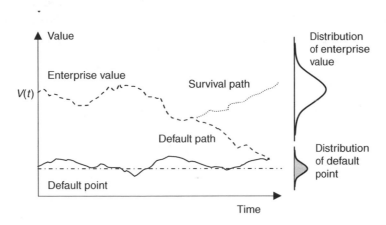

This model provides a reasonable account of the default process. It is also reasonably easy to calibrate: only two volatilities are needed, and the processes driving enterprise value and the default point are usually assumed to be uncorrelated, so there is no need to estimate a correlation. Moreover, the model still has at its heart the normal distribution—albeit in bivariate form—so implementation is straightforward.

11.4.2.2 Equity smiles in capital structure models

Both the simpler Merton-style model and the stochastic barrier capital structure model predict an equity skew: as enterprise value decreases, leverage increases, and so a fixed volatility of enterprise value produces a larger volatility of equity. Unfortunately, the shape of the smile generated is usually too flat. Nevertheless, it suggests an intriguing possibility that the smile could emerge as a property of a capital structure model rather than being imposed upon it *post hoc* as calibration.

11.4.3 Equity/Credit Optionality and Hybrid Security Modelling

A different approach to equity/credit modelling is to work directly with the stock process. Remember that for a local volatility model we had

$$\frac{dS}{S} = \mu(t)\,dt + \sigma(S,t)\,dW$$

---

* This style of model has been introduced by the CreditGrades consortium, and so is sometimes known as a CreditGrades™ model: see *The CreditGrades Technical Document* available from www.creditgrades.com for more details.

A simple modification to this would allow us to account for the possibility of default. Suppose N is a jump process representing a default with intensity $\lambda(S, t)$. Then we could write

$$\frac{\mathrm{d}S}{S} = \mu(t)\,\mathrm{d}t + \sigma(S, t)\,\mathrm{d}W + \mathrm{d}N$$

Much of the standard Black–Scholes theory of equity derivatives can be extended into this setting: an analogue of the Black–Scholes PDE is obtained, with a second equation for the dynamics of λ.* Here there are two sources of uncertainty, W and N, so the hedge of a derivative will require positions in two instruments in general: even a straight call has a credit delta.

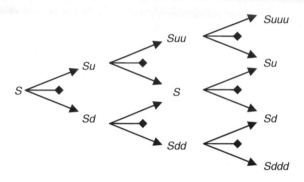

In tree terms, the stock model with a jump to default could just involve adding a default jump at every node as shown in the illustration. Calibration would be to both the default swap spread and the equity volatility, the difficult part being the selection of the functional form of λ.

This form of model is particularly useful for looking at highly credit-sensitive equity-linked securities: next therefore we look at an examination of this type of security.

11.4.3.1 Example
Perpetual convertibles offer an interesting equity/credit problem. Here we have a security which is convertible for a fixed period as usual, but at the expiry of the conversion option, instead of receiving principal, the investor is left holding a perpetual note. To remove significant interest rate sensitivity from the problem, let us assume that this note is floating rate. Therefore, we have the following structure:

Perpetual Convertible Callable Note		Issue Price	£1000
Underlying	ABC plc common stock	Conversion period	5 years
Converts into	200 shares	Coupon during conversion period	1%
Share price at issue	£4.00	FRN spread after conversion period	200 bps
Issuer call	Callable at par on each coupon date after 2 years		

---

* See Ayache, Henrotte, Nassar and Wang's *Can Anyone Solve the Smile Problem?* (available from www.ito33.com) for more details.

There are three outcomes for an investor in this note:

- The equity goes above the £5.00 strike and the note is converted.
- The equity does not go above the strike, but the issuer's cost of funds at the end of the conversion period is less than Libor + 2%, so the note is called then.
- The equity does not go above the strike, and the issuer's cost of funds at the end of the conversion period is more than Libor + 2%. Here the rational decision by the issuer will be not to call the note.

The only time the investor gets the FRN, in other words, is when the issuer's credit spread is high, and this is likely to happen when the stock price has fallen significantly.

Any of the models discussed in this section could be used to estimate the impact of this short credit put feature of the security. We could even get a quick idea of the magnitude of the issue using a modification of the blended discount factor model [of section 11.3.3]:

11.4.3.2 The 1½-factor model

Consider the previous expression for blended discount factors:

$$\text{DF}_i(t) = p_i^{\text{conversion}}(t) \times \exp(-r\Delta t) + (1 - p_i^{\text{conversion}}(t)) \times \exp(-(r+s)\Delta t)$$

This could be generalised by introducing a stock price-dependent risky discount factor. For instance, we could just set

$$s(S) = k\left(\frac{S}{S_0}\right)^{\alpha}$$

Here k and α are constants and S_0 is the share price at issue. This could be calibrated using similar credits or a full capital structure model to get an estimate for α then fixing k to recover the current CDS spread for ABC plc. The model gives a risky discount factor that varies node by node with the equity price at that point in the tree. Since this model uses the equity price to deterministically imply a credit spread, it is sometimes called a *1½ factor model*.

The point is not that this is a wonderfully insightful model of a wide range of hybrid securities—it is not—but rather that it is simple, easy to build and easy to calibrate and gives some intuition into the particular sensitivities of our example perpetual convertible. There is a place for models like this: they might not be industry strength, but they do allow us to get a sense for the magnitude of a problem quickly. In this case, the difference between a standard 1-factor model price for the perpetual convertible and the 1½-factor valuation should scare us enough to improve the model.

11.4.4 Credit Copulas and Credit Event Association

The prices of the standard tranches for credit index products [discussed in section 5.3.4] provide us with a useful calibration of a portfolio credit risk model. We know the individual credit spreads of the names in the index: can these be put together in such a way that the tranche prices are recovered?

11.4.4.1 Another copula problem

Following the analysis earlier in this chapter, we know that this is a copula problem. The credit spreads of the individual names tell us (at least once we fix a model of default) what the univariate distributions are. The tranche prices depend on the multivariate distribution since [as we saw in section 5.3.5] the more credit events move together, the more valuable the senior tranche is and the less valuable the junior tranche is.

Market practice in this area has *not* been to try to find a copula that allows all the tranche prices to be recovered simultaneously. Rather some dealers use the Gaussian copula, but rather than calibrating it with a single correlation, instead speak of an *implied correlation*. In analogy to implied volatility, this is also the wrong input to put into the wrong model to get the right price, in this case the correlation needed in the Gaussian copula to recover a given tranche price. The implied correlation for the traded index tranches depends on the tranche concerned, and tends to show the form illustrated.

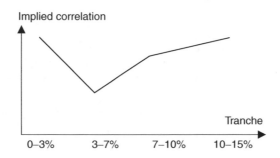

This approach is perfectly reasonable for traded tranches where the implied correlation can be readily backed out from market quotations. It is less useful for other baskets or for the general portfolio credit risk problem where implied correlations are not available.

It is worth noting here that just because an implied correlation is reasonably stable for a while, it does not mean that it will continue to be so. A good example at the moment comes in the tranched MBS market: lower-credit-quality mortgages (such as second liens) used to trade on a fairly low implied correlation. Now, however, the U.S. housing market is in retreat, so there is a systematic effect which has dramatically increased implied correlation lowering the value of the senior and supersenior tranches dramatically.

Exercise. How much difference does the choice of copula make for:

— The price of the first-, second- and third-to-default notes on a basket of six emerging market bonds?

— The price of the standard index tranches using a single dependency measure (i.e., one correlation for the Gaussian copula rather than different correlations per tranche)?

— The price of the tranches of a large SME loan securitisation?

11.5 NEW PRODUCTS

New products have repeatedly given risk to significant issues for firms. We discuss some of the controls that might assist in allowing innovation to proceed in a controlled fashion.

11.5.1 The New Product Approval Process

The *new product approval process* is designed to ensure that before a new product is traded, all areas of the bank can handle it. Typically, firms use a process whereby representatives of all the functional areas consider the issues around any new product for their area. This often happens in a committee process and the committee's sign-off is needed before trading. The committee acts as the gatekeeper, ensuring that new products do not pose unacceptable market, credit, operational, reputational or strategic risk.

The committee also often acts as a forum, so concerns can be aired and members can understand each other's issues. In addition to a flat prohibition of a proposed new product, the committee might also impose volume or size limits, or require further development before a new product can be traded in unlimited amounts. Some of the issues the committee might consider can be grouped under the following headings:

11.5.1.1 Management

- It should be clear who is responsible for trading the product and in which business group that responsibility arises. This is particularly an issue for cross-asset trading where there might be more than one claim on a product.
- The business should state the anticipated volume of trading and estimate the likely P/L per trade.
- An economic capital calculation should be available for the product, and a projected ROE should be estimated. An anticipated effect on earnings volatility might also be estimated.
- If the product conveys ongoing responsibilities on the firm such as a requirement to make a secondary market, these should be stated.
- The new product application should also consider competitors and their motivations in trading the product. This may be important reputationally: even if the firm's own conscience is clear, it does not want to be drawn into a wider controversy.

11.5.1.2 Booking and market risk

- New risk classes introduced by the product must be identified, and any new limits required must be proposed. This discussion should identify any cross-optionality or contingent risks and consider if they are properly captured and priced in.
- If the product requires the extension of trader mandates, this should be discussed.
- A backtest analysis of the product in ordinary and stress conditions should be presented.

- The system used for booking the product should be identified and any issues with valuation or risk measurement should be identified. An extensive discussion will be required for trades not booked in a main trading system.*

- Any manual processes should also be highlighted as this can lead to higher product service costs, less reliable data feeds and increased operational risk.

11.5.1.3 Credit risk

- Any form of credit extended to clients should be stated including a presentation of a PFCE calculation for the product.

- If trading the product requires an extension of credit limits, this should be quantified.

- The credit terms of the product (ISDA master, prime brokerage agreement or other enforceable netting agreement) should be discussed together with any requirements for the firm to post collateral and margin. The operational support for such credit mitigation should be outlined.

- Any country risk issues should be detailed.

11.5.1.4 Operational, legal and reputational risk

- If the product is not correctly and completely handled in a straight-through processing environment, all resulting issues should be addressed.

- The operational risk loss data gathering process for the product should be discussed, and any known operational risk issues should be commented upon.

- The process for providing client valuations should be discussed if these are necessary. Particular attention should be given to any situation where different valuations are provided to clients as those on the firm's own books and records.

- The legal entity involved in trading the product should be identified, and any particular legal, reputational or operational issues involved in using this entity should be highlighted.

- The intended clients for the product should be identified, and a discussion should be held on why the product is suitable for them. The worst case(s) for clients should be identified, and any disclosures and disclaimers made should be stated. Circumstances under which clients can lose more than their initial investment may be separately identified, and the client's legal right to transact the product should be discussed.

---

* Products which are not booked in the main trading systems are sometimes known as *not-in-system* trades (NIS). These pose considerable challenges to a risk management infrastructure and should be avoided if at all possible. Some trading systems have a sufficiently open architecture so that trades can be booked in them without a full valuation model being present, so at least the system has a place holder for the trade and it is clear what is not known about it.

- Any impact of the product on a listed or regulated entity should be identified including a discussion of additional disclosures required or any other visible impact on the consolidated group's reporting.

11.5.1.5 Valuation and reserving

- The valuation model for the product should be discussed together with the results of model review. The performance of both the intended hedge strategy and P/L explanation for the product should be considered.
- In particular, any questionable assumptions by the valuation model should be highlighted, and any need for mark adjustments or model risk provisions should be outlined. The impact of the product on portfolio mark adjustments must also be considered.
- The implications of trading the product on the market should be considered if the anticipated market impact is large.
- Any liquidity assumptions should be highlighted, and the implications of their failure should be discussed.
- The accounting of the product should be considered, and any accounting policy issues should be discussed.
- Any new legal entities required such as SPVs must be discussed in detail.

11.5.1.6 Liquidity and capital

- The issuance of debt by any of the firm's consolidated legal entities or by any entity guaranteed or otherwise credit enhanced by a group company should be detailed.
- Repo or other secured funding assumptions made in pricing or managing the product should be detailed: unsecured funding needs must also be discussed. The TP of any funding or liability issuance should be outlined.
- The currencies involved in trading the product and its hedges should be listed and any convertibility issues discussed.
- Any impact on the firm's liquidity stress planning must be considered together with any modifications necessary to the disaster recovery plan.
- The product's impact on the firm's balance sheet over time should be estimated.
- The product's regulatory capital usage should be estimated and any communications needed with regulators or ratings agencies should be considered in detail.

This list is not exhaustive and is intended simply to give a flavour of some of the issues. Any functional area—including finance, operations, law and technology—might potentially have concerns with the product, so many firms have a detailed checklist for each area.

Finally, senior management and legal staff should consider any additional reputational and strategic risks involved in a new product: just because we *can* trade something and make money does not mean that we *should*.

11.5.2 Managing Product Complexity

Firms typically only make money if risk is taken. Therefore, one of risk management's jobs is to facilitate risk taking within the firm's risk appetite. An effective risk manager therefore does not often absolutely prohibit a trade: rather the answer is typically 'yes, but …', with the qualification giving the risk taker an understanding of what needs to be done to allow the deal to be done given the firm's risk appetite.

New products give rise to particular challenges here: history suggests that it often takes several iterations before an effective product design is discovered, and managing suboptimal features of new products in the meantime can involve some heavy machinery.

Product innovation is intense in many markets, with significant fractions of some institutions' profits coming from new product areas. Certainly, spreads are wide, and large per-deal profits are possible in some new business lines. Therefore, there is a great temptation to get into novel products fast. This is not necessarily a bad thing if the firm has an appetite for product development risk. Firms sometimes choose to feel their way forward, allowing some new products even within a sub-optimal control environment, but tightly controlling the size of trading and working hard to develop systems and processes as demand rises. This can be pragmatic: not every new product becomes a business line, and it may not be worth developing infrastructure for a small number of products.

The real risk here can be in the growth of a product type rather than in the first trade: one $10M transaction is unlikely to cause significant loss (unless there is reputational or legal risk), but 20 or 50 such trades in a control infrastructure designed for one might be a problem. New product approval might therefore include a requirement to revisit the committee after a certain number of trades or after a certain risk position has developed.

11.5.2.1 Dialogue

The head of risk management, Dr. R. Careful, is on his way to lunch when he meets Rather Long, the head of U.S. Equity Derivatives. Long uses the chance meeting as an opportunity to raise some concerns.

'I want to trade La Cienga options'.

'You can't. The system can't price them'.

'I know. But Esmeralda is convinced we can sell half a yard retail and we'll have a pricing model soon. The quants are working on it'.

'Be serious. You can't trade five hundred million dollars of an exotic option to retail clients'.

'No one else on the street is offering them. We'll have the first La Cienga retail note and we'll make a fortune. Ten or twelve bucks, easy. And we'll dominate trading in them'.

'What trading? There is no trading because no one has any confidence in what they are worth. Selling stuff neither we nor our client understand isn't trading, it's suicide'.

'But we know they aren't worth more than those Bermudans I showed you the other day, you remember, the ones with the Melville feature'.

'Listen. I'll let you do one trade. But no more than twenty million dollars notional, you have to mark it as if it was a Melville, you don't get to show any P/L on day 1, and you can only do it with a professional counterparty. And once you can convince me you understand it, we'll talk about doing some more'.

'That will do for now'.

11.5.3 Hedge Fund Risk Management

The management of proprietary market risk taking activities has always been an important task for risk managers. Recently hedge funds—either as separate entities or as internal hedge funds within investment banks—have become significant market risk takers so in this section we examine some of the particular challenges they pose.

11.5.3.1 Leverage in investment management

Hedge funds are often unregulated and thus unconstrained by regulatory capital. Many are also unrated. These freedoms mean that the only constraint on many funds' leverage is provided by their contractual arrangements. Typically a fund will have one or more *prime brokers* who lend it money, act as its counterparty on securities and derivatives transactions, and possibly provide other services such as custody or record keeping. The prime broker is typically an investment bank who will act in this regard for a number of funds. It usually demands that each fund posts *margin* against its exposures on a portfolio basis so it is this margin or collateral requirement that constrains a fund's leverage: the lower the margin for a given risk, the more the fund can leverage its capital, and so the higher the potential return. Since many fund managers' compensation depends on return there may be a temptation to over leverage. If this happens and the market moves against the fund, the prime broker will try to close out the fund's positions before its capital is depleted to protect itself. Therefore hedge fund's risk management often includes the necessity of understanding how portfolio margin requirements change as the market moves.

11.5.3.2 Two key concepts: alpha and beta

One measure of a portfolio's leverage is *beta*. If we plot a fund's return versus the broad market return over time, beta is the gradient of the best fit straight line through the data. Thus beta greater than one implies a leveraged portfolio.

Investors are usually unwilling to pay funds just for being leveraged. They could do that themselves, for instance by buying the underlying investments on margin. Instead investors want *alpha*: excess returns that do not come from excess risk.

A fund's alpha might come from an arbitrage strategy that profits from market mispricings without taking (much) overall risk, or from fundamental stock or bond analysis. Unfortunately what appears to be alpha—returns without correlation to the broad market— is often really just beta on a different market risk factor. For instance a CB arbitrage fund that buys convertibles and delta hedges the embedded equity option might appear to add alpha in ordinary markets as its return is uncorrelated with the return on equity indices. In fact it has significant volatility risk (and perhaps some credit spread risk), i.e., *alternative beta*. Similarly the strategy of selling short dated out of the money puts will show no beta against the market provided that there is not a significant fall—it might look like a strategy with added alpha—but of course if the market does fall, beta will be significant. This phenomenon—low fund vs. market return correlation provided there is not a crash but high if there is—is known as *asymmetric beta*, and is a common feature of hedge fund risk taking.

11.5.3.3 Disclosure and performance attribution

Sometimes it is necessary to assess a hedge fund's risk profile without complete knowledge of its positions:

- Funds typically make only limited and infrequent disclosures to investors;
- A fund may have more than one prime broker, and then none of them can see the complete position;
- A *fund of funds* manager can often extract more information from funds than an ordinary investor can but they still may not have complete information.

All of these parties are therefore interested in where the fund's returns come from. P/L explanation for investment strategies is sometimes called *performance attribution*. Here in addition to decomposing the P/L into the effect of various risk factors, it is also common to examine:

- The effect of *asset selection*. Given its net position over a period, did the fund make less or more than an investment in the broad market (or the fund's benchmark) would have done?
- The effect of *market timing*. Is the difference between the P/L due to a risk position and the P/L that would have resulted had that position been held for the whole of the period positive or negative? In other words, does the fund show good or bad market timing?

If a manager can show that what appears to be alpha is due to consistently good asset selection or market timing then investors should be impressed. If it comes simply from taking risk on sensitivities other than broad market delta, we have alternative or asymmetric beta instead.

11.5.3.4 Crash risk

As we have seen earlier, in a crash:

- Market prices of risky assets fall;
- Both realised and implied volatilities;
- Asset return correlations tend towards either +1 or –1;
- There is a flight to quality increasing liquidity premiums.

These phenomena mean that margin requirements—including portfolio margin requirements calculated using VAR—typically increase in a crash. Moreover many hedge fund positions will lose money regardless of their normal market beta. To see this in more detail consider the common thread connecting the following strategies:

- Long an illiquid/risky/structured asset, short a related by more liquid/less risky/simpler one;
- The yen carry trade with the proceeds invested in a risky asset;
- Any mean reversion or trend following strategy;
- Buying the senior tranche of securitisations or selling tail risk insurance.

All of these will suffer in a crash, and *even if the strategy eventually makes money* the combination of leverage and increasing margin requirements poses a threat to the fund's ability to continue operating.

Managing proprietary risk taking activities, then, involves not just an assessment of the final outcome of a strategy but also a consideration of the circumstances which could result in positions being forcibly closed out.

11.5.3.5 The RMBS difficulties of 2007

As this book goes to press in summer 2007, various markets are, if not in crisis, then certainly creating distress for some players. The rough chronology of events was:

- There was a real estate bubble in a number of economies, prominently the U.S., until 2006;
- The combination of rising real estate prices and a benign real economy made lending against real estate appear relatively low risk even to borrowers who would hitherto not have qualified for a mortgage;
- Partly as a result of the securitisation treadmill [discussed in section 10.3.2], many institutions lent significant sums to subprime borrowers, lent using mortgage

structures which were more exposed to falls in collateral value than traditional (i.e., max 80% LTV) loans, or both;

- These mortgages were packaged into PTs, tranched, and/or included in CDOs of RMBS, thus transmitting the risk to the security buyers;

- When the real estate bubble burst, it slowly became clear that these securities were significantly riskier than some investors (and arguably the ratings agencies) had appreciated.

By this stage the problems were more or less confined to the RMBS market. The issues became broader when it became clear that the value of many of these RMBS PTs, CDO tranches and related securities was uncertain. Liquidity disappeared from the market and it became more or less impossible to sell an RMBS much more complicated than a conforming agency PT. This then had knock-on effects:

- Mutual funds and especially hedge funds investing in ABS had to suspend redemptions as the value of their assets could not be determined;

- Firms who relied on the securitisation market for funding suffered higher cost of funds and in some cases sufficient liquidity risk to threaten their ability to continue in business;

- The shares of companies directly exposed to real estate (builders and mortgage banks) fell sharply.

This in turn led to a wider credit crisis: spreads went out, credit was rationed, and so debt-financed PE deals could not be completed. Equity market volatility increased significantly with the VIX going from the low teens in the spring to over thirty in August. At the same time short end swap spreads went out signficantly in most currencies. Central banks then intervened, pumping hundreds of billions of dollars of extra liquidity into the financial system by opening the window for repo of a range of securities including RMBS presumably as they were worried about the systematic risk implications of a severe credit crunch.

One key driver in these events has been the interaction between *price falls* and *liquidity*. Without liquidity black holes in the ABS market the impact of the events would have been less severe: funds, banks and broker/dealers could have liquidated their positions, taken their losses, and moved on. But as matters stand, they have a large and uncertain mark-to-market loss *and* they still have the risk position. The next few months will be interesting.

Concluding Remarks

The book ends with a discussion of some overarching issues: the risk management process in general and some of the factors which determine its effectiveness.

THE RISK MANAGEMENT PROCESS

The same pattern has repeated itself a number of times in this book:

- We begin by understanding the *products* and how they behave, together with the processes, systems and infrastructure that are used to support them.
- Some assumptions are made about the *dynamics* of the variables of interest to produce a risk model.
- The model is *calibrated* on the basis of available data.
- Lower-level *risk metrics* are designed to capture the likelihood and impact of possible situations in which money could be lost.
- Next, higher-level risk metrics are produced by *aggregating* lower-level measures from different areas of the firm.
- The risk model is *documented* and *validated*.
- The firm's *risk appetite* is defined at various levels and risk reporting is designed on the basis of metrics available.
- The output of the risk measurement process is *monitored* regularly, and action to modify the firm's risk profile is taken as needed on the basis of its articulated risk appetite.
- *Capital allocation* and *performance measures* are derived from the risk measures.
- Finally, the performance of the whole framework is monitored; it is regularly audited and enhanced as needed.

There should be no pretence that this process offers a uniquely correct and perfectly accurate measurement of risk: rather we wish to measure the significant risks accurately enough to give stakeholders—including management, shareholders, supervisors and ratings agencies—reasonable confidence that risk is being taken in a controlled manner.

SUFFICIENTLY GOOD RISK MANAGEMENT

Like any other area of a firm's business, risk management can always be improved. The real issue is whether this would add value to the firm. Effective risk management is often about using an insightful model rather than merely a sophisticated one, and about using relevant, reasonably accurate data rather than serried rows of irrelevant or out-of-date information. Hence the suggestion in the introduction that a good risk manager is an artisan: certainly getting a good enough answer in limited time with limited resources is a craft skill that most good risk managers value. As in any other facet of the firm, a risk process or model should be good enough and no better. In this context that means it should roughly quantify the potential for loss or to suggest the need for further study, but not aspire to an accuracy which is unnecessary, costly to achieve and potentially spurious given the complexities of financial return distributions.

INCENTIVE STRUCTURES

Most policies and processes within a firm set up an incentive structure. In particular, any method of measuring performance defines important, behaviour-altering incentives. This means that economic capital allocation is typically not just a question of deciding how much capital is needed to support various risks: rather it is about creating a set of rules which encourage behaviour that is in the firm's best interests. Therefore, a key test of any risk management policy or process—and especially the firm's capital allocation methodology—is whether it encourages risk takers to do the right thing. Another key characteristic of good risk managers is that they look beyond the theory of risk to see what behaviour a given policy will produce and shape their views accordingly.

Risk measures are created by people, for people. It means that considerations of status, advantage and appearance are sometimes present. I define a new measure partly because it needs to be done, but partly to impress my boss. You report a systems failure because you are required to as part of the firm's operational risk loss data collection process, but also to make me look bad. This is so obvious that its mention is perhaps jejune. But it does mean that the effective risk manager has to work with these tendencies rather than opposing them: political adeptness is an important skill too.

CULTURE AND ORGANISATION

Just as capital is no substitute for risk management, quantitative risk management is no substitute for a bad risk culture. If a trader is permitted to ignore or evade a risk manager's questions, if limits are only respected by the naïve, or if reserves are manipulated to smooth the P/L, then no amount of detailed risk reporting or modelling will bring the firm back under control. Risk management starts with behaviour and culture, not with processes or models.

FURTHER STEPS

This book has tried to give some intuition about a number of areas of risk management and provided pointers towards understanding others. There are a number of further steps depending on the reader's inclinations and needs. For instance, nothing is better than spending time on a trading desk to develop your understanding of the market.

The references in the footnotes provide some suggestions for those interested in developing their understanding of various topics, hopefully guided by some of the intuitions presented here.

The focus for most of this book has been financial risk management as practised in banks or broker/dealers. The foundations are the same for other uses—for instance, in investment management or insurance—but some of the methods used and the dominant risk classes can be very different. Readers with an interest in these areas may therefore wish to move on to develop their understanding by further examining the application of the ideas discussed here.

UNDERSTANDING RISK

This book focuses on understanding risk: developing an intuition for what can lose us money. Some of the discussion has been quantitative as we need to be able to estimate the impact of financial risks, but much of it has been more informal, reflecting the importance of understanding the behaviours of markets and portfolios before modelling them. Good risk management often involves a blend of the mathematical and the practical approaches: of modelling and market knowledge, of process and paranoia, of systems and savvy.

Figures

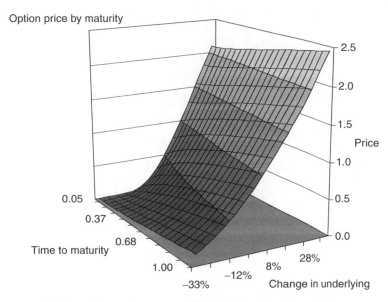

FIGURE 1 The payoff of a call as a function of underlying level and time to maturity.

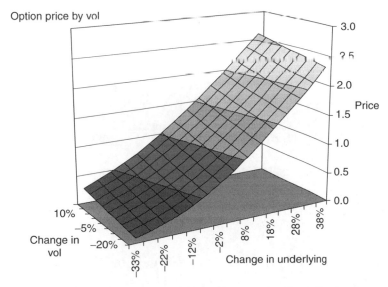

FIGURE 2 The payoff of a call as a function of underlying level and volatility.

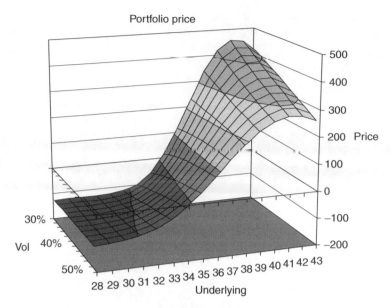

FIGURE 3 Illustration of a scenario for an options position.

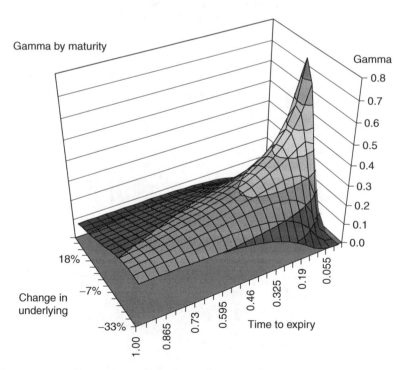

FIGURE 4 The gamma of a call as a function of spot and time to maturity.

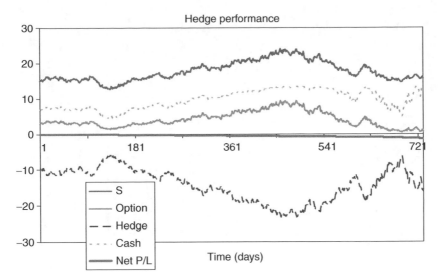

FIGURE 5 Behaviour of an option and its hedges for one path of the underlying with daily delta rebalances.

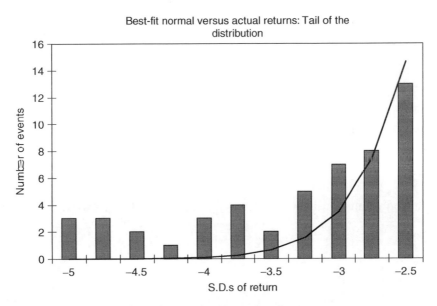

FIGURE 6 The best-fit normal distribution in the tails of CAC 40 returns.

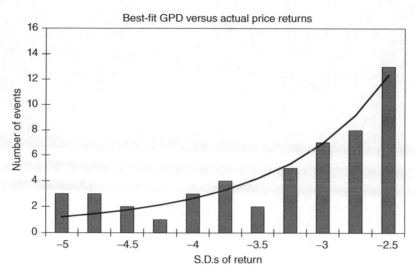

FIGURE 7 A generalised Pareto distribution (GPD) fit to the tails of CAC 40 returns.

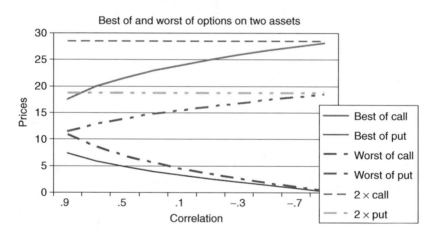

FIGURE 8 Rainbow and vanilla option prices as a function of correlation.

Index